VOLUME ONE HUNDRED AND THIRTEEN

ADVANCES IN
COMPUTERS

VOLUME ONE HUNDRED AND THIRTEEN

ADVANCES IN
COMPUTERS

Edited by

ATIF M. MEMON
College Park, MD,
United States

ELSEVIER

ACADEMIC PRESS
An imprint of Elsevier

Academic Press is an imprint of Elsevier
50 Hampshire Street, 5th Floor, Cambridge, MA 02139, United States
525 B Street, Suite 1650, San Diego, CA 92101, United States
The Boulevard, Langford Lane, Kidlington, Oxford OX5 1GB, United Kingdom
125 London Wall, London, EC2Y 5AS, United Kingdom

First edition 2019

ISBN: 978-0-12-816070-1
ISSN: 0065-2458

For information on all Academic Press publications
visit our website at https://www.elsevier.com/books-and-journals

Working together
to grow libraries in
developing countries

www.elsevier.com • www.bookaid.org

Publisher: Zoe Kruze
Acquisition Editor: Zoe Kruze
Editorial Project Manager: Shellie Bryant
Production Project Manager: James Selvam
Cover Designer: Matthew Limbert

Typeset by SPi Global, India

CONTENTS

Preface   *ix*

1. A Survey on Regression Test-Case Prioritization   1
Yiling Lou, Junjie Chen, Lingming Zhang, and Dan Hao

  1. Introduction   2
  2. Framework   5
  3. Criterion   9
  4. Prioritization Algorithm   12
  5. Measurement   17
  6. Constraint   20
  7. Application Scenario   23
  8. Empirical Study   25
  9. Some Discussions   28
  10. Conclusion   33
  Acknowledgments   33
  References   33
  About the Authors   45

2. Model-Based Test Cases Reuse and Optimization   47
Mohamed Mussa and Ferhat Khendek

  1. Introduction   48
  2. Overall MBT Framework   49
  3. Integration Test Generation   55
  4. Acceptance Test Optimization   67
  5. A Case Study: A Library Management System   70
  6. Related Work   71
  7. Conclusion   75
  Acknowledgments   76
  Appendix A: Properties of the Integration Test Generation Approach   77
  References   84
  About the Authors   87

3. Three Open Problems in the Context of E2E Web Testing and a Vision: NEONATE   89
Filippo Ricca, Maurizio Leotta, and Andrea Stocco

  1. Introduction   90
  2. The Three Open Problems in the Context of E2E Web Testing   92

3. State of the Art on the Three Open Problems   97
4. Overcoming the Three Open Problems: The NEONATE Vision   104
5. Architecture of the NEONATE Integrated Testing Environment   107
6. NEONATE's Examples of Use   119
7. NEONATE's Long-Term Impact   124
8. Conclusions   126
References   128
About the Authors   132

4. Advances in Using Agile and Lean Processes for Software Development   135

Pilar Rodríguez, Mika Mäntylä, Markku Oivo, Lucy Ellen Lwakatare, Pertti Seppänen, and Pasi Kuvaja

1. Introduction   136
2. Trends on Agile, Lean, and Rapid Software Development   142
3. A Walk Through the Roots of Agile and Lean Thinking   148
4. Agile and Lean in Software Development   158
5. Beyond Agile and Lean: Toward Rapid Software Development, Continuous
    Delivery, and CD   182
6. DevOps   190
7. The Lean Startup Movement   197
8. Miscellany   202
9. Conclusions and Future Directions   209
References   212
About the Authors   221

5. Advances in Symbolic Execution   225

Guowei Yang, Antonio Filieri, Mateus Borges, Donato Clun, and Junye Wen

1.   Introduction   226
2.   Background   228
3.   Constraint Solving   230
4.   Path Explosion   236
5.   Compositional Analysis   244
6.   Memory Modeling   246
7.   Concurrency   250
8.   Test Generation   253
9.   Security   259
10. Probabilistic Symbolic Execution   264
11. Tools Support   269

12. Conclusion 271
References 271
About the Authors 286

6. Symbolic Execution and Recent Applications to Worst-Case Execution, Load Testing, and Security Analysis 289

Corina S. Păsăreanu, Rody Kersten, Kasper Luckow, and Quoc-Sang Phan

1. Introduction 290
2. Symbolic Execution 290
3. Tools and Scalability Challenges 293
4. Applications 296
5. Conclusion 305
References 305
About the Authors 313

7. Experiences With Replicable Experiments and Replication Kits for Software Engineering Research 315

Steffen Herbold, Fabian Trautsch, Patrick Harms, Verena Herbold,

and Jens Grabowski

1. Introduction 316
2. What Is Replication 318
3. Replication Kits 319
4. Experience Reports 320
5. Discussion 330
6. Conclusion 339
References 340
About the Authors 342

PREFACE

This volume of Advances in Computers is the 113th in this series. This series, which has been continuously published since 1960, presents in each volume four to seven chapters describing new developments in software, hardware, or uses of computers. For each volume, I invite leaders in their respective fields of computing to contribute a chapter about recent advances.

Volume 113 focuses on three main topics. Chapters 1–4 discuss specific advances in software development; Chapters 5 and 6 discuss advances in symbolic execution, an important technique for checking programs; and Chapter 7 discusses important advances in replicating experiments in software engineering research. More specifically, in Chapter 1, entitled "A survey on regression test-case prioritization," by Yiling Lou et al., the authors posit that regression testing is crucial for ensuring the quality of modern software systems, but can be extremely costly in practice. Test-case prioritization has been proposed to improve the effectiveness of regression testing by scheduling the execution order of test cases to detect regression bugs faster. Since its first proposal, test-case prioritization has been intensively studied in the literature. This chapter presents an extensive survey and analysis on existing test-case prioritization techniques, as well as pointing out future directions for test-case prioritization. The survey includes 191 papers on test-case prioritization from 1997 to 2016. These papers are studied from six aspects: algorithms, criteria, measurements, constraints, empirical studies, and scenarios. For each of the six aspects, the existing work and the trend during the evolution of test-case prioritization is examined. Current limitations/issues in test-case prioritization research, as well as potential future directions on test-case prioritization are also presented.

In Chapter 2, entitled "Model based test cases reuse and optimization," authors Mussa and Khendek summarize a subset of test generation techniques that have been proposed in the literature. These techniques target separately specific levels of testing without relating them to each other in order to avoid redundancy and enable reuse and optimization. This chapter looks into connecting different levels of testing and proposes a model based testing framework that enables reusability and optimization across different levels of testing. Test cases at one level are reused to generate test cases of subsequent levels of testing. Furthermore, test cases at one level are optimized by relating them to test cases of preceding testing levels and removed if they are found redundant.

In Chapter 3, "Three open problems in the context of E2E web testing and a vision: NEONATE," authors Ricca, Leotta, and Stocco discuss web applications, which are critical assets of society and thus assuring their quality is of undeniable importance. Despite the advances in software testing, the ever-increasing technological complexity makes it difficult to prevent errors. The chapter provides a thorough description of three open problems hindering web test automation: fragility, strong coupling, and low cohesion and incompleteness. The authors call for a major breakthrough in test automation because the problems are closely correlated, and hence need to be attacked together rather than separately. They describe Neonate, a novel integrated testing environment specifically designed to empower the web tester.

Chapter 4, entitled "Advances in using agile and lean processes for software development," authored by Rodríguez et al., discusses that software development processes have evolved according to market needs. Fast-changing conditions that characterize current software markets have favored methods advocating speed and flexibility. Agile and Lean software development are in the forefront of these methods. The chapter presents a unified view of Agile software development, lean software development, and most recent advances towards rapid releases, then introduces the area and explains the reasons why the software development industry begun to move into this direction in the late 1990s.

In Chapter 5, "Advances in symbolic execution," authors Guowei Yang et al. assert that symbolic execution is a systematic technique for checking programs, and forms a basis for various software testing and verification techniques. It provides a powerful analysis in principle but remains challenging to scale and generalize symbolic execution in practice. The chapter reviews the cutting-edge research accomplishments in addressing these challenges in the last five years, including advances in addressing the scalability challenges such as constraint solving and path explosion, as well as advances in applying symbolic execution in testing, security, and probabilistic program analysis.

Chapter 6, "Symbolic Execution and Recent Applications to Worst-Case Execution, Load Testing, and Security Analysis," authored by Păsăreanu et al. discusses symbolic execution, which is a systematic program analysis technique that executes programs on symbolic inputs, representing multiple concrete inputs, and represents the program behavior using mathematical constraints over the symbolic inputs. Solving the constraints with off-the-shelf solvers yields inputs that exercise different program paths. Typical applications of the technique include test input generation and error detection. This chapter reviews symbolic execution and associated tools, and describes some of the main challenges in applying symbolic execution in practice: handling of programs with complex inputs, coping with path explosion and ameliorating the cost of constraint solving. The chapter also surveys applications of the technique that go beyond checking functional properties of programs. These include finding worst-case execution time in programs, load testing and security analysis, via combinations of symbolic execution with fuzzing.

Finally, in Chapter 7, "Experiences with replicable experiments and replication kits for software engineering research," authors Herbold et al. discuss recent advances in replication in experimentation. Replicable experiments are currently gaining traction in the software engineering research community. The authors have made an effort in recent years to make their research accessible to other researchers, through the provision of replication kits that allow rerunning experiments. This chapter presents their experiences with replication kits, including contents required, how to structure them, how to document them, and also how to best share them with other researchers. While this sounds very straightforward, there are many small potential mistakes, which may have a strong negative impact on the usefulness and long-term availability of replication kits. The chapter discusses best practices for the content and the sharing of replication kits.

I hope that you find these articles of interest. If you have any suggestions of topics for future chapters, or if you wish to be considered as an author for a chapter, please do reach out to the series editor.

PROF. ATIF M. MEMON, PH.D.
College Park, MD, USA

CHAPTER ONE

A Survey on Regression Test-Case Prioritization

Yiling Lou*,†, Junjie Chen*,†, Lingming Zhang‡, Dan Hao*,†

*Key Laboratory of High Confidence Software Technologies, Peking University, Ministry of Education, Beijing, China
†Institute of Software, EECS, Peking University, Beijing, China
‡Department of Computer Science, University of Texas at Dallas, Richardson, TX, United States

Contents

1. Introduction   2
2. Framework   5
3. Criterion   9
   3.1 Structural Criterion   9
   3.2 Model-Level Criterion   10
   3.3 Fault-Related Criterion   10
   3.4 Test Input-Based Criterion   11
   3.5 Change Impact-Based Criterion   11
   3.6 Other Criteria   12
4. Prioritization Algorithm   12
   4.1 Greedy Algorithm   13
   4.2 Search-Based Algorithm   14
   4.3 Integrate-Linear-Programming-Based Algorithm   15
   4.4 Information-Retrieval-Based Algorithm   15
   4.5 Machine-Learning-Based Algorithm   16
5. Measurement   17
   5.1 APFD   17
   5.2 AFPD$_C$ 17
   5.3 APXC   18
   5.4 WGFD   19
   5.5 HMFD   19
   5.6 NAPFD and RAPFD   20
6. Constraint   20
   6.1 Time Constraint   20
   6.2 Fault Severity   22
   6.3 Other Constraints   22
7. Application Scenario   23
   7.1 General Test-Case Prioritization   24
   7.2 Version-Specific Test-Case Prioritization   24

Advances in Computers, Volume 113
ISSN 0065-2458
https://doi.org/10.1016/bs.adcom.2018.10.001

1

8. Empirical Study   25
  8.1 Studies on Traditional Dynamic Prioritization   25
  8.2 Comparison With Traditional Dynamic Techniques   26
9. Some Discussions   28
  9.1 Existing Issues   29
  9.2 Other Challenging Problems   31
10. Conclusion   33
Acknowledgments   33
References   33
About the Authors   45

Abstract

Regression testing is crucial for ensuring the quality of modern software systems, but can be extremely costly in practice. Test-case prioritization has been proposed to improve the effectiveness of regression testing by scheduling the execution order of test cases to detect regression bugs faster. Since its first proposal, test-case prioritization has been intensively studied in the literature. In this chapter, we perform an extensive survey and analysis on existing test-case prioritization techniques, as well as pointing out future directions for test-case prioritization. More specifically, we collect 191 papers on test-case prioritization from 1997 to 2016 and conduct a detailed survey to systematically investigate these work from six aspects, i.e., algorithms, criteria, measurements, constraints, empirical studies, and scenarios. For each of the six aspects, we discuss the existing work and the trend during the evolution of test-case prioritization. Furthermore, we discuss the current limitations/issues in test-case prioritization research, as well as potential future directions on test-case prioritization. Our analyses provide the evidence that test-case prioritization topic is attracting increasing interests, while the need for practical test-case prioritization tools remains.

1. INTRODUCTION

Modern software systems keep evolving to refine software functionality and maintainability, as well as fixing software flaws. Regression testing has been widely used during software evolution to ensure that software changes do not bring new regression faults. Although crucial, regression testing can be extremely costly [1–3]. In the research literature, it has been reported to consume 80% of the testing cost [4]. Furthermore, modern industry companies also suffer from regression testing cost due to the large number of accumulated test cases during software evolution. For example, Google engineers have witnessed a quadratic increase in their regression testing time, and the number of tests executed each day within Google already exceeds 100 million [5–7].

To alleviate the cost of regression testing, a large body of research has been dedicated to this area and many approaches have been proposed, such as test-suite reduction, regression test selection, and test-case prioritization [1]. Test-suite reduction (also denoted as test-suite minimization) [2, 8–16] aims at reducing the number of test cases by excluding redundant test cases. Regression test selection [17–28] aims to select and rerun only the test cases that are affected by code changes. Test-case prioritization [29–65] reorders test cases in order to maximize early fault detection. Among the three areas, both test-suite reduction and regression test selection exclude some test executions and may suffer from *unsafe* test execution (i.e., missing regression faults). In contrast, test-case prioritization, the target area of this work, simply reorders test executions and does not discard any test case. Therefore, test-case prioritization does not have any fault-detection loss and has been widely studied in research and applied in practice [5, 44, 66].

Test-case prioritization was first proposed in regression testing to deal with the trade-off between what ideal regression testing should do and what is affordable by scheduling the execution order of test cases [67]. However, test-case prioritization is not the focus of that work. Later, Rothermel et al. [29] presented a widely known industrial case to show the necessity of test-case prioritization. Show in that work, the industry case has a product with about 20,000 lines of code consuming 7 weeks on running the entire test suite. Furthermore, that work also proposed various basic test-case prioritization techniques, including the total and additional techniques, which are usually taken as the control techniques in the evaluation of novel test-case prioritization techniques, and still represent state-of-the-art test-case prioritization according to three recent studies [68–70]. These two pieces of work witness the beginning of test-case prioritization, and a large amount of work has been proposed in the following two decades.

Briefly speaking, test-case prioritization aims to schedule the execution order of test cases so as to satisfy some testing requirements. Formally, test-case prioritization is defined as the following process: given any test suite T, test-case prioritization is to find a permutation T' of T satisfying $f(T') \geq f(PT)$, where PT represents any permutation of T and f is a function defined to map permutations of T to real numbers representing the prioritization goal [32]. Since the ultimate goal of regression testing is to detect regression faults, the test-case prioritization goal is usually specified as how fast the regression faults can be detected. That is, test-case prioritization is usually regarded as scheduling test cases to detect more faults earlier.

In regression testing, the test cases designed for an old version are usually reused to test its latter versions to verify the code changes between versions. That is, to reveal faults in the latter versions as early as possible, the reused test cases should be executed in some specified order, which is the aim of test-case prioritization. In other words, regression test-case prioritization (usually abbreviated as RTP) targets at scheduling the execution order of test cases designed for an old version so as to detect faults in its latter versions as early as possible. Besides regression testing, test-case prioritization is also applied to other testing scenarios where test cases are not designed for an old version but for the current version, which is called initial testing [71]. That is, test-case prioritization in initial testing (abbreviated as ITP in this chapter) targets at scheduling the execution order of test cases designed for the current version so as to detect faults in the current version as early as possible. Due to the characteristics of ITP (e.g., does not rely on old version information), its techniques are usually applicable to regression testing, whereas the techniques of the latter may not be applicable for the former.

As the ultimate goal of test-case prioritization, detecting more faults early is usually infeasible, because we can hardly know whether a test case detects faults without running the test case. Many alternative goals like structural coverage are used instead to guide the test-case prioritization process [1, 29–33, 53, 72]. However, due to the inherent difference between alternative goals and the ultimate goal, test-case prioritization becomes more difficult. Furthermore, even taking these alternative goals, test-case prioritization is also an NP-hard problem [73]. Therefore, test-case prioritization suffers from both the effectiveness and the efficiency issues.

To promote the long-term development of the test-case prioritization topic, it is necessary to review and summarize it systematically. However, the existing surveys either summarized this topic at a high level together with other topics (e.g., test-case selection and test-suite reduction) [1], or just reviewed test-case prioritization techniques before 2013 [74, 75]. During the recent years, researchers have still been making obvious achievements on this topic. For example, based on the papers collected for this survey (details shown in Section 2), recent 3 years witness another upsurge in test-case prioritization paper publications due to the popularity of continuous integration. Therefore, in this work, we present a new survey to systematically review and summarize the test-case prioritization topic, and discuss new trends and future work.

2. FRAMEWORK

In this section, we analyze the papers considered in this survey and present the analysis framework of this survey.

To conduct an extensive survey, it is necessary for us to collect a sufficient number of test-case prioritization papers, which represent the past and current status of test-case prioritization. To achieve this goal, we collected representative papers through two steps. First, we used keywords "test," "prioritiz," and "prioritis" to obtain an initial set of related papers. Second, we manually checked the initial set of papers to keep the most representative papers. Finally, we have a set of 191 papers on test-case prioritization in total. To the best of our knowledge, this is the most comprehensive study on test-case prioritization in the literature.

Fig. 1 shows the number of analyzed papers on test-case prioritization from 1997 to 2016. X-axis represents the year and Y-axis represents the number of papers. From Fig. 1, we observe that the number of test-case prioritization papers overall has a clear increasing trend since the first proposal of test-case prioritization. The reason is that software systems grow larger and larger during the last two decades (e.g., the Debian OS system [76] increased from 55 million LoC to 419 million LoC between 2000 and 2012), and more

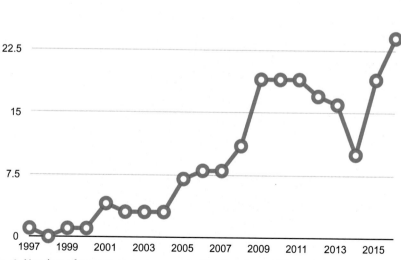

Fig. 1 Number of papers on test-case prioritization from 1997 to 2016.

and more regression test cases are also accumulated during the process, thus stimulating the development of efficient regression testing techniques including test-case prioritization. In addition, we also see several upsurges during the development of test-case prioritization in 2004–2005, 2008–2009, and 2014–2015. We looked into the phenomenon, and found the potential reasons for that. During 2004–2005, the modern distributed version control systems including Git [77] and Mercurial [78] were being proposed. With the advanced version control systems, more and more projects are hosted in code repositories, bringing regression testing techniques to the attention of the developers to test code revisions. During 2008–2009, the GitHub [79] open-source project hosting service (the largest source-code hosting website to date, with 20 million users and 57 million code repositories as of April 2017) was initially released, and the hosted projects usually use regression testing to validate code revisions. Another potential reason for the 2008–2009 resurge is that the financial crisis increased the graduate student population. Finally, we think that the resurge during 2014–2015 may be due to the recent development of mature Continuous Integration (CI) services, such as Travis [80] and Jenkins [81], which extensively use regression testing to provide fast quality feedback.

Following prior work on test-case prioritization [82], we also classify the existing test-case prioritization work according to the following aspects: algorithms, criteria, measurements, scenarios, constraints, and empirical studies. Fig. 2 shows the percentage of papers related to each aspect. Note that some

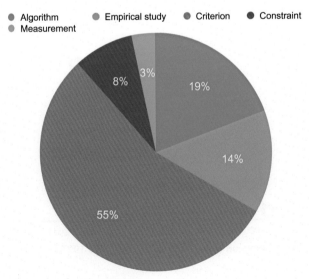

Fig. 2 Ratio of papers of each category.

papers cover multiple of those aspects, thus we categorize each paper based on the main contribution of the work. Also note that this figure does not show the percentage of papers on scenarios, because each test-case prioritization technique has to be evaluated on some specific scenario, such as version-specific test-case prioritization or general test-case prioritization (details shown in Section 7).

According to Fig. 2, more than half of papers focus on investigating criteria for prioritization, followed by the papers proposing prioritization algorithms and the papers on empirical studies. Accessing the fault-detection capability of each test case is always a big challenge and the key for prioritization problem, which is hard to obtain in practice. Fault-detection capability interacts with many other capability such as coverage capability, mutant-killing capability. Thus researchers always keep figuring out many different ways to represent or simulate the fault-detection capability, and plenty of test criteria are newly proposed each year. Since prioritization problem is an NP-hard problem, the algorithm to find the optimal solution among the solution space also matters. Many advanced algorithms in other field can also be adopted to solve the test-case prioritization problem, thus there are also a large number of papers investigating prioritization algorithms. Naturally, due to the large number of test-case prioritization approaches, the comparison between these approaches is also crucial for providing practical guidelines in regression testing, leading to the large number of empirical studies.

To further analyze the trend of each category in test-case prioritization, Fig. 3 further shows the number of papers belonging to each category per year. Consistent with the ratio results of Fig. 2, most of the papers published each year work on investing effective test criteria for test-case prioritization, indicating the researchers' effort in finding optimal test criteria to simulate the fault-detection capabilities of tests a cross the last two decades. Besides, since 2009, prioritization algorithms and empirical studies also attracted increasing attentions, indicating the switch of research interests in this area. We suspect the reason to be as follows. In the initial stage of test-case prioritization (i.e., in 1997), there were not many works in this area, and thus the researchers mainly focused on the core problem of finding suitable surrogates (i.e., various test criteria) for real fault-detection capabilities. Later on, when the test-case prioritization area became more mature since 2009, researchers began to spend more efforts on designing new prioritization algorithms. Meanwhile, due to the large number of emerging papers on test-case prioritization, practitioners often found it hard to find the optimal technique. Therefore, a large body of research has also been dedicated to empirically evaluating and comparing various test-case prioritization techniques.

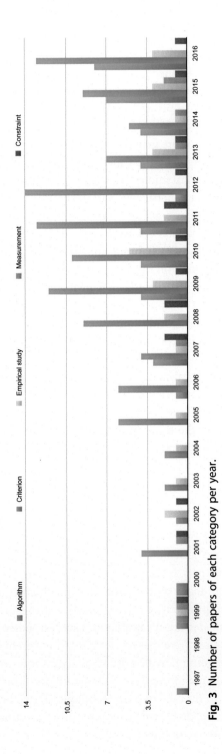

Fig. 3 Number of papers of each category per year.

In this survey, we discuss the development and future directions for each aspect of test-case prioritization in details. The remaining of this chapter is organized as follows. Sections 3–8 review test-case prioritization from the aforementioned six aspects, i.e., coverage criterion, prioritization algorithm, measurement, constraint, scenario, and empirical study. Section 9 discusses the challenges, issues, and future work in test-case prioritization, and finally Section 10 concludes this chapter.

3. CRITERION

Since it is hard to obtain the fault-detection capability of each test case directly in practice, various criteria are proposed to assess test-case fault-detection capability in prioritization. Besides test-case prioritization, criteria are also widely used in test generation, selection, and minimization [8, 17, 83].

Usually, large coverage criterion value means large probability of exposing faults in a program, and thus maximizing criterion values can be an intermediate goal of test-case prioritization. That is, criteria are actually used to guide the prioritization process. For example, branch-coverage-based prioritization [84] schedules the execution order of test cases based on the branch coverage of these test cases. Due to the importance of criteria, many of the existing work [34, 38, 60, 84–135] investigates their influence in the evaluation.

3.1 Structural Criterion

Among all criteria, structural coverage is the mostly used one. In particular, a structural coverage criterion is defined as the percentage of structural units covered by a test case [136–142]. For example, the widely used structural coverage criterion is statement coverage, which measures to what extent a test case covers statements during test-case execution. Higher statement coverage indicates larger fault-detection capability because without covering faulty statements a test case cannot reveal the corresponding faults.

Besides statement coverage [29], some other structural units like functions/methods [30], blocks [67], and modified condition/decision [32] have also been considered as a type of structural coverage criterion.

Interestingly, the experiment results in the work of Rothermel et al. [29, 30] showed that in most cases, branch coverage outperformed statement coverage using a set of C programs. But in more recent work of Lu et al. [68], statement coverage usually performs best among these coverage criteria on a set of real-world Java programs. One potential reason could be that branches are less prevalent for the object-oriented Java programs than the procedural C programs, making branch coverage ineffective for Java.

3.2 Model-Level Criterion

Though structural criteria are widely used, sometimes, structural coverage can be unavailable for black-box or can be quite expensive to obtain for large systems. System models can capture the different behaviors of a system and there are some modeling languages proposed to model state-based software systems. Recently, model-based techniques have been adopted in software testing, such as test-case generation [143, 144], test-suite reduction [145], and test-case prioritization [117, 146–150]

Korel et al. [147] presented a novel test-case prioritization based on state-based models which execution information of the original and modified models is used for retesting the modified software system. Furthermore, Korel et al. [146] further proposed several model-based test-case prioritization heuristics and empirically investigated the improvements of these heuristics strategies. Xu and Ding [117] proposed an aspect-related test-case prioritization based on the incremental testing paradigm. Aspects are incremental modifications to the base classes, thus the tests targeting the aspects would be selected to execute first for they are more likely to detect the failures.

3.3 Fault-Related Criterion

As testing criteria are usually used to measure the fault-detection capability of a test case or a test suite, some researchers presented some fault-detection criteria directly because the preceding code-based coverage criteria cannot sufficiently assess the capability of a test case or a test suite [151–156].

In particular, Rothermel et al. [84] introduced mutation score to represent each test case's fault-exposing-potential, which regarded mutation-killing-capability as fault-detection-capability. Elbaum et al. [157] used the fault-index to estimate the fault proneness for each program unit, which had been proved effective in previous work [158, 159]. The calculation process of fault-index was as follows: (1), each function was associated with some measurable attributes; (2), all attribute values were standardized according to a group of baseline values; (3), the set would be reduced to a smaller one by principal components analysis [160]; (4), the left values were represented by a linear function which could generate one fault-index for each function in the program.

Ma and Zhao [125] proposed a new prioritization index called testing-importance of module (TIM), which consisted of two factors: fault proneness and importance of module, which acted as a new metric to measure the severe fault proneness for module covered by test cases. Lou et al. [95] seeded mutation on the changed code between versions to imitate the real faults introduced during software evolution. Therefore, the capability of killing these mutants can represent the capability of detecting real faults to some extent.

3.4 Test Input-Based Criterion

Since the structural coverage, model information and mutation analysis can be costly to obtain, recently researchers started to measure the fault-detection capability of test cases based on the input data alone rather than the execution information of test cases. That is, this type of criteria measures the fault-detection capability by calculating the difference between test input data, which are usually regarded as strings or vectors.

In particular, Ledru [161] proposed a prioritization approach which compared the string distance between test cases with a greedy algorithm. Chen et al. [162] proposed a test-vector-based approach to prioritizing test programs for compilers by analyzing the extracted features of test programs to solve the efficiency problem of compiler testing [163]. Recently, Chen et al. [164] proposed to predict the bug-revealing probabilities per unit time of test programs for compilers via machine learning, and schedule the execution order of these test programs based on the descending order of these bug-revealing probabilities per unit time. Chen et al. [89] transformed test cases into a form of vectors for clustering. Jiang et al. [165] proposed a novel family of input-based prioritization techniques, which calculates the difference between test cases by three types of distance functions.

3.5 Change Impact-Based Criterion

Change information are also used very frequently in prioritization criteria [166, 167]. For example, the modified condition mentioned in structural unit level criteria also used the change information during program evolution. However, there is a category of criteria analyzing the change code in a more specific way, so this section introduce these criteria individually.

Haraty et al. [168] proposed a clustering prioritization approach based on code change relevance, which mainly prioritized clusters of test cases based on their relevance to code changes. Alves et al. [169] proposed a refactoring-based approach to prioritizing tests for detecting refactoring bugs. The approach first collected the change edits between two versions of a program and then analyzed the change impact based on a number of refactoring fault models to determine the execution order of test cases.

Panda et al. [170] presented a static analysis approach to prioritizing test cases based on affected component coupling of object-oriented programs. It first constructed affected slice graph whose nodes had different fault-proneness and then scheduled execution order based on the nodes covered by each test case.

3.6 Other Criteria

Besides, some studies are hard to categorize into aforementioned categories.

3.6.1 Risk

Hettiarachchi et al. [171] proposed a risk-based test case prioritization approach, which applied a fuzzy expert system to estimate the risks systematically for requirements and prioritized test cases based on the risks they involved.

3.6.2 Similarity

Fang et al. [100] proposed a similarity-based test case prioritization which transformed test case's execution profile into an ordered sequence of program entities and compared distance of the sequence of each test case.

3.6.3 Service History

Srikanth et al. [172] prioritized building acceptance test cases based on the service history data from several months, i.e., service interaction and historically failing services.

3.6.4 Requirement

Arafeen et al. [173] proposed a test-case prioritization approach which clustered test cases according to the requirement similarities in order to utilize requirements information.

4. PRIORITIZATION ALGORITHM

In this section, we introduce the algorithms used to guide test-case prioritization. Specifically, we classify the existing prioritization algorithms into several groups, i.e., greedy algorithm, search-based algorithm, information-retrieval-based algorithm, integrate-linear-programming-based algorithm, machine-learning-based algorithm. Moreover, when introducing the prioritization algorithms, we take the statement coverage criterion as the representative, although many following algorithms can be applied to various criteria, e.g., method coverage, branch coverage, and even advanced data-flow coverage criteria [84, 85].

4.1 Greedy Algorithm

Greedy algorithms are widely used to address the test-case prioritization problem, which focus on always selecting the current "best" test case during test-case prioritization. The greedy algorithms can be classified into two groups. The first group aims to select tests covering more statements, whereas the second group aims to select tests that is farthest from the selected tests.

Regarding to the first group, the most popular greedy algorithms are the total and additional algorithms. In particular, the total algorithm prioritizes test cases based on the descendent order of statements covered by each test case, whereas the additional algorithm prioritizes test cases based on the descendent order of statements that are covered by each unselected test case but uncovered by the existing selected test cases. As the total and additional algorithms can have best performance in different cases, Zhang et al. [174, 175] proposed a unified prioritization model, which uses a probabilistic model to bridge the gap between the total and additional algorithms so that the total and additional algorithms can be regarded as its two extreme instances. Moreover, this model yields a spectrum of specific prioritization algorithms between the total and additional algorithms. Besides, Li et al. [72] proposed the 2-optimal strategy which was based on K-optimal algorithm [176] where $K = 2$. Different from approaches mentioned above, the 2-optimal approach tries to select "next two best" test cases according to coverage ability of each pair of test cases.

Regarding to the second group, the typical greedy algorithm is adaptive random test-case prioritization [177], which is proposed based on adaptive random testing [178, 179]. In particular, it first iteratively generates a candidate set of test cases and selects one test case based on a selecting algorithm. The selecting algorithm aims to select a test case that is the farthest from the already selected test cases based on a distance definition function $f1$ and a farthest selection function $f2$. In particular, this work proposed to use Jaccard distance to define $f1$ and defined three types of selection function $f2$.

The greedy algorithms focus on searching the local optimal solution to prioritization, and thus their prioritization results may not be the optimal solution.

4.2 Search-Based Algorithm

Since the prioritization problem is an NP-hard problem, greedy algorithms can not always obtain the optimal solution within the solution space. Therefore, some search-based algorithms are applied to solve the prioritization problem, aiming to achieve better prioritization results with acceptable computation cost.

In particular, Li et al. [72] applied meta-heuristic search-based algorithms to test-case prioritization. That is, they applied steepest ascent hill-climbing and genetic algorithms. In particular, steepest ascent hill climbing is a local search algorithm, where each test permutation is regarded as a state. This algorithm iteratively switches to best state among all neighbors of the current state. The genetic algorithm [72] is based on the processes of natural selection according to Darwinian theory of biological evolution. In this algorithm, each test sequence is encoded in an N-sized array representing an instance of chromosome. In the initial step, a group of test sequences is generated randomly as the initial individuals. Iteratively, a new generation is generated by combining selected individuals guided by the fitness function. The whole search process will be terminated until certain requirement is satisfied.

Besides the traditional single-objective test-case prioritization, there is another form of test-case prioritization problem, called multiobjective test-case prioritization. Given a test suite T, the set of $T's$ permutations PT, and a vector of M objective functions, $f_i(i = 1, 2, ..., M)$, multiobjective test-case prioritization aims at finding $T' \subset PT$ such that T' is a Pareto-optimal permutation set with respect to the objective functions, $f_i(i = 1, 2, ..., M)$. The objective functions usually are some important prioritization criteria. Pareto-optimal means that strategy A improves strategy B without making things worse.

Epitropakis et al. [180] investigated multiobjective test-case prioritization through three objectives: average percentage of coverage, average percentage of coverage of changed code, and average percentage of past fault coverage and evaluated the fault-detection capabilities in the experiment.

Solving multiobjective problem in software engineering by multiobjective evolutionary algorithms usually faces with the challenge of scalability problem due to the population size and iterations. Therefore, Li et al. [181] proposed a novel GPU-based parallel fitness evaluation algorithm for test-case prioritization, which implemented the fitness evaluation and crossover computation by graphic processing units on GPU.

Overall, the characteristics of search-based prioritization algorithms lie in searching for the optimal solution guided by the predefined fitness function within the searching space.

4.3 Integrate-Linear-Programming-Based Algorithm

Integrated linear programming (abbreviated as ILP) is a mathematical optimization or feasibility program where all the variables, objective functions, and constraints are linear, which is an NP-hard problem. Recently, researchers applied ILP to describe the problem of test-case prioritization and thus the solutions to the ILP formula are the prioritization results. That is, the problem of test-case prioritization is transformed into formula construction and solving process.

In particular, Zhang et al. [82] firstly applied ILP to solve time-aware test-case prioritization. In particular, this approach first selects a set of test cases by solving the ILP formula describing time-aware test-case prioritization, and then prioritizes selected test cases through some greedy strategies. Recently, to investigate the bound of coverage-based test-case prioritization, Hao et al. [182] used ILP to represent coverage-based test-case prioritization so as to learn the performance of optimal coverage-based test-case prioritization techniques.

4.4 Information-Retrieval-Based Algorithm

Information retrieval (abbreviated as IR) techniques [183] aim to obtain information needed from a collection of information resources, which have been fully studied in the last 40 years and applied to various domains, including software engineering. The main idea of information-retrieval-based algorithm is as follows: (1) it uses test case information such as execution information or source code of each test case to construct the corresponding document collection for each test case, namely, each document represents one test case; (2) it uses source code information (usually the changed part of source code) serving as the input query of IR, and IR will return a ranked list of the documents constructed in the first step, which in fact is a ranked list of test cases by the relevance to the input information.

In particular, Nguyen et al. [184] proposed an IR-based approach to prioritizing test cases for web services, which used the identifier documents extracted from the execution trace to represent each test case and used the web service change description as the input query of IR.

Kwon et al. [185] proposed an IR-based approach which adapted term frequency (TF) and inverse document frequency (IDF) to prioritize test cases. This approach considers not only code coverage information but also how many times a coverage element is executed by a test case (TF) and source code elements are tested by few test cases (IDF). Linear regression model is applied to weigh the value of the information.

Later on, Saha et al. [186] proposed an IR-based approach to prioritize JUnit test cases. Their approach used the test source code to construct the relative document for each test case and used the changed code of the program under test as the input query to get a ranked list of test cases by their relevance to program changes.

4.5 Machine-Learning-Based Algorithm

Machine learning is a data-analysis technique that builds a model from sample input to make prediction for new data. Typically, machine learning techniques consist of supervised learning and unsupervised learning (called clustering as well).

Tonella et al. [133] presented a machine-learning-based test-case prioritization approach which incorporated user knowledge by case-based ranking model. This approach used the indicator of priority, which was defined by user cases, and test case information such as coverage and fault proneness metrics as features to train a model to predict the priority of test cases. Chen et al. [162] proposed a test-vector based approach to prioritizing test programs for compilers, which did not need to collect coverage information but only analyze necessary features from each test program itself to prioritize test programs for compilers. More recently, Chen et al. [164] developed LET (short for learning to test), which learned from existing test programs to accelerate future test execution. LET first designed and extracted a lot of features from the source code of test programs (e.g., address features and pointer comparison features). Then, LET trained a capability model to predict the bug-revealing probability of each new test program, and a time model to predict the execution time of each new test program, based on these features. Finally, LET prioritized new test programs as the descending order of their bug-revealing probabilities in unit time.

5. MEASUREMENT

To access the performance of test-case prioritization techniques, it is necessary to propose a measurement for test-case prioritization, including efficiency and effectiveness.

With regard to the efficiency of test-case prioritization, researchers usually use the complexity analysis of a prioritization algorithm to measure its cost. For example, Elbaum et al. [157] analyzed that the complexity of the statement-coverage-based total prioritization technique is $O(mn + mlogm)$ and the complexity of the statement-coverage-based additional prioritization technique is $O(m^2n)$, where m represents the number of test cases and n represents the number of statements in a program.

With regard to the effectiveness of test-case prioritization, most of the existing work uses the average of percentage of faults detected (abbreviated as APFD). Besides, as this measurement suffers from the widely known problems, e.g., ignoring the impact of testing time and fault severities, many researchers further improved this measurement accordingly. In the following, we briefly introduce the measurements used in test-case prioritization.

5.1 APFD

Rothermel et al. [29] proposed the first measurement for assessing the effectiveness of test-case prioritization, which is called weighted average of the percentage of faults detected (APFD). APFD measures how rapidly a prioritized test suite detects faults. Higher APFD values mean faster fault-detection rates. Formula (1) shows how to calculate APFD values for a test-case prioritization technique. In this formula, TF_j refers to the first test case in prioritized test suite that detects the jth fault, n refers to the number of test cases, and m refers to the number of faults detected by the test suite. APFD has already become one of the most widely used measurements for assessing the performance of test-case prioritization in the literature [71].

$$\text{APFD} = 1 - \frac{\sum_{j=1}^{m} TF_j}{nm} + \frac{1}{2n} \tag{1}$$

5.2 AFPD$_C$

Actually, APFD does not reflect the practical performance of test-case prioritization, since it ignores the influence of test execution costs and fault severity. Therefore, Elbaum et al. [33] further proposed another measurement to measure the practical performance of test-case prioritization by considering the influence of the two factors, which is called cost-cognizant weighted average percentage of faults detected ($APFD_C$). $APFD_C$ is actually adapted from APFD, which is defined as Formula (2). In this formula, f_i refers to the severity of the ith fault detected by the prioritized test suite, and t_j refers to the test cost of the jth test case in the prioritized test suite.

$$APFD_C = \frac{\sum_{j=1}^{m}\left(f_i * \left(\sum_{i=TF_j}^{n} t_i - \frac{1}{2} * t_{TF_j}\right)\right)}{\sum_{j=1}^{n} t_j * \sum_{j=1}^{m} f_j} \tag{2}$$

In practice, it tends to be quite difficult to know the severity of each fault in advance. Therefore, a simplified $APFD_C$ is usually used to measure the performance of test-case prioritization by treating all faults as sharing the same severity [180]. The simplified $APFD_C$ is shown as Formula (3).

$$APFD_C(\text{simplified}) = \frac{\sum_{j=1}^{m}\left(\sum_{i=TF_j}^{n} t_i - \frac{1}{2} * t_{TF_j}\right)}{\sum_{j=1}^{n} t_j * m} \tag{3}$$

5.3 APXC

In order to measure the performance of test-case prioritization before test-case execution, researchers [72, 182] also proposed to leverage the average percentage of some structural coverage (abbreviated as APXC) as a measurement. APXC has the similar formula with APFD. For APXC, TF_j in Formula (1) refers to the first test case in prioritized test suite that covers structural units (e.g., statement and block) j, and m refers to the total number of structural units covered by the test suite. In particular, higher APXC values mean faster coverage rates.

According to the general definition of APXC, we may have APBC to measure the rate at which a prioritized test suite covers the blocks, APSC to measure the rate at which a prioritized test suite covers the statements. Such measurements are defined based on structural units, which are not the ultimate goal. Therefore, they are actually widely used as an intermediate goal (e.g., fitness function) during search-based test-case prioritization to guide test-case prioritization, rather than as a measurement for the performance of test-case prioritization.

5.4 WGFD

Higher APFD values mean faster fault detection. However, the problem is how to define "fastness." In different testing scenarios, "fastness" tends to have different definitions. Therefore, Lv et al. [187] proposed a new generalized measurement from a control theory viewpoint, which is called the weighted gain of faults detected (WGFD). The basic idea is to weight and sum fault-detection rates of different test cases so as to define "fastness" in different testing scenarios. That is, since the number of test cases detected at different time should have different impact on measuring the performance (fastness) of a prioritization technique, different weights should be assigned to the fault-detection rates of different test cases. In particular, WGFD is defined in Formula (4), where n refers to the number of test cases in the test suite, $r(i)$ refers to the fault-detection rate of test case i, and $w(i)$ refers to the assigned weight to the fault-detection rate of test case i.

$$\mathrm{WGFD} = \sum_{i=1}^{n} w(i) * r(i) \tag{4}$$

5.5 HMFD

According Formula (1), the APFD measure increases as the size of the test suite increases. In other words, APFD is affected by the size of a given test suite. To relieve this issue of APFD, Zhai et al. [99] proposed a new measurement to measure how quickly a prioritized test suite can detect faults, which is independent from the size of a given test suite. The new measurement is called the harmonic mean of the rate of fault detection (HMFD). In particular, HMFD is defined as Formula (5), where TF_j refers to the first test case in the prioritized test suite that detects the ith fault, and m is the number of faults detected by the test suite. Note that low HMFD values mean better performance of test-case prioritization.

$$\mathrm{HMFD} = \frac{m}{\sum_{j=1}^{m} \frac{1}{TF_j}} \tag{5}$$

5.6 NAPFD and RAPFD

In practice, there may be various constraints in test-case prioritization. Due to the existence of practical constraints in test-case prioritization, not all of the faults can be detected by a given test suite. Moreover, we may not execute the same number of test cases. Walcott et al. [188] proposed to assign a penalty to the missing faults so as to solve the first problem. In addition, Qu et al. [189] proposed normalized APFD (abbreviated as NAPFD) to measure the performance of test-case prioritization in order to solve the two problems. In particular, NAPFD is defined as Formula (6), where p refers to the value that is calculated by dividing the number of faults detected by the prioritized test suite by the number of faults detected by the full test suite. To further improve these measurements, Wang and Chen [190] proposed the relative average percent of faults detected (RAPFD) by considering the given testing resource constraint, which determines how many test cases could be run. Furthermore, Do and Rothermel [191, 192] further proposed many improved cost–benefit models for assessing regression testing methodologies (including test-case prioritization). In particular, these models incorporate context factors (e.g., the costs of some essential testing activities such as test setup and obsolete test identification) and lifecycle factors (e.g., the costs and benefits for techniques across system lifetimes).

$$\text{NAPFD} = p - \frac{\sum_{j=1}^{m} TF_j}{nm} + \frac{p}{2n} \tag{6}$$

6. CONSTRAINT

As a practical problem, test-case prioritization tends to suffer from various practical constraints. Therefore, many studies investigated how to prioritize test cases when considering practical constraints [88, 121, 127, 188, 193].

6.1 Time Constraint

The mostly studied constraint in test-case prioritization is the time constraint, also called time budget [188]. Ideally, all the test cases in the prioritized test suite are expected to be executed during the process of software testing, so as to avoid fault-detection capability loss of the test suite. However, under the practical environment of software testing, the allowed testing time may not be quite sufficient, which causes that the prioritized test suite may not be totally executed. For example, in some companies, software testing is just allowed in night [188, 194], and thus if the time of executing the whole test suite is more than one night, some prioritized test cases will not be executed. Besides, new software development processes, e.g., extreme programming, also advocate a short testing cycle. Therefore, on this occasion, the time constraint is quite necessary to be considered when prioritizing test cases.

To make test-case prioritization more effective given the allowed testing time, various approaches have been proposed to select only a subset of test cases and schedule their execution order rather than all the test cases. Walcott et al. [188] proposed time-aware test-case prioritization. More specifically, they used a genetic algorithm to prioritize test cases in order to achieve two goals. The first goal is to ensure that the prioritized test cases can be executed within the given testing time. The second goal is to make the prioritized test cases achieve the largest fault-detection capability. To achieve the same goals, Alspaugh et al. [195] proposed to use 0/1 knapsack solvers to prioritize test cases, including greedy, dynamic programming, and the core algorithms. Zhang et al. [82] identified that time-aware test-case prioritization implied to select a subset of test cases from the test suite for prioritization. Therefore, they proposed to combine test-case selection and test-case prioritization to achieve the goals of time-aware test-case prioritization. More specifically, they first used integer linear programming [196] to select a subset of test cases that can achieve the maximum test coverage within the time budget, and then applied traditional test-case prioritization techniques to schedule the execution order of the selected test cases. Note that, in this way, the traditional total technique and the traditional additional technique are both adapted to be time-aware total technique and time-aware additional technique. Later on, Suri et al. [197] also applied ant colony optimization to prioritize test cases in the time constraint environment.

Based on the existing research [188, 191], considering the time constraint in test-case prioritization may influence the costs and benefits of test-case prioritization techniques. Do et al. [198] conducted a series of experiments to investigate such influence. Their experimental results demonstrated that the time constraint indeed has a significant influence on the cost-effectiveness of test-case prioritization techniques. Furthermore, You et al. [199] conducted an empirical study to investigate whether the time cost of each test case influences the effectiveness of time-aware test-case prioritization. Their experimental results showed that the effectiveness of the prioritization techniques considering the time cost of each test case has no significant difference with that of the prioritization techniques omitting the time cost of each test case.

That is, it tends to be not worth considering the time cost of each test case for time-aware test-case prioritization. In addition, Marijan [38] proposed a framework for optimal test-case prioritization in the time constraint environment by integrating three different perspectives, including business perspective, performance perspective, and test design perspective. More specifically, from a business perspective, failure impact is regarded as an important factor influencing test effectiveness; from a performance perspective, test execution time is regarded as an obvious factor of test effectiveness; from a technical perspective, both failure frequency and cross-functionality are regarded as important factors of test effectiveness. In particular, failure frequency refers to a measure of how often test cases detect failures, and cross-functionality refers to a measure of how much the functionality of the system under test is covered by a test case.

6.2 Fault Severity

Another widely studied constraint in test-case prioritization is fault severity. The fault severity reflects the costs or resources required if a fault persists in and influences the users/organization/developers. The existing test-case prioritization is based on the assumption that the severity of all the faults are considered equally. However, the assumption may not hold in practice, and thus the fault severity is also a practical constraint for test-case prioritization.

Elbaum et al. [33] firstly considered the fault severity constraint when measuring the effectiveness of test-case prioritization techniques. Park et al. [200] proposed to prioritize test cases by considering fault severity. In particular, they estimate the current fault severity using history information. Actually, their approach has an assumption, i.e., test costs and fault severities are not largely changed from one version to a later version. Malishevsky et al. [201] adapted traditional test-case prioritization (e.g., the total technique and the additional technique) to cost-cognizant test-case prioritization by considering the fault severity constraint and the time cost of each test case. Huang et al. [108] also proposed a history-based cost-cognizant test-case prioritization. More specifically, their approach collected the historical records from the latest regression testing and then used a genetic algorithm to schedule the most effective execution order of test cases.

6.3 Other Constraints

Besides, resource (e.g., hardware resource) is also a constraint in test-case prioritization. Kim and Porter [193] proposed a test-case prioritization based on history information by considering the resource constraint and time constraint. That is, they assigned a selection probability for each test case based on history information, and selected a test case to run based on these probabilities until testing time is exhausted. More specifically, their utilized history information contains the execution history of each test case, the corresponding fault detection, and/or the covered program entities. Wang et al. [88] proposed a resource-aware multiobjective optimization solution to produce an optimal execution order of test cases by considering the resource constraint and the time constraint. In the multiobjective optimization solution, they defined a fitness function based on four cost-effectiveness measures, including (1) minimizing the time for executing prioritized test cases and allocating relevant test resources; (2) maximizing the number of test cases to be executed; (3) maximizing the usage of available test resources; and (4) maximizing fault-detection achieved by prioritized test cases.

Furthermore, there are some other constraints, e.g., testing requirement priorities and the request quotas of web service. To prioritize test cases by considering testing requirement priorities, Zhang et al. [127] proposed to utilize test history information to evaluate the priorities of test cases so as to prioritize test cases based on them. Here various types of code elements can be regarded as testing requirements, e.g., statements, basic blocks, methods; or features and attributions of system; or faults in system. About the constraint of the request quotas of web service (e.g., the upper limit of the number of requests that a user can send to a Web Service during a certain time range), Hou et al. [121] proposed quota-constrained test-case prioritization for service-centric systems by maximize testing requirement coverage. More specifically, they first divided the testing time into time slots, and then selected and prioritized test cases for each slot by using integer linear programming.

7. APPLICATION SCENARIO

Test-case prioritization aims to speed up fault detection for the new software version during software evolution. Balancing the overhead and effectiveness, two different application scenarios have been explored—(1) *general* test-case prioritization and (2) *version-specific* test-case prioritization.

General test-case prioritization techniques [68–70, 84, 202] usually compute the optimal test order once for one revision, and then reuse that test order for a number of subsequent revisions. On the contrary, version-specific test-case prioritization techniques [66, 95, 146, 186] compute the optimal test order right before each revision in order to achieve effective test-case prioritization. While version-specific test-case prioritization may achieve more precise results, it may incur higher overhead due to the frequent test-case prioritization runs. In this section, we discuss the details for such two application scenarios.

7.1 General Test-Case Prioritization

Given a program P and its corresponding test suite T, general test-case prioritization [68–70, 84, 202] computes test execution order valid for a number of subsequent modified revisions of P. Therefore, they are usually based on general program/test information shared by various revisions, e.g., the set of program elements covered by each test.

For example, if test t_1 covers more program elements than t_2 on one program revision, the same may still hold for later program revisions. Therefore, traditional test-case prioritization techniques based on coverage information, e.g., the *total/additional* [84, 157], *adaptive-random-testing-based* [177], and *search-based* techniques [72], can all be directly utilized for general test-case prioritization.

When prioritizing using coverage information obtained from historical revisions, software changes and test additions could make test-case prioritization techniques ineffective since coverage information can be obsolete (due to software changes) or absent (for newly added tests) during the software evolution. To study the impacts of software changes and test additions for general test-case prioritization, Lu et al. [68] recently performed a study on real-world evolving GitHub projects. The study results demonstrate that software changes do not impact general test-case prioritization much, whereas test additions, which incur tests without coverage information, may significantly impact the effectiveness of general test-case prioritization. The study provides practical guidelines for determining the intervals of applying general test-case prioritization—general test-case prioritization should be reapplied whenever there are nontrivial number of added tests.

7.2 Version-Specific Test-Case Prioritization

Given a program P and its corresponding test suite T, version-specific test-case prioritization [66, 95, 146, 186] computes optimal test execution orderings specifically for P', the next revision of P. Version-specific test-case prioritization is performed after changes have been made to P and prior to regression testing of P'. The prioritized test suite may be more effective for testing P' than that computed by general test-case prioritization, but may be inferior on average on a succession of subsequent releases of P.

In the literature, researchers have also applied traditional coverage-based test-case prioritization techniques to the version-specific scenario. Furthermore, since regression faults are mainly due to software changes, researchers have also proposed various version-specific test-case prioritization techniques [66, 95, 146, 186] based on the detailed change information during software revision for more effective test-case prioritization. For example, Srivastava and Thiagarajan [66] analyzed the binary-level basic block changes to execute tests covering more changes earlier for faster regression fault detection. Korel et al. [146] analyzed the system models and computed model-level modifications for precise version-specific test-case prioritization. Lou et al. [95] presented a mutation-based version-specific test-case prioritization technique, which simulates faults occurred in software evolution by mutants on the change and prioritizes test cases based on their killing information on these simulation faults. Recently, Saha et al. [186] transformed the version-specific test-case prioritization problem into an information retrieval problem by treating source-code level changes as queries and test-case source code as documents. Then, the tests with more textual similarities with software changes are executed earlier to detect regression bugs faster.

8. EMPIRICAL STUDY

Due to the large number of existing test-case prioritization techniques, it can be hard to make the right/optimal choices in practice. Therefore, researchers have also performed various studies on test-case prioritization techniques to provide practical guidelines for test-case prioritization.

8.1 Studies on Traditional Dynamic Prioritization

Due to the dominant position of traditional dynamic test-case prioritization techniques, the vast majority of studies explore various factors around these techniques.

Rothermel et al. [84] empirically compared various dynamic test-case prioritization techniques (including coverage-based and mutation-based techniques) against unordered or randomized test suites on a suite of C programs. Later on, Elbaum et al. [33] further studied the impacts of fault severities and test execution time on test-case prioritization. Elbaum et al. [157] also investigated the impacts of program versions, program types, and different coverage granularities on test-case prioritization on C programs. Do et al. [203]

performed the first study of test-case prioritization on JUnit tests for Java programs. The study demonstrated the effectiveness of dynamic test-case prioritization on Java programs besides C programs, and also revealed divergent behaviors of test-case prioritization on Java and C programs. Do et al. [198] also investigated the effect of time constraints on the cost-effectiveness of test-case prioritization, as well as demonstrating the validity of using mutation faults for test-case prioritization experiments [156, 204]. Recently, Lu et al. [68] investigated the impacts of real-world software evolution on test-case prioritization and found that code changes do not impact the effectiveness of test-case prioritization much while test additions can significantly lower the effectiveness of traditional dynamic test-case prioritization.

In terms of effectiveness, various studies have confirmed that the traditional additional [84] and search-based [72] test-case prioritization techniques represent the state of the art [68, 72, 174, 177].

8.2 Comparison With Traditional Dynamic Techniques

Besides the traditional dynamic test-case prioritization techniques, researchers have also proposed various other test-case prioritization techniques. In the next, we present two recent but important studies comparing traditional dynamic test-case prioritization with other static or black-box techniques.

8.2.1 Dynamic vs Static

Traditional dynamic test-case prioritization techniques [72, 84, 177] mainly reply on dynamic execution information (e.g., statement or method coverage) to prioritize tests. Although effective, they may not be suitable for all the cases. For some software systems, it may not be possible to collect dynamic execution information via code instrumentation, e.g., code instrumentation may interrupt normal test run for real-time systems. For some software systems, dynamic execution information may not be always precise, e.g., code with concurrency and randomness. Even it is possible to collect precise dynamic execution for some software systems, dynamic instrumentation may incur high overhead, e.g., even the coarse file/class-level dynamic information may incur $8\times$ slowdown for commons-math [205]. Finally, the dynamic execution information may not always be available on the old version [137, 206]. Therefore, Zhang et al. [137] firstly proposed to use static analysis to simulate the dynamic execution information. More specifically, they used the static call graph information of each test to simulate the method-level coverage of the test, since the static call graph is always a superset of the actual method coverage. Later on, Mei et al. [206] further extended the call-graph-based test prioritization techniques via considering the method body information. Ledru et al. [161] directly treated each test (e.g., test source code or test input) as a string and prioritized tests to maximum string distances of the executed tests. The main insight is that executing more diverse tests may have higher probability to detect unknown regression bugs. Thomas et al. [148] found that simply treating each test as a string may include useless terms while missing important latent terms of the test. Therefore, they proposed to further use topic model to infer the latent semantic representation of each test. Then, they computed the string distances between test semantic representations, and prioritized tests to execute more diverse tests.

Although various static test-case prioritization techniques have been proposed, there lack extensive studies comparing different static techniques as well as comparing static techniques against dynamic techniques. For example, the call-graph-based techniques [137, 206] were not compared against other static techniques since there were no other static techniques before, while the more recent topic-model-based technique was only evaluated using only two subject systems. Therefore, recently, Luo et al. [70] performed an extensive study on state-of-the-art static and dynamic test-case prioritization techniques using 30 modern real-world GitHub projects. The study results show that the call-graph-based techniques outperform all the studied dynamic and static techniques at the test-class level, while the topic-model-based technique performs better than other static techniques but worse than two dynamic techniques at the test-method level. The call-graph-based techniques have also been shown to incur the lowest prioritization overhead among all the static techniques. Overall, while almost all techniques perform better at the test-method level, the static techniques perform comparatively worse to dynamic techniques at the test method level as opposed to the test class level. Finally, the study results show that there is minimal overlap between the detected faults by the static and dynamic techniques, e.g., top 10% prioritized tests only share less than 30% of detected faults, indicating a promising future for applying static and dynamic test-case prioritization in tandem.

8.2.2 Block-Box vs White-Box

Since the first proposal of test-case prioritization two decades ago [29, 67], white-box test-case prioritization techniques have been intensively studied. Such white-box techniques rely on the source code or dynamic execution information (obtained via code instrumentation) of the program under test to perform effective test-case prioritization. However, such techniques may not be applicable when the program source code and dynamic execution information are not accessible or available. Furthermore, white-box techniques can be expensive due to the collection of dynamic execution information [137, 206]. Therefore, researchers have also proposed black-box test-case prioritization techniques which do not require accessing source code or performing code instrumentation. Bryce and Colbourn [130, 134] proposed the first black-box test-case prioritization technique inspired by combinatorial interaction testing (CIT). Based on the test input information, they adopted a "one-test-at-a-time" greedy approach to prioritize test cases to achieve high pair-wise interactions of the test inputs faster. Bryce et al. [115, 207] later used t-wise interaction from CIT to prioritize test cases for GUI applications. Qu et al. [208, 209] also used the notion of CIT to prioritize tests for the highly configurable software systems (e.g., software product lines). Henard et al. [210] recently proposed a search-based technique to prioritize the configurations for testing highly configurable software systems based on CIT.

Due to the presence of various black-box and white-box test-case prioritization techniques, it can be hard for the developers or testers to choose the right technique. Therefore, recently, Henard et al. [69] systematically studied and compared the existing white-box and black-box test-case prioritization techniques. They studied 20 state-of-the-art test-case prioritization techniques, including 10 white-box techniques and 10 black-box techniques. The study was performed on six real-world C programs, widely used in prior work on test-case prioritization. The study results reveal a number of practical guidelines. First, the CIT and diversity-based techniques perform the best among all studied black-box test-case prioritization techniques. Second, although white-box techniques outperform black-box techniques for the majority of the cases, surprisingly, the performance (in terms of APFD) difference between white-box and black-box techniques is negligible, e.g., at most 4% APFD difference. Third, the overlap between the faults detected by the black-box and white-box techniques tend to be high: the first 10% prioritized tests agree on over 60% of the detected faults. Overall, the study provides practical guidelines that the developers or testers who may not have source code information available can use black-box test-case prioritization as a reliable substitute of white-box test-case prioritization.

9. SOME DISCUSSIONS

In this section, we first discuss existing issues in test-case prioritization following the previous classification and then point out some other challenge problems and potential future work in test-case prioritization.

9.1 Existing Issues

In this section, we discuss the existing issues in test-case prioritization through three aspects—criteria, measurements, and empirical studies.

9.1.1 Criteria

Testing criteria are used to guide the selection of test cases in test-case prioritization. Most of the widely used testing criteria can be classified into two categories, structural coverage-based criteria and mutation-based criteria. However, these two types of testing criteria are either less precise or costly. In particular, the structural coverage criteria (e.g., statement coverage or branch coverage) actually measure the percentage of code elements (e.g., statements or branches) covered by a test case or a test suite. That is, these coverage criteria measure the effectiveness of only test input, ignoring test oracle [211, 212]. Therefore, such a type of criteria is less precise. On the contradictory, mutation-based criteria tend to measure the effectiveness of a test case or a test suite based on the output of the program. Therefore, mutation-based criteria consider both test input and test oracle, which seem to have higher precision than coverage-based criteria. However, mutation testing suffers from the widely known cost issue. To sum up, neither structural coverage-based criteria nor mutation-based criteria are good enough serving as testing criteria, and thus another precise but less costly testing criterion is needed. Recently, Zhang et al. [213] proposed predictive mutation testing (PMT). The approach built predictive models based on a series of lightweight features related to mutants and tests, and predicts mutant execution results without executing the mutants. It greatly reduces the cost of mutation testing while incurring only minor loss of accuracy, which may provide effective but efficient supports for future test-case prioritization.

9.1.2 Measurement

First, the effectiveness measurement taken by the existing work has obvious flaws. In the past, most of the existing work evaluated test-case prioritization techniques based on APFD [84]. However, APFD assumes that all the tests have the same execution time and treats them equivalently, which is usually not true in practice. For example, for project MapDB [214], the test with the longest running time spends $8.8*10^5 X$ more time than that with the shortest running time. To address this measurement issue, Elbaum et al. [33] proposed a cost-cognizant version of APFD, $APFD_C$, which considers different test costs and fault severities. Since fault severities can be hard to determine in practice, Epitropakis et al. [180] simplified this measurement by assuming all faults have the same severity. We encourage researchers to evaluate future test-case prioritization work using $APFD_C$ or simplified $APFD_C$ to explicitly consider test execution time. Meanwhile, $APFD_C$ may also not be suitable for all cases, since its values are influenced by various factors like the number of tests, the number of faults. Therefore, it is hard to use the values of such measurements to explain the effectiveness of a prioritization technique in different cases. To illustrate, we can hardly tell whether a prioritization technique whose $APFD_C$ value is 0.7800 is good or not for a particular test suite. Furthermore, such measurements do not explicitly consider the actual switching costs between test executions (e.g., time to load and schedule the next test). In the future, we suggest researchers to also consider measuring test-case prioritization techniques based on the actual time spent on fault detection, e.g., TTFF (time to detect the first fault) and TTLF (time to detect the last fault), since such measurements precisely measure the actual time cost during regression testing.

Second, the efficiency measurement is mostly ignored in test-case prioritization, although its results influence the usage of test-case prioritization techniques. In the past, the efficiency of test-case prioritization is mostly evaluated through complexity analysis rather than the actual prioritization time. However, the complexity of some prioritization algorithms (e.g., genetic algorithm [72]) can be hard to estimate. Furthermore, although the time complexity of some algorithms (e.g., integer linear programming-based algorithm [82]) is large, their actual prioritization time may be acceptable since the test-case prioritization process is usually performed *offline* beforehand, i.e., before the new version is ready. On the other hand, the efficiency of test-case prioritization can also be crucial for some cases (e.g., version-specific test-case prioritization). In such cases, test-case prioritization is usually performed *online* (e.g., after the new version is ready), making it unbearable when the prioritization time is close to the time spent on test-case execution. Therefore, it is necessary to study the end-to-end testing time (i.e., including the prioritization time and the test execution time) for the online test-case prioritization techniques.

Finally, besides the prioritization cost, it is also important to measure the cost on collecting the necessary data required by test-case prioritization techniques. Most prioritization techniques require extra information besides test cases (e.g., structural coverage) for test-case prioritization. Apparently, obtaining such information may occur extra cost. However, many studies simply take the information as given and do not report the collection cost. In particular, some prioritization techniques require structural coverage

[84], static coverage [206], or mutation execution information on some early version [95]. Although such information is usually collected offline, i.e., before test-case prioritization, it still consumes computing resources and should be measured to provide practical guidelines.

9.1.3 Empirical Studies

In the literature, existing empirical studies investigated the various factors (e.g., programming languages [203], coverage granularity and type [203, 206], fault type [156, 204], test granularity [95, 206], and constraints [192]) that may influence the effectiveness and efficiency of test-case prioritization. Besides these factors, it is also important to investigate the following (but not limited to) factors.

Some experimental factors have been recognized as threats in the past, e.g., subjects, faults, and test cases, but they are seldom studied. For example, subjects are a widely recognized external factor, but the early work of test-case prioritization (especially the papers published around 2000) mostly used the seven small programs (whose number of lines of code is smaller than 600) in Siemens as the subjects. Fortunately, this threat is reduced to some extent after 2000, because researchers started to use larger projects, e.g., *grep* and *gzip* whose number of lines of code is about 10,000. Furthermore, most prior work uses mutation faults or seeded faults, which may be a nonnegligible threat, since there might be some gap between mutants and real faults during software evolution. In other words, we suggest considering using real regression faults in test-case prioritization.

Besides these well-recognized threats, researchers started to notice the difference between practice and existing experimental setup of test-case prioritization. For example, recently Lu et al. [68] identified another one important flawed setting in the existing evaluation, evolution of source code and test cases. That is, previous work on test-case prioritization is usually evaluated based on the source code and test cases with artificial changes simulated via mutation testing, which do not represent real software evolution. Lu et al. [68] investigated the influence of this factor on the effectiveness of many existing general prioritization techniques, and found that changes on source code do not have much influence on the effectiveness of test-case prioritization, but changes on test code (e.g., test additions) do have.

9.2 Other Challenging Problems

Besides these issues in the current work, test-case prioritization, test-case prioritization also suffers from other challenging problems.

9.2.1 Intermediate/Ultimate Goal

Test-case prioritization has been studied for long, and a large number of prioritization techniques have been proposed and investigated in the literature. However, most of the prioritization techniques are less effective than the simple greedy algorithm, such as the additional algorithm, resulting from the difference between the ultimate goal and the intermediate goal of test-case prioritization. In particular, as the ultimate goal of test-case prioritization can hardly serve to guide prioritization, existing prioritization techniques actually use an intermediate goal instead, and thus these "well-designed" prioritization techniques do not optimize the execution order of test cases in terms of the ultimate goal. In recent years, researchers in test-case prioritization started to notice this fact [72] and investigated this fact [182]. Unfortunately, no work in the literature actually solves this problem, and it becomes a fundamental challenge for test-case prioritization. In the future, researchers should investigate other intermediate goals (e.g., detection of mutation faults or detection of similar real faults), which have closer relationship with the ultimate goal rather than the existing intermediate goals (e.g., structural coverage).

9.2.2 Practical Values

Test-case prioritization is a practical problem raised from industry, and thus it is important to study test-case prioritization in practice.

Test-case prioritization aims to facilitate fault detection in software testing, and thus it brings more benefits when the time spent on test-case execution is not ignorable (e.g., several days or months). In other words, when the total execution time of all test cases is small (e.g., several minutes), it does not matter so much whether a fault is detected by the first test case or the last test case. However, to our knowledge, most of the existing research work is actually evaluated on the subjects whose total execution time of test cases is not large at all. That is, the existing techniques are not evaluated in its most possible application scenario. In other word, to facilitate practical usage, it is necessary to investigate test case prioritization in a proper practical scenario.

Besides, test-case prioritization may have variants besides its default setting. Traditionally, test-case prioritization aims to address the test effectiveness problem when the total execution time of test cases are long. However, in practice, it may be costly to run an individual test case. In particular, a test suite may consist of only several test cases, each of which consumes long execution time. Therefore, it is also interesting to study how to optimize the execution of an individual test case, e.g., transferring a test case with long execution time to several test cases with short execution time by modifying its components (e.g., test input data). Apparently, this problem is different from the existing prioritization problem, and thus a totally new method for this problem is needed.

Furthermore, surprisingly, to the best of our knowledge, although test-case prioritization techniques have been studied for decades, there still lack practical test-case prioritization tools that are effective and easy to use. For example, despite the large number of papers on JUnit test-case prioritization, there is no practical test prioritization technique fully integrated with JUnit. To demonstrate the practical value of test-case prioritization, we encourage the researchers to provide practical tool supports on test-case prioritization in the near future.

In summary, although test-case prioritization has been studied for decades, it is yet not fully explored and evaluated, leaving many future work in this promising area. In addition, to gain practical impacts, we encourage researchers to investigate this problem in real practical scenarios and provide practical tool supports.

10. CONCLUSION

To alleviate the cost of regression testing, test-case prioritization is proposed, which aims to achieve some testing requirements by scheduling the execution order of test cases. This domain has been studied for decades and dedicated efforts have been made accordingly. In this work, we conduct a survey to systematically investigate the existing work on test-case prioritization.

More specifically, in this survey, we review the existing work by classifying them into six categories: algorithms, criteria, measurements, constraints, scenarios, and empirical studies. Based on these analyses, we further discuss challenges, issues, and future opportunities in test-case prioritization.

ACKNOWLEDGMENTS

This work is supported in part by NSF Grant No. CCF-1566589, UT Dallas faculty start-up fund, Google Faculty Research Award, Samsung GRO Award, and generous supports from Huawei, the National Key Research and Development Program 2016YFB1000801, and the National Natural Science Foundation of China under Grant No. 61522201.

REFERENCES

[1] S. Yoo, M. Harman, Regression testing minimization, selection and prioritization: a survey, Softw. Test. Verification Reliab. 22 (2) (2012) 67–120.
[2] G. Rothermel, M.J. Harrold, J. Von Ronne, C. Hong, Empirical studies of test-suite reduction, Softw. Test. Verification Reliab. 12 (4) (2002) 219–249.

[3] J. Zhang, Y. Lou, L. Zhang, D. Hao, L. Zhang, H. Mei, Isomorphic regression testing: executing uncovered branches without test augmentation, in: Proceedings of the 2016 24th ACM SIGSOFT International Symposium on Foundations of Software Engineering, ESE, 2016, pp. 883–894.

[4] P.K. Chittimalli, M.J. Harrold, Recomputing coverage information to assist regression testing, IEEE Trans. Softw. Eng. 35 (4) (2009) 452–469.

[5] S. Elbaum, G. Rothermel, J. Penix, Techniques for improving regression testing in continuous integration development environments, in: FSE, 2014, pp. 235–245.

[6] Testing at the speed and scale of Google, 2011, http://goo.gl/2B5cyl.

[7] Tools for Continuous Integration at Google Scale, 2011, https://goo.gl/Gqj7uL.

[8] M.J. Harrold, R. Gupta, M.L. Soffa, A methodology for controlling the size of a test suite, ACM Trans. Softw. Eng. Methodol. 2 (3) (1993) 270–285.

[9] J. Pan, L.T. Center, Procedures for reducing the size of coverage-based test sets, in: Proceedings of International Conference on Testing Computer Software, 1995.

[10] J. Black, E. Melachrinoudis, D. Kaeli, Bi-criteria models for all-uses test suite reduction, in: Proceedings of the 26th International Conference on Software Engineering, IEEE Computer Society, 2004, pp. 106–115.

[11] J. Chen, Y. Bai, D. Hao, L. Zhang, L. Zhang, B. Xie, How do assertions impact coverage-based test-suite reduction? Proceedings of the 10th International Conference on Software Testing, Verification and Validation, 2017, pp. 418–423.

[12] T.Y. Chen, M.F. Lau, A new heuristic for test suite reduction, Inf. Softw. Technol. 40 (5–6) (1998) 347–354.

[13] D. Jeffrey, N. Gupta, Test suite reduction with selective redundancy, in: IEEE International Conference on Software Maintenance, IEEE, 2005, pp. 549–558.

[14] S. Sprenkle, S. Sampath, E. Gibson, L. Pollock, A. Souter, An empirical comparison of test suite reduction techniques for user-session-based testing of web applications, in: IEEE International Conference on Software Maintenance, IEEE, 2005, pp. 587–596.

[15] H. Zhong, L. Zhang, H. Mei, An experimental study of four typical test suite reduction techniques, Inf. Softw. Technol. 50 (6) (2008) 534–546.

[16] G. Fraser, F. Wotawa, Redundancy based test-suite reduction, Fundam. Approaches Softw. Eng. (2007) 291–305.

[17] K.F. Fischer, A test case selection method for the validation of software maintenance modifications, in: Computer Software and Applications Conference, vol. 77, 1977, pp. 421–426.

[18] K. Fischer, F. Raji, A. Chruscicki, A methodology for retesting modified software, in: Proceedings of the National Telecommunications Conference B-6-3, 1981, pp. 1–6.

[19] G. Rothermel, M.J. Harrold, Selecting tests and identifying test coverage requirements for modified software, in: Proceedings of the 1994 ACM SIGSOFT International Symposium on Software Testing and Analysis, ACM, 1994, pp. 169–184.

[20] G. Rothermel, M.J. Harrold, A safe, efficient regression test selection technique, ACM Trans. Softw. Eng. Methodol. 6 (2) (1997) 173–210.

[21] S. Yoo, M. Harman, Pareto efficient multi-objective test case selection, in: International Symposium on Software Testing and Analysis, ACM, 2007, pp. 140–150.

[22] M. Grindal, B. Lindström, J. Offutt, S.F. Andler, An evaluation of combination strategies for test case selection, Empir. Softw. Eng. 11 (4) (2006) 583–611.

[23] S. Fujiwara, G.V. Bochmann, F. Khendek, M. Amalou, A. Ghedamsi, Test selection based on finite state models, IEEE Trans. Softw. Eng. 17 (6) (1991) 591–603.

[24] T.L. Graves, M.J. Harrold, J.-M. Kim, A. Porter, G. Rothermel, An empirical study of regression test selection techniques, ACM Trans. Softw. Eng. Methodol. 10 (2) (2001) 184–208.

[25] Y. Chen, R.L. Probert, D.P. Sims, Specification-based regression test selection with risk analysis, in: Conference of the Centre for Advanced Studies on Collaborative Research, IBM Press, 2002, p. 1.

[26] G. Rothermel, M.J. Harrold, Analyzing regression test selection techniques, IEEE Trans. Soft. Eng. 22 (8) (1996) 529–551.

[27] L.C. Briand, Y. Labiche, S. He, Automating regression test selection based on UML designs, Inf. Softw. Technol. 51 (1) (2009) 16–30.

[28] L. Zhang, Hybrid regression test selection, in: ICSE, 2018, pp. 199–209. (to appear).

[29] G. Rothermel, R.H. Untch, C. Chu, M.J. Harrold, Test case prioritization: an empirical study, in: IEEE International Conference on Software Maintenance, IEEE, 1999, pp. 179–188.

[30] G. Rothermel, R.H. Untch, C. Chu, M.J. Harrold, Prioritizing test cases for regression testing, IEEE Trans. Softw. Eng. 27 (10) (2001) 929–948.

[31] S. Elbaum, D. Gable, G. Rothermel, Understanding and measuring the sources of variation in the prioritization of regression test suites, in: Proceedings. Seventh International Software Metrics Symposium, 2001, IEEE, 2001, pp. 169–179.

[32] J.A. Jones, M.J. Harrold, Test-suite reduction and prioritization for modified condition/decision coverage, IEEE Trans. Softw. Eng. 29 (3) (2003) 195–209.

[33] S. Elbaum, A. Malishevsky, G. Rothermel, Incorporating varying test costs and fault severities into test case prioritization, in: Proceedings of the 23rd International Conference on Software Engineering, IEEE Computer Society, 2001, pp. 329–338.

[34] Z.-W. He, C.-G. Bai, GUI test case prioritization by state-coverage criterion, in: Proceedings of the 10th International Workshop on Automation of Software Test, IEEE Press, 2015, pp. 18–22.

[35] H. Hemmati, Z. Fang, M.V. Mantyla, Prioritizing manual test cases in traditional and rapid release environments, in: 2015 IEEE 8th International Conference on Software Testing, Verification and Validation, IEEE, 2015, pp. 1–10.

[36] E.J. Rapos, J. Dingel, Using fuzzy logic and symbolic execution to prioritize UML-RT test cases, in: 2015 IEEE 8th International Conference on Software Testing, Verification and Validation, IEEE, 2015, pp. 1–10.

[37] B. Jiang, W.K. Chan, T.H. Tse, PORA: proportion-oriented randomized algorithm for test case prioritization, in: 2015 IEEE International Conference on Software Quality, Reliability and Security, IEEE, 2015, pp. 131–140.

[38] D. Marijan, Multi-perspective regression test prioritization for time-constrained environments, in: 2015 IEEE International Conference on Software Quality, Reliability and Security, IEEE, 2015, pp. 157–162.

[39] D. Di Nucci, A. Panichella, A. Zaidman, A. De Lucia, Hypervolume-based search for test case prioritization, in: International Symposium on Search Based Software Engineering, Springer, 2015, pp. 157–172.

[40] F. Yuan, Y. Bian, Z. Li, R. Zhao, Epistatic genetic algorithm for test case prioritization, in: International Symposium on Search Based Software Engineering, Springer, 2015, pp. 109–124.

[41] Y. Bian, S. Kirbas, M. Harman, Y. Jia, Z. Li, Regression test case prioritisation for guava, in: International Symposium on Search Based Software Engineering, Springer, 2015, pp. 221–227.

[42] C. Jia, L. Mei, W.K. Chan, Y.-T. Yu, T.H. Tse, Is XML-based test case prioritization for validating WS-BPEL evolution effective in both average and adverse scenarios? in: 2014 IEEE International Conference on Web Services, IEEE, 2014, pp. 233–240.

[43] X. Zhang, T. Chen, H. Liu, An application of adaptive random sequence in test case prioritization, in: 26th International Conference on Software Engineering and Knowledge Engineering, Knowledge Systems Institute Graduate School, 2014, pp. 126–131.

[44] D. Marijan, A. Gotlieb, S. Sen, Test case prioritization for continuous regression testing: an industrial case study, in: IEEE International Conference on Software Maintenance, IEEE, 2013, pp. 540–543.

[45] R. Malhotra, D. Tiwari, Development of a framework for test case prioritization using genetic algorithm, ACM SIGSOFT Softw. Eng. Notes 38 (3) (2013) 1–6.

[46] W. Sun, Z. Gao, W. Yang, C. Fang, Z. Chen, Multi-objective test case prioritization for GUI applications, in: Proceedings of the 28th Annual ACM Symposium on Applied Computing, ACM, 2013, pp. 1074–1079.

[47] D. Garg, A. Datta, T. French, A novel bipartite graph approach for selection and prioritisation of test cases, ACM SIGSOFT Softw. Eng. Notes 38 (6) (2013) 1–6.

[48] D. Di Nardo, N. Alshahwan, L. Briand, Y. Labiche, Coverage-based test case prioritisation: an industrial case study, in: 2013 IEEE Sixth International Conference on Software Testing, Verification and Validation, IEEE, 2013, pp. 302–311.

[49] H. Srikanth, S. Banerjee, Improving test efficiency through system test prioritization, J. Syst. Softw. 85 (5) (2012) 1176–1187.

[50] M.M. Islam, A. Marchetto, A. Susi, F.B. Kessler, G. Scanniello, MOTCP: a tool for the prioritization of test cases based on a sorting genetic algorithm and Latent Semantic Indexing, in: IEEE International Conference on Software Maintenance, IEEE, 2012, pp. 654–657.

[51] C. Malz, N. Jazdi, P. Gohner, Prioritization of test cases using software agents and fuzzy logic, in: 2012 IEEE Fifth International Conference on Software Testing, Verification and Validation, IEEE, 2012, pp. 483–486.

[52] P. de Alca´ntara dos Santos Neto, R. Britto, T. Soares, W. Ayala, J. Cruz, R.A.L. Rabelo, Regression testing prioritization based on fuzzy inference systems, in: International Conference on Software Engineering and Knowledge Engineering, 2012, pp. 273–278.

[53] H. Yoon, B. Choi, A test case prioritization based on degree of risk exposure and its empirical study, Int. J. Softw. Eng. Knowl. Eng. 21 (2) (2011) 191–209.

[54] E. Shihab, Z.M. Jiang, B. Adams, A.E. Hassan, R. Bowerman, Prioritizing the creation of unit tests in legacy software systems, Softw. Pract. Exp. 41 (10) (2011) 1027–1048.

[55] R. Carlson, H. Do, A. Denton, A clustering approach to improving test case prioritization: an industrial case study, in: IEEE International Conference on Software Maintenance, IEEE, 2011, pp. 382–391.

[56] S. Sampath, R.C. Bryce, S. Jain, S. Manchester, A tool for combination-based prioritization and reduction of user-session-based test suites, in: IEEE International Conference on Software Maintenance, IEEE, 2011, pp. 574–577.

[57] J. Czerwonka, R. Das, N. Nagappan, A. Tarvo, A. Teterev, Crane: failure prediction, change analysis and test prioritization in practice-experiences from windows, in: 2011 IEEE Fourth International Conference on Software Testing, Verification and Validation, IEEE, 2011, pp. 357–366.

[58] E. Engström, P. Runeson, A. Ljung, Improving regression testing transparency and efficiency with history-based prioritization-an industrial case study, in: 2011 IEEE Fourth International Conference on Software Testing, Verification and Validation, IEEE, 2011, pp. 367–376.

[59] N. Kaushik, M. Salehie, L. Tahvildari, S. Li, M. Moore, Dynamic prioritization in regression testing, in: 2011 IEEE Fourth International Conference on Software Testing, Verification and Validation Workshops, IEEE, 2011, pp. 135–138.

[60] C. Malz, P. Göhner, Agent-based test case prioritization, in: 2011 IEEE Fourth International Conference on Software Testing, Verification and Validation Workshops, IEEE, 2011, pp. 149–152.

[61] E. Salecker, R. Reicherdt, S. Glesner, Calculating prioritized interaction test sets with constraints using binary decision diagrams, in: 2011 IEEE Fourth International Conference on Software Testing, Verification and Validation Workshops, IEEE, 2011, pp. 278–285.

[62] B. Jiang, W.K. Chan, On the integration of test adequacy, test case prioritization, and statistical fault localization, in: 2010 10th International Conference on Quality Software, IEEE, 2010, pp. 377–384.

[63] E. Shihab, Z.M. Jiang, B. Adams, A.E. Hassan, R. Bowerman, Prioritizing unit test creation for test-driven maintenance of legacy systems, in: 2010 10th International Conference on Quality Software, IEEE, 2010, pp. 132–141.

[64] L. Chen, Z. Wang, L. Xu, H. Lu, B. Xu, Test case prioritization for web service regression testing, in: 2010 Fifth IEEE International Symposium on Service Oriented System Engineering, IEEE, 2010, pp. 173–178.

[65] C.L.B. Maia, R.A.F. do Carmo, F.G. de Freitas, G.A.L. de Campos, J.T. de Souza, Automated test case prioritization with reactive GRASP, Adv. Softw. Eng. 2010 (2010).

[66] A. Srivastava, J. Thiagarajan, Effectively prioritizing tests in development environment, in: ACM SIGSOFT Software Engineering Notes, vol. 27, ACM, 2002, pp. 97–106.

[67] W.E. Wong, J.R. Horgan, S. London, H. Agrawal, A study of effective regression testing in practice, in: International Symposium on Software Reliability Engineering, 1997, pp. 264–274.

[68] Y. Lu, Y. Lou, S. Cheng, L. Zhang, D. Hao, Y. Zhou, L. Zhang, How does regression test prioritization perform in real-world software evolution? in: Proceedings of the 38th International Conference on Software Engineering, ACM, 2016, pp. 535–546.

[69] C. Henard, M. Papadakis, M. Harman, Y. Jia, Y. Le Traon, Comparing white-box and black-box test prioritization, in: Proceedings of the 38th International Conference on Software Engineering, ACM, 2016, pp. 523–534.

[70] Q. Luo, K. Moran, D. Poshyvanyk, A large-scale empirical comparison of static and dynamic test case prioritization techniques, in: Proceedings of the 2016 24th ACM SIGSOFT International Symposium on Foundations of Software Engineering, ACM, 2016, pp. 559–570.

[71] D. Hao, L. Zhang, H. Mei, Test-case prioritization: achievements and challenges, Front. Comp. Sci. 10 (5) (2016) 769–777.

[72] Z. Li, M. Harman, R.M. Hierons, Search algorithms for regression test case prioritization, IEEE Trans. Softw. Eng. 33 (4) (2007) 225–237.

[73] S. Li, N. Bian, Z. Chen, D. You, Y. He, A simulation study on some search algorithms for regression test case prioritization, in: 2010 10th International Conference on Quality Software (QSIC), IEEE, 2010, pp. 72–81.

[74] Y. Singh, A. Kaur, B. Suri, S. Singhal, Systematic literature review on regression test prioritization techniques, Informatica (Slovenia) 36 (4) (2012) 379–408.

[75] C. Catal, D. Mishra, Test case prioritization: a systematic mapping study, Softw. Qual. J. 21 (3) (2013) 445–478.

[76] Debian https://www.debian.org/, n.d.

[77] Git, https://git-scm.com/, n.d.

[78] Mercurial, https://www.mercurial-scm.org/, n.d.

[79] GitHub, https://github.com/, n.d.

[80] Travis CI, https://travis-ci.org/, n.d.

[81] Jenkins CI, https://jenkins.io/, n.d.

[82] L. Zhang, S.-S. Hou, C. Guo, T. Xie, H. Mei, Time-aware test-case prioritization using integer linear programming, in: Proceedings of the Eighteenth International Symposium on Software Testing and Analysis, ACM, 2009, pp. 213–224.

[83] G. Fraser, A. Arcuri, Evosuite: automatic test suite generation for object-oriented software, in: Proceedings of the 19th ACM SIGSOFT Symposium and the 13th European Conference on Foundations of Software Engineering, ACM, 2011, pp. 416–419.

[84] G. Rothermel, R.H. Untch, C. Chu, M.J. Harrold, Test case prioritization: an empirical study, in: IEEE International Conference on Software Maintenance, 1999, pp. 179–188.

[85] G. Rothermel, R.H. Untch, C. Chu, M.J. Harrold, Prioritizing test cases for regression testing, IEEE Trans. Softw. Eng. 27 (10) (2001) 929–948.

[86] J. Qian, D. Zhou, Prioritizing test cases for memory leaks in android applications, J. Comput. Sci. Technol. 31 (5) (2016) 869–882.

[87] R. Huang, W. Zong, J. Chen, D. Towey, Y. Zhou, D. Chen, Prioritizing interaction test suites using repeated base choice coverage, in: Computer Software and Applications Conference, vol. 1, IEEE, 2016, pp. 174–184.

[88] S. Wang, S. Ali, T. Yue, Ø. Bakkeli, M. Liaaen, Enhancing test case prioritization in an industrial setting with resource awareness and multi-objective search, in: Proceedings of the 38th International Conference on Software Engineering Companion, ACM, 2016, pp. 182–191.

[89] J. Chen, L. Zhu, T.Y. Chen, R. Huang, D. Towey, F.-C. Kuo, Y. Guo, An adaptive sequence approach for OOS test case prioritization, in: International Symposium on Software Reliability Engineering Wokshops, IEEE, 2016, pp. 205–212.

[90] P.E. Strandberg, D. Sundmark, W. Afzal, T.J. Ostrand, E.J. Weyuker, Experience report: automated system level regression test prioritization using multiple factors, in: International Symposium on Software Reliability Engineering, IEEE, 2016, pp. 12–23.

[91] X. Zhang, X. Xie, T.Y. Chen, Test case prioritization using adaptive random sequence with category-partition-based distance, in: 2016 IEEE International Conference on Software Quality, Reliability and Security (QRS), IEEE, 2016, pp. 374–385.

[92] H. Wang, J. Xing, Q. Yang, D. Han, X. Zhang, Modification impact analysis based test case prioritization for regression testing of service-oriented workflow applications, in: Computer Software and Applications Conference, vol. 2, IEEE, 2015, pp. 288–297.

[93] R. Huang, J. Chen, D. Towey, A.T.S. Chan, Y. Lu, Aggregate-strength interaction test suite prioritization, J. Syst. Softw. 99 (2015) 36–51.

[94] L. Mei, Y. Cai, C. Jia, B. Jiang, W.K. Chan, Z. Zhang, T.H. Tse, A subsumption hierarchy of test case prioritization for composite services, IEEE Trans. Serv. Comput. 8 (5) (2015) 658–673.

[95] Y. Lou, D. Hao, L. Zhang, Mutation-based test-case prioritization in software evolution, in: International Symposium on Software Reliability Engineering, IEEE, 2015, pp. 46–57.

[96] T.B. Noor, H. Hemmati, A similarity-based approach for test case prioritization using historical failure data, in: International Symposium on Software Reliability Engineering, IEEE, 2015, pp. 58–68.

[97] R. Wang, S. Jiang, D. Chen, Similarity-based regression test case prioritization, in: International Conference on Software Engineering and Knowledge Engineering, 2015, pp. 358–363.

[98] H. Srikanth, S. Banerjee, L. Williams, J. Osborne, Towards the prioritization of system test cases, Softw. Test. Verification Reliab. 24 (4) (2014) 320–337.

[99] K. Zhai, B. Jiang, W.K. Chan, Prioritizing test cases for regression testing of location-based services: metrics, techniques, and case study, IEEE Trans. Serv. Comput. 7 (1) (2014) 54–67.

[100] C. Fang, Z. Chen, K. Wu, Z. Zhao, Similarity-based test case prioritization using ordered sequences of program entities, Softw. Qual. J. 22 (2) (2014) 335–361.

[101] R. Huang, J. Chen, R. Wang, D. Chen, How to do tie-breaking in prioritization of interaction test suites? in: International Conference on Software Engineering and Knowledge Engineering, 2014, pp. 121–125.

[102] R. Huang, X. Xie, D. Towey, T.Y. Chen, Y. Lu, J. Chen, Prioritization of combinatorial test cases by incremental interaction coverage, Int. J. Softw. Eng. Knowl. Eng. 23 (10) (2013) 1427–1457.

[103] T. Miller, et al., Using dependency structures for prioritization of functional test suites, IEEE Trans. Softw. Eng. 39 (2) (2013) 258–275.

[104] D. Hao, X. Zhao, L. Zhang, Adaptive test-case prioritization guided by output inspection, in: Computer Software and Applications Conference, IEEE, 2013, pp. 169–179.

[105] E.L.G. Alves, P.D.L. Machado, T. Massoni, S.T.C. Santos, A refactoring-based approach for test case selection and prioritization, in: 2013 8th International Workshop on Automation of Software Test, IEEE, 2013, pp. 93–99.

[106] J.F.S. Ouriques, Strategies for prioritizing test cases generated through model-based testing approaches, in: 2015 IEEE/ACM 37th IEEE International Conference on Software Engineering, vol. 2, IEEE, 2015, pp. 879–882.

[107] L. Mei, Y. Cai, C. Jia, B. Jiang, W.K. Chan, Prioritizing structurally complex test pairs for validating WS-BPEL evolutions, in: 2013 IEEE 20th International Conference on Web Services (ICWS), IEEE, 2013, pp. 147–154.

[108] Y.-C. Huang, K.-L. Peng, C.-Y. Huang, A history-based cost-cognizant test case prioritization technique in regression testing, J. Syst. Softw. 85 (3) (2012) 626–637.

[109] Y. Ledru, A. Petrenko, S. Boroday, N. Mandran, Prioritizing test cases with string distances, Autom. Softw. Eng. 19 (1) (2012) 65–95.

[110] K. Wu, C. Fang, Z. Chen, Z. Zhao, Test case prioritization incorporating ordered sequence of program elements, in: Proceedings of the 7th International Workshop on Automation of Software Test, IEEE Press, 2012, pp. 124–130.

[111] P. Caliebe, T. Herpel, R. German, Dependency-based test case selection and prioritization in embedded systems, in: 2012 IEEE Fifth International Conference on Software Testing, Verification and Validation, IEEE, 2012, pp. 731–735.

[112] M. Staats, P. Loyola, G. Rothermel, Oracle-centric test case prioritization, in: International Symposium on Software Reliability Engineering, IEEE, 2012, pp. 311–320.

[113] L. Mei, W.K. Chan, T.H. Tse, R.G. Merkel, XML-manipulating test case prioritization for XML-manipulating services, J. Syst. Softw. 84 (4) (2011) 603–619.

[114] A. Gonzalez-Sanchez, É. Piel, R. Abreu, H.-G. Gross, A.J.C. van Gemund, Prioritizing tests for software fault diagnosis, Softw. Pract. Exp. 41 (10) (2011) 1105–1129.

[115] R.C. Bryce, S. Sampath, A.M. Memon, Developing a single model and test prioritization strategies for event-driven software, IEEE Trans. Softw. Eng. 37 (1) (2011) 48–64.

[116] Y.-C. Huang, C.-Y. Huang, J.-R. Chang, T.-Y. Chen, Design and analysis of cost-cognizant test case prioritization using genetic algorithm with test history, in: Computer Software and Applications Conference, IEEE, 2010, pp. 413–418.

[117] D. Xu, J. Ding, Prioritizing state-based aspect tests, in: 2010 Third International Conference on Software Testing, Verification and Validation, IEEE, 2010, pp. 265–274.

[118] R. Krishnamoorthi, S.A. Sahaaya Arul Mary, Requirement based system test case prioritization of new and regression test cases, Int. J. Softw. Eng. Knowl. Eng. 19 (3) (2009) 453–475.

[119] X. Qu, Configuration aware prioritization techniques in regression testing, in: 31st International Conference on Software Engineering-Companion Volume, 2009. ICSE-Companion 2009, IEEE, 2009, pp. 375–378.

[120] I.-C. Yoon, A. Sussman, A. Memon, A. Porter, Prioritizing component compatibility tests via user preferences, in: IEEE International Conference on Software Maintenance, IEEE, 2009, pp. 29–38.

[121] S.-S. Hou, L. Zhang, T. Xie, J.-S. Sun, Quota-constrained test-case prioritization for regression testing of service-centric systems, in: IEEE International Conference on Software Maintenance, IEEE, 2008, pp. 257–266.

[122] S. Sampath, R.C. Bryce, G. Viswanath, V. Kandimalla, A.G. Koru, Prioritizing user-session-based test cases for web applications testing, in: 2008 1st International Conference on Software Testing, Verification, and Validation, IEEE, 2008, pp. 141–150.

[123] D. Jeffrey, N. Gupta, Experiments with test case prioritization using relevant slices, J. Syst. Softw. 81 (2) (2008) 196–221.

[124] P.R. Srivastva, K. Kumar, G. Raghurama, Test case prioritization based on requirements and risk factors, ACM SIGSOFT Softw. Eng. Notes 33 (4) (2008) 7.

[125] Z. Ma, J. Zhao, Test case prioritization based on analysis of program structure, in: 2008. 15th Asia-Pacific Software Engineering Conference, IEEE, 2008, pp. 471–478.

[126] H. Stallbaum, A. Metzger, K. Pohl, An automated technique for risk-based test case generation and prioritization, in: Proceedings of the 3rd International Workshop on Automation of Software Test, ACM, 2008, pp. 67–70.

[127] X. Zhang, C. Nie, B. Xu, B. Qu, Test case prioritization based on varying testing requirement priorities and test case costs, in: Seventh International Conference on Quality Software, 2007, IEEE, 2007, pp. 15–24.

[128] M. Sherriff, M. Lake, L. Williams, Prioritization of regression tests using singular value decomposition with empirical change records, in: International Symposium on Software Reliability Engineering, IEEE, 2007, pp. 81–90.

[129] A.M. Smith, J. Geiger, G.M. Kapfhammer, M.L. Soffa, Test suite reduction and prioritization with call trees, in: Proceedings of the Twenty-Second IEEE/ACM international conference on Automated Software Engineering, ACM, 2007, pp. 539–540.

[130] R.C. Bryce, C.J. Colbourn, Prioritized interaction testing for pair-wise coverage with seeding and constraints, Inf. Softw. Technol. 48 (10) (2006) 960–970.

[131] D. Jeffrey, N. Gupta, Test case prioritization using relevant slices, in: Computer Software and Applications Conference, vol. 1, IEEE, 2006, pp. 411–420.

[132] P.L. Li, J. Herbsleb, M. Shaw, B. Robinson, Experiences and results from initiating field defect prediction and product test prioritization efforts at ABB Inc, in: Proceedings of the 28th International Conference on Software Engineering, ACM, 2006, pp. 413–422.

[133] P. Tonella, P. Avesani, A. Susi, Using the case-based ranking methodology for test case prioritization, in: IEEE International Conference on Software Maintenance, IEEE, 2006, pp. 123–133.

[134] R.C. Bryce, C.J. Colbourn, Test prioritization for pairwise interaction coverage, in: ACM SIGSOFT Software Engineering Notes, vol. 30, ACM, 2005, pp. 1–7.

[135] H. Srikanth, L. Williams, On the economics of requirements-based test case prioritization, in: ACM SIGSOFT Software Engineering Notes, vol. 30, ACM, 2005, pp. 1–3.

[136] A. Beszédes, T. Gergely, L. Schrettner, J. Jász, L. Langó, T. Gyimóthy, Code coverage-based regression test selection and prioritization in WebKit, in: IEEE International Conference on Software Maintenance, IEEE, 2012, pp. 46–55.

[137] L. Zhang, J. Zhou, D. Hao, L. Zhang, H. Mei, Prioritizing JUnit test cases in absence of coverage information, in: IEEE International Conference on Software Maintenance, IEEE, 2009, pp. 19–28.

[138] B. Korel, G. Koutsogiannakis, Experimental comparison of code-based and model-based test prioritization, in: International Conference on Software Testing, Verification and Validation Workshops, 2009, IEEE, 2009, pp. 77–84.

[139] W. Masri, M. El-Ghali, Test case filtering and prioritization based on coverage of combinations of program elements, in: Proceedings of the Seventh International Workshop on Dynamic Analysis, ACM, 2009, pp. 29–34.

[140] J.J. Li, D. Weiss, H. Yee, Code-coverage guided prioritized test generation, Inf. Softw. Technol. 48 (12) (2006) 1187–1198.

[141] D.C. Episkopos, J.J. Li, H.S. Yee, D.M. Weiss, Prioritize code for testing to improve code coverage of complex software, 2011. US Patent 7,886,272.

[142] K.K. Aggrawal, Y. Singh, A. Kaur, Code coverage based technique for prioritizing test cases for regression testing, ACM SIGSOFT Softw. Eng. Notes 29 (5) (2004) 1–4.

[143] K.T. Cheng, A.S. Krishnakumar, Automatic functional test generation using the extended finite state machine model, in: Proceedings of the 30th International Design Automation Conference, ACM, 1993, pp. 86–91.

[144] J. Dick, A. Faivre, Automating the generation and sequencing of test cases from model-based specifications, in: FME'93: Industrial-Strength Formal Methods, Springer, 1993, pp. 268–284.

[145] B. Korel, L.H. Tahat, B. Vaysburg, Model based regression test reduction using dependence analysis, in: Proceedings. International Conference on Software Maintenance, 2002, IEEE, 2002, pp. 214–223.

[146] B. Korel, G. Koutsogiannakis, L.H. Tahat, Application of system models in regression test suite prioritization, in: IEEE International Conference on Software Maintenance, IEEE, 2008, pp. 247–256.

[147] B. Korel, L.H. Tahat, M. Harman, Test prioritization using system models, in: IEEE International Conference on Software Maintenance, IEEE, 2005, pp. 559–568.

[148] S.W. Thomas, H. Hemmati, A.E. Hassan, D. Blostein, Static test case prioritization using topic models, Empir. Softw. Eng. 19 (1) (2014) 182–212.

[149] L. Tahat, B. Korel, M. Harman, H. Ural, Regression test suite prioritization using system models, Software Testing, Verification and Reliability 22 (7) (2012) 481–506.

[150] C.R. Panigrahi, R. Mall, Model-based regression test case prioritization, ACM SIGSOFT Softw. Eng. Notes 35 (6) (2010) 1–7.

[151] Y.T. Yu, M.F. Lau, Fault-based test suite prioritization for specification-based testing, Inf. Softw. Technol. 54 (2) (2012) 179–202.

[152] R. Just, G.M. Kapfhammer, F. Schweiggert, Using non-redundant mutation operators and test suite prioritization to achieve efficient and scalable mutation analysis, in: International Symposium on Software Reliability Engineering, IEEE, 2012, pp. 11–20.

[153] S.A.S.A. Mary, R. Krishnamoorthi, Time-aware and weighted fault severity based metrics for test case prioritization, Int. J. Softw. Eng. Knowl. Eng. 21 (1) (2011) 129–142.

[154] S. Kim, J. Baik, An effective fault aware test case prioritization by incorporating a fault localization technique, in: Proceedings of the 2010 ACM-IEEE International Symposium on Empirical Software Engineering and Measurement, ACM, 2010, p. 5.

[155] C. Simons, E.C. Paraiso, Regression test cases prioritization using failure pursuit sampling, in: 2010 10th International Conference on Intelligent Systems Design and Applications, IEEE, 2010, pp. 923–928.

[156] H. Do, G. Rothermel, On the use of mutation faults in empirical assessments of test case prioritization techniques, IEEE Trans. Softw. Eng. 32 (9) (2006) 733–752.

[157] S. Elbaum, A.G. Malishevsky, G. Rothermel, Test case prioritization: a family of empirical studies, IEEE Trans. Softw. Eng. 28 (2) (2002) 159–182.

[158] J.C. Munson, S.G. Elbaum, Code churn: a measure for estimating the impact of code change, in: Proceedings. International Conference on Software Maintenance, 1998, IEEE, 1998, pp. 24–31.

[159] A. Nikora, J. Munson, Software evolution and the fault process, 1998. Tech. rep.

[160] I.H. Bernstein, Applied Multivariate Analysis, Springer Science & Business Media, 2012.

[161] Y. Ledru, A. Petrenko, S. Boroday, Using string distances for test case prioritisation, in: Proceedings of the 2009 IEEE/ACM International Conference on Automated Software Engineering, IEEE Computer Society, 2009, pp. 510–514.

[162] J. Chen, Y. Bai, D. Hao, Y. Xiong, H. Zhang, L. Zhang, B. Xie, Test case prioritization for compilers: a text-vector based approach, in: 2016 IEEE International Conference on Software Testing, Verification and Validation, IEEE, 2016, pp. 266–277.

[163] J. Chen, W. Hu, D. Hao, Y. Xiong, H. Zhang, L. Zhang, B. Xie, An empirical comparison of compiler testing techniques, in: Proceedings of the 38th International Conference on Software Engineering, ACM, 2016, pp. 180–190.

[164] J. Chen, Y. Bai, D. Hao, Y. Xiong, H. Zhang, B. Xie, Learning to prioritize test pro-grams for compiler testing, in: Proceedings of the 39th International Conference on Software Engineering, IEEE Press, 2017, pp. 700–711.

[165] B. Jiang, W.K. Chan, Input-based adaptive randomized test case prioritization: a local beam search approach, J. Syst. Softw. 105 (2015) 91–106.

[166] D. Garg, A. Datta, Test case prioritization due to database changes in web applications, in: 2012 IEEE Fifth International Conference on Software Testing, Verification and Validation, IEEE, 2012, pp. 726–730.

[167] C.D. Nguyen, A. Marchetto, P. Tonella, Change sensitivity based prioritization for audit testing of webservice compositions, in: 2011 IEEE Fourth International Conference on Software Testing, Verification and Validation Workshops, IEEE, 2011, pp. 357–365.

[168] R.A. Haraty, N. Mansour, L. Moukahal, I. Khalil, Regression test cases prioritization using clustering and code change relevance, Int. J. Softw. Eng. Knowl. Eng. 26 (5) (2016) 733–768.

[169] E.L.G. Alves, P.D.L. Machado, T. Massoni, M. Kim, Prioritizing test cases for early detection of refactoring faults, Softw. Test. Verification Reliab. 26 (5) (2016) 402–426.

[170] S. Panda, D. Munjal, D.P. Mohapatra, A slice-based change impact analysis for regres-sion test case prioritization of object-oriented programs, Adv. Soft. Eng. 2016 (2016) 1.

[171] C. Hettiarachchi, H. Do, B. Choi, Risk-based test case prioritization using a fuzzy expert system, Inf. Softw. Technol. 69 (2016) 1–15.

[172] H. Srikanth, M. Cashman, M.B. Cohen, Test case prioritization of build acceptance tests for an enterprise cloud application: an industrial case study, J. Syst. Softw. 119 (2016) 122–135.

[173] M.J. Arafeen, H. Do, Test case prioritization using requirements-based clustering, in: 2013 IEEE Sixth International Conference on Software Testing, Verification and Validation, IEEE, 2013, pp. 312–321.

[174] L. Zhang, D. Hao, L. Zhang, G. Rothermel, H. Mei, Bridging the gap between the total and additional test-case prioritization strategies, in: Proceedings of the 2013 International Conference on Software Engineering, IEEE Press, 2013, pp. 192–201.

[175] D. Hao, L. Zhang, L. Zhang, G. Rothermel, H. Mei, A unified test case prioritization approach, ACM Trans. Softw. Eng. Methodol. 24 (2) (2014) 10.

[176] S. Lin, Computer solutions of the traveling salesman problem, Bell Syst. Tech. J. 44 (10) (1965) 2245–2269.

[177] B. Jiang, Z. Zhang, W.K. Chan, T.H. Tse, Adaptive random test case prioritization, in: Proceedings of the 2009 IEEE/ACM International Conference on Automated Software Engineering, IEEE Computer Society, 2009, pp. 233–244.

[178] T.Y. Chen, F.-C. Kuo, R.G. Merkel, T.H. Tse, Adaptive random testing: the art of test case diversity, J. Syst. Softw. 83 (1) (2010) 60–66.

[179] T.Y. Chen, H. Leung, I.K. Mak, Adaptive random testing, in: ASIAN, vol. 4, Springer, 2004, pp. 320–329.

[180] M.G. Epitropakis, S. Yoo, M. Harman, E.K. Burke, Empirical evaluation of pareto efficient multi-objective regression test case prioritisation, in: Proceedings of the 2015 International Symposium on Software Testing and Analysis, ACM, 2015, pp. 234–245.

[181] Z. Li, Y. Bian, R. Zhao, J. Cheng, A fine-grained parallel multi-objective test case prioritization on GPU, in: International Symposium on Search Based Software Engi-neering, Springer, 2013, pp. 111–125.

[182] D. Hao, L. Zhang, L. Zang, Y. Wang, X. Wu, T. Xie, To be optimal or not in test-case prioritization, IEEE Trans. Softw. Eng. 42 (5) (2016) 490–505.

[183] A. Singhal, Modern information retrieval: a brief overview, IEEE Data Eng. Bull. 24 (4) (2001) 35–43.

[184] C.D. Nguyen, A. Marchetto, P. Tonella, Test case prioritization for audit testing of evolving web services using information retrieval techniques, in: 2011 IEEE International Conference on Web Services, IEEE, 2011, pp. 636–643.

[185] J.-H. Kwon, I.-Y. Ko, G. Rothermel, M. Staats, Test case prioritization based on information retrieval concepts, in: 2014 21st Asia-Pacific Software Engineering Conference (APSEC), vol. 1, IEEE, 2014, pp. 19–26.

[186] R.K. Saha, L. Zhang, S. Khurshid, D.E. Perry, An information retrieval approach for regression test prioritization based on program changes, in: 2015 IEEE/ACM 37th IEEE International Conference on Software Engineering, vol. 1, IEEE, 2015, pp. 268–279.

[187] J. Lv, B. Yin, K.-Y. Cai, On the gain of measuring test case prioritization, in: Computer Software and Applications Conference, IEEE, 2013, pp. 627–632.

[188] K.R. Walcott, M.L. Soffa, G.M. Kapfhammer, R.S. Roos, Time-aware test suite prioritization, in: Proceedings of the 2006 International Symposium on Software Testing and Analysis, ACM, 2006, pp. 1–12.

[189] X. Qu, M.B. Cohen, K.M. Woolf, Combinatorial interaction regression testing: a study of test case generation and prioritization, in: IEEE International Conference on Software Maintenance, IEEE, 2007, pp. 255–264.

[190] Z. Wang, L. Chen, Improved metrics for non-classic test prioritization problems, in: International Conference on Software Engineering and Knowledge Engineering, 2015, pp. 562–566.

[191] H. Do, G. Rothermel, An empirical study of regression testing techniques incorporating context and lifetime factors and improved cost-benefit models, in: ACM SIGSOFT International Symposium on Foundations of Software Engineering, 2006, pp. 141–151.

[192] H. Do, G. Rothermel, Using sensitivity analysis to create simplified economic models for regression testing, in: International Symposium on Software Testing and Analysis, 2008, pp. 51–62.

[193] J.-M. Kim, A. Porter, A history-based test prioritization technique for regression testing in resource constrained environments, in: Proceedings of the 24th International Conference on Software Engineering, ACM, 2002, pp. 119–129.

[194] Y. Chen, A. Groce, C. Zhang, W.-K. Wong, X. Fern, E. Eide, J. Regehr, Taming compiler fuzzers, in: ACM SIGPLAN Conference on Programming Language Design and Implementation, vol. 48, 2013, pp. 197–208.

[195] S. Alspaugh, K.R. Walcott, M. Belanich, G.M. Kapfhammer, M.L. Soffa, Efficient time-aware prioritization with knapsack solvers, in: WEASELTech, 2007, pp. 13–18.

[196] H.P. Williams, Model Building in Mathematical Programming, Wiley, 1999.

[197] B. Suri, S. Singhal, Analyzing test case selection & prioritization using ACO, ACM SIGSOFT Softw. Eng. Notes 36 (6) (2011) 1–5.

[198] H. Do, S. Mirarab, L. Tahvildari, G. Rothermel, The effects of time constraints on test case prioritization: a series of controlled experiments, IEEE Trans. Softw. Eng. 36 (5) (2010) 593–617.

[199] D. You, Z. Chen, B. Xu, B. Luo, C. Zhang, An empirical study on the effectiveness of time-aware test case prioritization techniques, in: Proceedings of the 2011 ACM Symposium on Applied Computing, ACM, 2011, pp. 1451–1456.

[200] H. Park, H. Ryu, J. Baik, Historical value-based approach for cost-cognizant test case prioritization to improve the effectiveness of regression testing, in: IEEE International Conference on Secure Software Integration and Reliability Improvement, 2008, pp. 39–46.

[201] A.G. Malishevsky, J.R. Ruthruff, G. Rothermel, S. Elbaum, Cost-cognizant test case prioritization, 2006. Technical Report.

[202] S. Eghbali, L. Tahvildari, Test case prioritization using lexicographical ordering, IEEE Trans. Softw. Eng. 42 (12) (2016) 1178–1195.

[203] H. Do, G. Rothermel, A. Kinneer, Prioritizing JUnit test cases: an empirical assessment and cost-benefits analysis, Empir. Softw. Eng. 11 (1) (2006) 33–70.

[204] H. Do, G. Rothermel, A controlled experiment assessing test case prioritization techniques via mutation faults, in: IEEE International Conference on Software Maintenance, IEEE, 2005, pp. 411–420.

[205] M. Gligoric, L. Eloussi, D. Marinov, Practical regression test selection with dynamic file dependencies, in: International Symposium on Software Testing and Analysis, 2015, pp. 211–222.

[206] H. Mei, D. Hao, L. Zhang, L. Zhang, J. Zhou, G. Rothermel, A static approach to prioritizing junit test cases, IEEE Trans. Softw. Eng. 38 (6) (2012) 1258–1275.

[207] R.C. Bryce, A.M. Memon, Test suite prioritization by interaction coverage, in: Workshop on Domain Specific Approaches to Software Test Automation: In Conjunction With the 6th ESEC/FSE Joint Meeting, ACM, 2007, pp. 1–7.

[208] X. Qu, M.B. Cohen, G. Rothermel, Configuration-aware regression testing: an empirical study of sampling and prioritization, in: Proceedings of the 2008 International Symposium on Software Testing and Analysis, ACM, 2008, pp. 75–86.

[209] X. Qu, M.B. Cohen, A study in prioritization for higher strength combinatorial testing, in: 2013 IEEE Sixth International Conference on Software Testing, Verification and Validation Workshops, IEEE, 2013, pp. 285–294.

[210] C. Henard, M. Papadakis, G. Perrouin, J. Klein, P. Heymans, Y. Le Traon, Bypassing the combinatorial explosion: using similarity to generate and prioritize t-wise test configurations for software product lines, IEEE Trans. Softw. Eng. 40 (7) (2014) 650–670.

[211] J. Zhang, J. Chen, D. Hao, Y. Xiong, B. Xie, L. Zhang, H. Mei, Search-based inference of polynomial metamorphic relations, in: Proceedings of the 29th ACM/IEEE International Conference on Automated Software Engineering, ASE '14, 2014, pp. 701–712.

[212] J. Chen, Y. Bai, D. Hao, L. Zhang, L. Zhang, B. Xie, H. Mei, Supporting oracle construction via static analysis, in: 2016 31st IEEE/ACM International Conference on Automated Software Engineering (ASE), IEEE, 2016, pp. 178–189.

[213] J. Zhang, Z. Wang, L. Zhang, D. Hao, L. Zang, S. Cheng, L. Zhang, Predictive mutation testing, in: Proceedings of the 25th International Symposium on Software Testing and Analysis, ACM, 2016, pp. 342–353.

[214] MapDB, http://www.mapdb.org/, n.d.

ABOUT THE AUTHORS

Yiling Lou received the B.S. degree in computer science and technology from Peking University. She is currently working toward the Ph.D. degree under the supervision of Professor Lu Zhang and Professor Dan Has at Peking University. Her research interests include software testing and debugging.

Junjie Chen is a Ph.D. candidate at the School of Electronics Engineering and Computer Science, Peking University. He received his B.S. degree from Beihang University. His research interests are software testing and debugging, mainly focusing on compiler testing, regression testing, automated debugging.

Lingming Zhang is an assistant professor in the Computer Science Department at the University of Texas at Dallas. He obtained his Ph.D. degree from the Department of Electrical and Computer Engineering in the University of Texas at Austin in May 2014. He received his M.S. degree and B.S. degree in Computer Science from Peking University (2010) and Nanjing University (2007), respectively. His research interests lie broadly in software engineering and programming languages, including automated software analysis, testing, debugging, and verification, as well as software evolution and mobile computing. He has authored over 40 papers in premier software engineering or programming language conferences and transactions. He has also served on the program/organization committee or artifact evaluation committee for various international conferences (including ICSE, ISSTA, FSE, ASE, ICST, ICSM, and OOPSLA). He has won the Google Faculty Research Award, his research is also being supported by NSF, Huawei, NVIDIA, and Samsung. More information available at: http://www.utdallas.edu/~lxz144130/.

Dan Hao received the B.S. degree in computer science from the Harbin Institute of Technology in 2002 and the Ph.D. degree in computer science from Peking University in 2008. She is an associate professor at the School of Electronics Engineering and Computer Science, Peking University, P.R. China. Her current research interests include software testing and debugging.

CHAPTER TWO

Model-Based Test Cases Reuse and Optimization

Mohamed Mussa, Ferhat Khendek
Electrical and Computer Engineering, Concordia University, Montréal, QC, Canada

Contents

1. Introduction   48
2. Overall MBT Framework   49
   2.1 Overall Framework   49
   2.2 Some Definitions   52
3. Integration Test Generation   55
   3.1 Identification Process   56
   3.2 Test Case Selection   57
   3.3 Test Behavior Generation   61
   3.4 Checking for Redundancy   63
   3.5 Test Architecture Generation   63
   3.6 Some Properties of the Integration Test Generation Approach   64
4. Acceptance Test Optimization   67
   4.1 Integration Test Case Selection   68
   4.2 Mapping Acceptance Test Cases to Integration Test Cases   69
5. A Case Study: A Library Management System   70
6. Related Work   71
7. Conclusion   75
Acknowledgments   76
Appendix A: Properties of the Integration Test Generation Approach   77
   A.1 System Specification   77
   A.2 Commutativity   80
   A.3 Associativity   82
References   84
About the Authors   87

Abstract

Several test generation techniques have been proposed in the literature. These techniques target separately specific levels of testing without relating them to each other in order to avoid redundancy and enable reuse and optimization. In this chapter, we look into connecting different levels of testing. We propose a model-based testing

Advances in Computers, Volume 113
ISSN 0065-2458
https://doi.org/10.1016/bs.adcom.2018.01.001

47

framework that enables reusability and optimization across different levels of testing. Test cases at one level are reused to generate test cases of subsequent levels of testing. Furthermore, test cases at one level are optimized by relating them to test cases of preceding testing levels and removed if they are found redundant.

1. INTRODUCTION

Testing aims at enhancing the quality of software products. Software goes through different levels of testing; the main ones are unit, component, integration, system, and acceptance level testing. For each level, engineers plan the test, design test cases, exercise the test cases on the target implementation, and evaluate the results. Different test approaches have been developed over the last 3 decades. They usually target specific levels of testing. The lack of clear and systematic connections between the software testing levels is a noteworthy problem in software testing [1,2].

The investigation of this issue is the main objective of this chapter. Moreover, we propose a model-based setting to tackle this problem. Model-driven engineering (MDE) [3] is gaining in maturity and popularity. The introduction of model-based testing (MBT) [4,5] has been an important progress in software testing. Several MBT approaches covering a wide spectrum of modeling languages and software domains, such as embedded systems and telecommunication [6], have been proposed. The unified modeling language (UML) [7] is now widely accepted. Recently, the object management group (OMG) [8] standardized a UML testing profile (UTP) [9]. The profile enables test concepts in UML models in order to create precise UML test models. The literature shows a growing interest in UTP-based approaches [1,10–15]. However, the focus is still on MBT approaches for specific software levels of testing [10–18].

In this chapter, we propose a model-based software testing framework to connect testing levels and enable reusability and optimization across different testing levels [19–22]. The framework is based on UTP. It is composed of two approaches for test generation and test optimization. The test generation approach generates test models for a target testing level by reusing test models from the previous level. We elaborate and discuss the specific case of generation of integration test models from component test models. For this purpose, we investigate the merging of test cases. While there has been a lot of research toward merging architectural models, only a little has been done for merging behavioral models [23–27]. The second approach optimizes test models by relating them to test models that have been already executed in the preceding testing levels. It aims at reducing test execution time without compromising the quality. For this purpose, we develop a model comparison process specific to test cases. We elaborate and discuss the optimization of acceptance test models by relating them to integration test models. Our framework has been implemented, and we have experienced with several case studies. In this chapter, we illustrate our approaches with a library management system case study and discuss the results.

The rest of this chapter is structured as following. Section 2 presents an overview of our MBT framework. We discuss the test generation approach in Section 3 and the test optimization approach in Section 4. In Section 5, we discuss a case study and its results. We review related work in Section 6 before concluding in Section 7.

2. OVERALL MBT FRAMEWORK

In this section, we introduce our overall MBT framework for linking testing levels and enabling reusability and optimization. We also provide formal definitions for some concepts used throughout the chapter.

2.1 Overall Framework

The framework enables reusability across testing levels during test design and enhances test execution through the optimization of test models. To conduct our research in rigorous settings, we use sequence diagrams, which have been formally investigated in [28–30], to model and design our test behaviors. Although UTP is used as the language for the framework, it can be replaced by any other language supporting test model description. Fig. 1 shows our overall MBT framework. The framework consists of

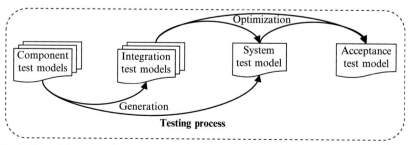

Fig. 1 Overall framework.

two approaches: a test generation approach and a test optimization approach. The generation approach links

- the component-level testing to the integration-level testing and
- the component-level testing to the system-level testing.

Component testing is a black box testing; tests are exercised on components through their interfaces. These interfaces can be internal, to communicate with other system components, or external, to communicate with the system environment, as shown in Fig. 2A. Hence, the same interface can be exercised by several test cases from different component test models. Each test case takes different perspective of the same interface as shown in Fig. 2B. Therefore, we can generate integration test cases from component test cases that examine internal interfaces and generate system test cases from component test cases that examine external interfaces.

In this chapter, we discuss the generation of integration test models from component test models. In our work, a component is defined as a self-coherent piece of software that provides one or more services and can interact with other components. In this framework, we also assume that component test models are generated from the design models using existing approaches such as [10–15].

As shown in Fig. 1, the framework enables also the optimization of test models by mapping them to the previously executed test models. The optimization approach links

- the integration-level testing to the system-level testing,
- the integration-level testing to the acceptance-level testing, and
- the system-level testing to the acceptance-level testing.

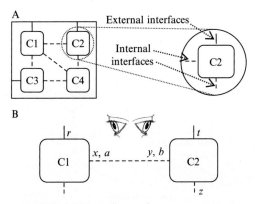

Fig. 2 Component interfaces: (A) interfaces of system components and (B) different views of the same interface.

Our framework is based on the idea of overlapping test cases. Test models are composed of a set of test cases. Test cases capture the test behavior that is exercised on the target implementation. Test behavior in general reflects the expected behavior of the implementation under test. We believe that the "collective" behavior of all test cases at any testing level captures the system behavior. In practice, some research activities migrate system behavior across different development stages using test cases since test cases are finite and precise compared to the system design models [31]. Therefore, we can intuitively conclude that test cases from different testing levels that examine the same portion of the system behavior are redundant if they meet the same test requirements of the subsequent testing level. We propose a test optimization approach that optimizes acceptance test cases by relating them to system and/or integration test cases, and optimize system test cases by relating them to integration test cases. In this chapter (Section 4), we discuss the optimization of acceptance test cases by avoiding redundancy with integration test cases.

We assume component test cases have the following characteristics:

- they are complete and cover all component interfaces,
- each test case for a component covers at least one service provided by the component, and
- there is consistency between the component test models since they describe different components of the same system. The names of the components, interfaces, and messages are used in a consistent manner in test models.

In the proposed approaches, we will compare test models. Comparing two different test models that have been built by different engineers with different views is a challenging task. It requires the identification of the similarities and differences among the elements of the two test models, and reconciling inconsistency between the two models. In the discipline of model comparison, there are two methods: three-way merging and two-way merging [32]. In addition to the merged models, the three-way merging, requires the existence of the base model or the changes logs. It compares each model to the base model to identify the changes of that model with respect to the base model. These changes can be classified into three categories: added, deleted, and/or modified. Based on this information, the method merges the models to generate a new base model. Comparing to three-way merging, the two-way merging is harder since it does not rely on any additional information beside the merged models. However, the two methods share the same assumption. They assume that the merged models are evolved from the same base model. Researchers build their approaches around certain model features, such as universally unique identifiers (UUID), to calculate the similarities and differences between the elements of their models. While the process of identifying similarities and differences between the models and detecting conflicts can be automated, reconciling conflicts require user interaction [32–34]. In the testing domain, component test models are usually generated independently from the corresponding design specifications. Hence, the assumption of the evolving of the merged models from the same source is not applicable in this domain. However, we can benefit from the characteristics of our domain since we are not developing a general merging approach. Test cases describe partial behavior of the system under test (SUT). They actually represent a partial view of the SUT that is the focus of the test designer. Thus, different test cases may describe the same system behavior from different angles. We focus on such test cases to build our integration approach. Our approach follows two-way merging. However, we do not assume that the test models evolved from the same source. We assume that test cases overlap since they describe the same system from different angles. Therefore, test models share elements of the system that can be identified through their names and attributes.

2.2 Some Definitions

We define here some concepts used throughout this chapter.

Definition 1. (Test Model)

A test model is expressed as a tuple $M = (P, T)$, where

- P is the test package and
- T is a set of test cases.

Definition 2. (Test Package)

A test package is expressed as a tuple $P = (tcn, tcm, sut)$, where

- tcn is the test control,
- tcm is a set of test components required to realize the test execution (test stubs), and
- sut is a set of components under test.

Definition 3. (Test Case)

A test case is expressed as a tuple $t = (I, E, R)$, where

- I is a set of instances,
- E is a set of events (defined further in Definition 4), and
- $R \subseteq (E \times E)$ is a partial order reflecting the transitive closure of the order relation between events on the same axis and the sending and receiving events of the same message.

We categorize events into three categories: message events, time events, and miscellaneous events. Message events, the sending event and the receiving event, represent the two ends of messages exchanged between two instances referred to as the sender and the receiver, respectively. In this chapter, messages are instances of an execution trace. Hence, they are unique throughout a single system execution. Time events represent events related to timers. Each timer is associated with one instance. We classify the rest of event types, such as instance termination and UTP verdict, into the third category. Notice that the association between events and instances is part of the event definition.

Definition 4. (Event)

1. A message event E_{msg} is a tuple *(ty, nm, owner, msg, oIns)*, where
   a. *ty* $\in$ {*send, receive*},
   b. *nm* is the event name,
   c. *owner* is the instance where the event belongs to. *owner = (nm, st)*, where
      i. *nm* is the instance name and
      ii. *st* is the UTP stereotype of the instance,
   d. *msg* is the message the event is related to,
   e. *oIns* is the other instance related to msg, *oIns = (nm, st)*, where
      i. *nm* is the instance name and
      ii. *st* is the UTP stereotype of the instance.
2. A time-related event E_{time} is a tuple *(ty, nm, tm, owner, pd)*, where
   a. *ty* $\in$ {*timeOutMessage, startTimerAction, stopTimerAction, readTimerAction, timerRunningAction*},
   b. *nm* is the event name,
   c. *tm is* the timer name,
   d. *owner* is the instance where the event belongs to, *owner = (nm, st)*, where
      i. *nm* is the instance name and
      ii. *st* is the UTP stereotype of the instance,
   f. *pd* is the timer value.
3. A miscellaneous event E_{misc} is a tuple *(ty, nm, v, owner)*, where
   a. *ty* $\in$ {*Action, Terminate, UTPverdict*},
   b. *nm* is the event name,
   c. *v* is the value associated with the event (this value can be *pass, fail, inconclusive, error* in case *ty = UTPverdict*),
   d. *owner* is the instance where the event belongs to, *owner = (nm, st)*, where

i. *nm* is the instance name, and

ii. *st* is the UTP stereotype of the instance.

We use the test model specified in Fig. 3 to illustrate our definitions. The test model is composed of a test package, *p*, represents the test architecture and two test cases, t_1 and t_2, represent the test behavior. To distinguish between the sending and receiving events of the same message, we suffix the message name with the first letter of the corresponding action. We represent this test model, *M*, as follows:

$M = (P, T)$, *with*

$P = (TC, \varnothing, \{CUT\})$,

$T = \{t_1, t_2\}$,

$t_1 = (\{tc, cut\}, \{m_{1s}, m_{2r}, m_{3s}, m_{4r}, ver, m_{1r}, m_{2s}, m_{3r}, m_{4s}\}, \{(m1s,m2r),(m2r,m3s),$
$(m3s,m4r),(m4r,ver),(m2s,m3r),(m3r,m4s),(m1s,m1r),(m2s,m2r),(m3s,m3r),(m4s,$
$m4r),(m1s,m3s),(m2r,m4r),(m2r,m3r),(m3s,ver),(m2s,m4s),(m3r,m4r),(m2s,m3s),$
$(m3s,m4s),(m4s,ver),(m1s,m4r),(m1s,m3r),(m2r,ver),(m2r,m4s),(m2s,m4r),$
$(m3r,ver),(m1s,ver), (m1s,m4s),(m2s,ver)\})$

$tc = (\text{``}tc\text{''}, TestContext)$,

$cut = (\text{``}cut\text{''}, SUT)$,

$m_{1s} = (send, \text{``}m_{1s}\text{''}, tc, m_1, cut)$,

$m_{2r} = (receive, \text{``} m_{2r}\text{''}, tc, m_2, cut)$,

$m_{3s} = (send, \text{``}m_{3s}\text{''}, tc, m_3, cut)$,

$m_{4r} = (receive, \text{``}m_{4r}\text{''}, tc, m_4, cut)$,

$ver = (UTPverdict, \text{``}ver\text{''}, \text{``}pass\text{''}, tc)$,

$m_{1r} = (receive, \text{``}m_{1r}\text{''}, cut, m_1, tc)$,

$m_{2s} = (send, \text{``}m_{2s}\text{''}, cut, m_2, tc)$,

$m_{3r} = (receive, \text{``}m_{3r}\text{''}, cut, m_3, tc)$,

$m_{4s} = (send, \text{``}m_{4s}\text{''}, cut, m_4, tc)$.

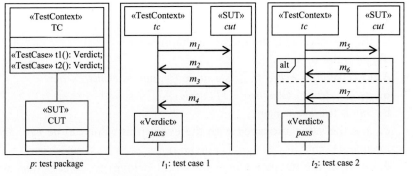

Fig. 3 Example of test model (*M*).

$t_2 = (\{tc, cut\}, \{m_{5s}, m_{6r}, m_{7r}, ver, m_{5r}, m_{6s}, m_{7s}\}, \{(m_{5s}, m_{6r}), (m_{5s}, m_{7r}), (m_{6r}, ver),$
$(m_{7r}, ver), (m_{5r}, m_{7s}), (m_{5s}, m_{5r}), (m_{6s}, m_{6r}), (m_{7s}, m_{7r}), (m_{5s}, ver), (m_{5r}, m_{7r}), (m_{5s}, m_{7s}),$
$(m_{6s}, ver), (m_{7s}, ver), (m_{5r}, ver)\}),$

$tc = (\text{``}tc\text{''}, \ TestContext),$

$cut = (\text{``}cut\text{''}, \ SUT),$

$m_{5s} = (send, \ \text{``}m_{5s}\text{''}, \ tc, \ m_5, \ cut),$

$m_{6r} = (receive, \ \text{``}m_{6r}\text{''}, \ tc, \ m_6, \ cut),$

$m_{7r} = (receive, \ \text{``}m_{7r}\text{''}, \ tc, \ m_7, \ cut),$

$ver = (UTPverdict, \ \text{``}ver\text{''}, \ \text{``}pass\text{''}, \ tc),$

$m_{5r} = (receive, \ \text{``}m_{5r}\text{''}, \ cut, \ m_5, \ tc),$

$m_{6s} = (send, \ \text{``}m_{6s}\text{''}, \ cut, \ m_6, \ tc),$

$m_{7s} = (send, \ \text{``}m_{7s}\text{''}, \ cut, \ m_7, \ tc).$

3. INTEGRATION TEST GENERATION

By definition integration-level testing puts emphasis on the interactions between the involved components. Hence, integration test cases can be generated from the component test cases that capture such interactions. These interactions can be direct between components or indirect through mediators. However, not all component test cases capture such interactions. Therefore, we need to analyze the available component test cases to identify and select the ones that capture interactions between the components.

The integration test generation approach supports incremental integration strategies, bottom–up, top–down, or ad hoc. With such strategies system integration is a recursive process which integrates components one by one until reaching the complete system. Our test generation approach supports such recursive process. Component test models are integrated incrementally to generate the integration test model for the current iteration as shown in Fig. 4. Eventually, the generated test model will be integrated with the

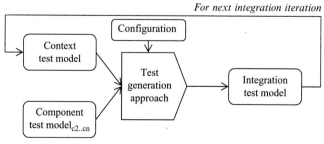

Fig. 4 Overall integration test generation approach.

component test model of the next integrated component to generate the next integration test model, and so on. In case of presence of complex mediators, test stubs, that are underspecified in the test behavior, a configuration model should be provided in order to reveal the behavior of the mediators; i.e., relate mediators' outputs to their corresponding inputs that is specified in the test behavior.

The approach is composed of four processes as shown in Fig. 5. The first two processes analyze the given test models to detect interactions between the integrated system components. The last two processes generate and optimize the output model. We elaborate more on these processes in the following subsections.

3.1 Identification Process

As mentioned earlier we adopted the two-way merging. In order to generate integration test cases from component test cases, we have to inspect the component test cases and select the ones that contain integration test scenarios. In order to inspect such test cases, we need to recognize the identities of the specified test objects. In this identification process, we aim at locating the declaration of one of the integrated components in the test model of the other integrated component or the existence of a shared test object that is specified in both test models. Test objects can be classified into three kinds: test control, implementation under test (IUT), and test stub. The IUT can be SUT, component under test (CUT) or any fragment of software under test. Using UTP stereotypes, the approach can easily recognize the test objects specified on the input test models as shown in Table 1.

However, the identification is not always straightforward. With the exception of the CUT, test objects can emulate the behavior of more than one system component and/or system environment. The most used pattern for test cases is composed of two test objects, the IUT and the test control.

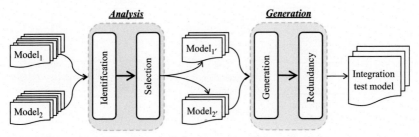

Fig. 5 Different processes (steps) of the test generation approach.

Table 1 UTP Stereotypes for Identifying Test Objects

UTP Stereotype	Test Objects Tagged to
«*TestContext*»	Test controls
«*SUT*»	System/Component under test
«*TestComponent*»	Test stubs, test environment

In this pattern, the test control emulates the test environment in addition to controlling the test case. In other words, the test control embeds the behavior of any system component and/or environment that is required to realize the test execution. Hence, the approach has to investigate the behavior of test objects stereotyped by *TestContext* or *TestComponent* to reveal the identity of the CUTs or shared test objects that may be embedded within these test objects. In order to achieve that, our approach maps the behavior of the test objects of one test model to the behavior of known test objects in the other test model. However, UTP stereotypes can be applied only on the test architecture. Up to UTP version 1.2, the behavior part has been left out from the UTP metamodel [10]. We have to rely on the UML specification to reveal the relations between the UML elements in the test architecture using UML class diagram, and the UML elements in the test behavior using UML sequence diagrams. Furthermore, we have two exceptions. First, there is no comparison between two test controls since both of them are unknown. Second, there is no comparison with test objects that are specified in both test models.

3.2 Test Case Selection

Based on the results of the identification process, the selection process analyzes the test cases in both test models and selects the ones that capture interactions between the integrated components. We investigated two patterns for such test cases. The first pattern comprises individual test cases while the second pattern comprises two test cases, one from each test model.

In the first pattern, we look for an individual test case that specifies both of the integrated components. One component is specified as CUT and the other is specified as a test stub or embedded in the behavior of a test object. In other words, we look for test cases that emulate the system component of the other test model. Furthermore, there must be an interaction between the integrated components captured by the selected test case with at least one exchanged message. In this case, we select such test cases to generate integration test cases, and we refer to such pattern as a *complete integration test scenario*. Fig. 6A illustrates such a pattern, where the other component *Comp2* is specified as a test stub. The figure shows two component test cases, one from each test model. By examining each test case individually, we cannot conclude that any one of them captures an integration test case. However, by mapping the two test cases using the information gathered from the identification process, we can make certain observations. The first observation is that the two test cases capture the same test scenario. The second observation is that the component of the second model, *Comp2*, is represented in the first model test case as a test stub. Hence, we conclude that component *Comp2* is emulated by the test case of the first model. The third observation is that there is an interaction between the integrated components, *Comp1* and *Comp2*, by exchanging messages *m2* and *m3*. Therefore, we conclude that the test case of the first test model captures a *complete integration test scenario*. Furthermore, we can observe that the test control *TC2* emulates the behavior of the component *Comp1*. Hence, we can reclaim this behavior and initiate a new instance for *Comp1* and select the test case as a *complete integration test scenario* too. However, one of the generated test cases will be removed later by the redundancy checking process.

In the second pattern, we investigate the existence of integration test scenarios that are split across two component test cases. Each part of such scenario is captured by one of the test cases of the two test models. The scenario must represent interaction among the integrated components. This interaction can be direct or indirect through other test objects. These test objects

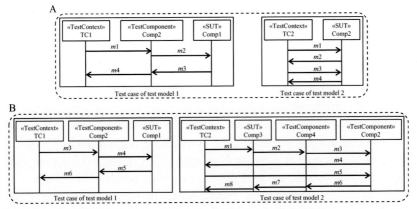

Fig. 6 Patterns of integration test scenarios: (A) Pattern 1: complete integration test scenario and (B) Pattern 2: complement integration test scenario.

can be other system components that have not been integrated yet or the system environment as in client/server applications. Fig. 6B illustrates this pattern. In this example, the integration is applied on components *Comp1* and *Comp3*. There is one shared test object which is explicitly specified in both test cases. In addition, test object *Comp4* is explicitly specified by an instance in one test case and implicitly specified in the other test case as partially emulated by the test control *TC1*. The next step in our process is to examine the existence of interaction between the integrated components with at least one exchanged message. The two test cases are selected to generate integration test case if an interaction is detected between the integrated components.

3.2.1 Interaction Detection

In order to detect interactions between the integrated components, we build the event dependency tree (EDT) as shown in Fig. 7. The EDT represents the order relation between the events of the involved test cases. Each node represents an event. As naming convention, the event name is composed of the message name followed by the first letter of the action name, *send* or *receive*. We build the EDT based on the given test cases. The approach builds the EDT in two or three steps depending on the selection pattern. In the first step it creates an EDT for each instance lifeline. Next, it merges the EDTs of the same test case by linking the nodes of the corresponding sending/receiving events of the same message. The process proceeds to the third step only in the case of the second selection pattern. In the third step, the process merges the two EDTs of the involved test cases by matching the shared events

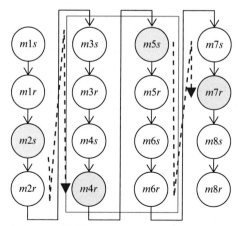

Fig. 7 Event dependency tree (EDT).

among the two test cases. Event matching, depending on the event type, is done according to Definition 5.

Definition 5. (Event Matching)

Let e_1 and e_2 be two events of the same kind from two different instances, then e_1 and e_2 match (and noted $e_1 = e_2$) if and only if:

1. $Match_{msg}(e_1, e_2) = \{e_1 \in E_{msg}, e_2 \in E_{msg} \mid (e_1.ty = e_2.ty) \wedge (e_1.msg = e_2.msg) \wedge ((e_1.nm = e_2.nm) \vee (((e_1.owner.nm = e_2.owner.nm) \vee (e1.owner.st \neq SUT) \vee (e2.owner.st \neq SUT) \wedge ((e_1.oIns.nm = e_2.oIns.nm) \vee (e1.oIns.st \neq SUT) \vee (e2.oIns.st \neq SUT))$, or

2. $Match_{ime}(e_1, e_2) = \{e_1 \in E_{time}, e_2 \in E_{time} \mid (e_1.ty = e_2.ty) \wedge (e_1.tm = e_2.tm) \wedge (e_1.pd = e_2.pd) \wedge ((e_1.nm = e_2.nm) \vee (e_1.owner.nm = e_2.owner.nm) \vee (e1.owner.st \neq SUT) \vee (e2.owner.st \neq SUT))\}$, or

3. $Match_{misc}(e_1, e_2) = \{e_1 \in E_{misc}, e_2 \in E_{misc} \mid (e_1.ty = e_2.ty) \wedge (e_1.v = e_2.v) \wedge ((e_1.nm = e_2.nm) \vee (e_1.owner.nm = e_2.owner.nm) \vee (e1.owner.st \neq SUT) \vee (e2.owner.st \neq SUT))\}$.

At the same time, the process takes into account the information gathered during the identification process to match instances that are syntactically different but one emulates the other, e.g., *TC1* and *Comp4*. The process examines two characteristics of the EDT. The first characteristic is the existence of overlapping between the EDTs of the two test cases in the final one. This characteristic is related to the second selection pattern and is evaluated during the third step based on the existence of shared events or not. Fig. 7 shows the EDT of the two test cases in Fig. 6B. The EDT of the first test case, surrounded by dotted rectangle, is overlapping completely with the EDT of the second test case, i.e., all of the events of the first test case are shared events. The second characteristic is the existence of interactions between the two CUTs. This characteristic is checked by:

1. locating a node that represents a sending event of one of the integrated components, then
2. searching the branches of such a node to locate a node that represents a receiving event of the other integrated components.

The process repeats these steps until a send and a receiving events located on the same path is found. From Fig. 7, there are two traces that satisfy this characteristic: (*m2s*, *m4r*) and (*m5s*, *m7r*). Therefore, the two component test cases are selected to generate integration test cases.

The process selects the involved test cases if the two characteristics are satisfied. Otherwise, it proceeds by examining other test cases from the given test models. The approach stops the current test integration generation if it does not select test cases from the given test models.

3.3 Test Behavior Generation

In this chapter, we first generate the test behavior, then we construct the test architecture. The process generates integration test cases corresponding to the two selection patterns of test cases that have been selected by the previous process.

In the first pattern, test objects of the selected test cases represent the integrated component of the other test model. These test objects can be test stubs or test controls. Furthermore, their instance can represent exclusively the integrated component or have additional test behavior to emulate other entities and/or provide test control. Hence, we have two different scenarios to handle in this pattern. In the first scenario where the instance of the test object represents exclusively the integrated component, the process generates integration test case by relating the instance to the integrated component. To illustrate this scenario, let us consider the example in Fig. 8 where the first integration test case is generated from the component test case of the first test model in Fig. 6A. In the case of the second scenario where the instance of the test object represents partially the integrated component, the process generates integration test case by creating a new instance that represents the integrated component and relocating the corresponding events to it. To illustrate this scenario, the second integration test case in Fig. 8 is generated from the component test case of the second test model in Fig. 6A. As one can notice the two integration test cases are identical. This is because the two component test cases capture the same test scenario. This redundancy is managed in the next step.

In the second pattern, pairs of test cases, one from each test model capture an integration test scenario. The two test cases, in each pair, have a shared test behavior and test objects. The process merges each pair of test cases to build integration test cases. During the merging, we have to align and merge the shared test behavior of identical instances, which are specified on both test cases.

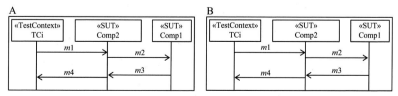

Fig. 8 Generated integration test cases: (A) integration test case 1 and (B) integration test case 2.

Definition 6. (Shared Events)

Let E_1 and E_2 be two sets of events of two different test cases. The set of shared events, *se*, is defined as follows:

$$se = \{(e_1, e_2): e_1 \in E_1 \text{ and } e_2 \in E_2 \mid e_1 = e_2\}$$

The process creates a new instance for each integration test case to represent the integration test control, and its behavior will be the sum of the behavior of the given test controls. At the same time, we have to maintain the specification of both test cases; e.g., if one test case specifies n instances of a test object and the other test case specifies m instances of the same test object, then the approach merges *min(n, m)* instances that define shared behavior. The merging operator is defined as follows:

Definition 7. (Merging Test Cases)

Let $t_1 = (I_1, E_1, R_1)$ and $t_2 = (I_2, E_2, R_2)$ be test two cases and se_{12} be the corresponding set of shared events. The generated integration test case is defined as follows:

$$t_{12} = t_1 + t_2$$
$$= (g(I_1) \cup g(I_2), f(E_1) \cup f(E_2), f(R_1) \cup f(R_2))$$

where

$g(I)$: $\{i: i \in I \text{ and } \forall\ i \text{ if } i.st = TestContext, \text{ then } i = tc_i\}$.

The function transforms component test controls to integration test control.

$f(E)$: $\{e: e \in E \text{ and if } (e_1, e_2) \in se_{12} \text{ and } e = e_1 \text{ then } e = e_2\}$.

The function replaces the first element of a pair in the shared events with the second element to eliminate the duplication of identical events. In other words, it relocates emulated events to their corresponding test objects.

Fig. 9 shows the integration test case generated from the merging of the two component test cases in Fig. 6B. At the end of the generation of the test behavior, the redundancy checking process removes duplicated test cases before the generation of the test architecture.

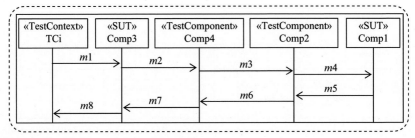

Fig. 9 Generated integration test case 3.

3.4 Checking for Redundancy

The generated test behavior may include redundant test cases which may be produced as a result of two situations. First, the same test scenario is specified in the two given test models as shown in Fig. 6A. The second case is when a test case is selected by the two selection patterns. In this case, in addition to the generated integration test case from the merging, the approach generates another integration test cases. However, the latter generated integration test case is identical to, or part of, the first generated integration test case. Hence, it should be removed from the generated test model. This case can be explained with the test cases in Fig. 6B. The two test cases contain a shared test object *Comp2* with shared behavior. The approach merges the two test cases to generate integration test case as shown in Fig. 9. On the other hand, the test control of the second test case *TC2* emulates the behavior of the CUT *Comp1*. Hence, the approach generates an integration test case by adding a new instance for *Comp1* and relocate the correspond events. The generated test case is similar to the one in Fig. 9. The second test case should be removed since it is redundant.

To remove redundancy among the generated integration test cases, we map the test cases against each other. We define test case inclusion as follows.
Definition 8. (Integration Test Case Inclusion)

Let $T_1 = (I_1, E_1, R_1)$ be an integration test case and $T_2 = (I_2, E_2, R_2)$ be another integration test case, then $T_1 \subseteq T_2$ if and only if the following conditions are satisfied:

$$I_1 \subseteq I_2$$
$$E_1 \subseteq E_2$$
$$R_1 \subseteq R_2$$

3.5 Test Architecture Generation

After generating the test behavior, we build the test architecture. The integration test architecture is created from the specification of the generated integration test behavior. The given test architectures of the component test models are used to relate test objects to their external models, if found. We use the UTP test package in this chapter. Table 2 summarizes the important mappings to generate test architecture from test behavior. The generation process traverses the test cases. It goes through the elements of each test case and creates the equivalent elements in the test architecture. Internal references between elements of the test behavior and the corresponding elements of the test architecture are built. After that, the process compares the generated test objects, UML classes, to their corresponding test objects in the given component test cases. In case where any test object has a reference to an external model, the process updates the corresponding generated test object with the same reference. The most important test object is the SUT, which is always externally referenced. Finally, the process adds a reference to the UTP to enable its stereotypes in the generated test model. Fig. 10 shows the generated test architecture for the generated test behavior in Fig. 9.

3.6 Some Properties of the Integration Test Generation Approach

During the development of the generation approach, we investigated the impact of the integration strategy on our approach. System components are integrated using different integration strategies, some of them are well known such as top-down, bottom-up, big-bang, and ad hoc. The overall generated test behavior for the same set of system components must be equivalent regardless of the applied integration strategy. Hence, we have investigated two properties of the generation approach: commutativity and associativity. More details are given in Appendix A.

Table 2 Mapping Test Behavior to Test Structure

Test Behavior	Test Architecture
UML lifeline	UML class
UML message	UML association
UML sequence diagram	UML operation

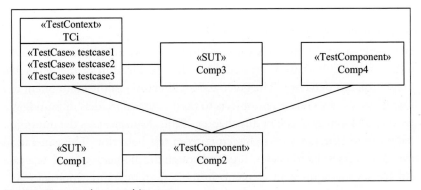

Fig. 10 Generated test architecture.

Furthermore, we have investigated the saving of test information from one integration iteration to the subsequent ones. Usually, there is a single component test model for each system component. The component test model holds all test information regarding the component. There is typically a test case or more for each targeted function. These test cases exercise the system component through its different interfaces. For each system integration, we need different set of test cases that capture test information related to the interfaces between the currently integrated components. Accordingly, integration test cases capture test information regarding the currently integrated components and neglect test information related to interfaces with system components that have not yet been integrated. Therefore, we need to carry-on test information of component test cases that is not captured by the generated integration test cases to be used in subsequent test integrations. We use the example in Fig. 11 to illustrate this point. The system is composed of four components, which are integrated according to the illustrated integration strategy. Usually, there is a component test model for each component that covers the corresponding component functionality through its interfaces, e.g., the component test model of the component A capture test information related to the interfaces ab and ad. The integration goes through three iterations: $(A + B)$, $((A + B) + C)$, and $(((A + B) + C) + D)$. In the first iteration, the approach analyzes the two test models A and B and uses test information related to the interface ab to generate the integration test model AB. In the second iteration, the approach analyzes the test models AB and C and uses test information related to the interface bc to generate the integration test model ABC. In the last iteration, the approach analyzes the test models ABC and D and uses test information related to the interfaces dc and ad to generate the integration test model $ABCD$. Here, we may encounter some issues during the second and third iterations. Let us take the second iteration to explain the issues. The integration test model AB capture test information related only to the interface ab; while the component test model

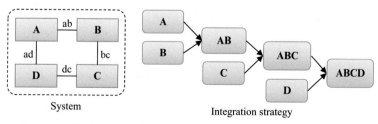

Fig. 11 Example of integration strategy.

C capture test information related to the interface *bc* and *dc*. Test information of component test model *B* related to the interface *bc* are probably ignored by the approach during the first iteration except if some test cases capture test information for both interfaces, *ab* and *bc*. The same situation applies to the test information related to the interface *ad*. The approach will probably ignore this information during the first iteration. Hence, the generated integration test model *AB* is probably missing some test information related to interfaces *bc* and *ad*. When the approach tries to generate the integration test model in the second iteration, it probably cannot identify and locate any shared test behavior between the two test models, *AB* and *C*. It will generate nothing. Hence, we need to save test information regarding interfaces *ad* and *bc* during the first iteration to be used in the subsequent iterations.

We have investigated two techniques, as shown in Fig. 12, to carry test information of component test models to subsequent integration iterations: selective and cumulative integration. The selective technique carries the component test models along with the generated integration test model to the subsequent integration iterations. In each integration iteration, the approach is applied several times to generate the corresponding integration test model. First, it uses the former integration test model and the component test model of the currently integrated system component to generate the integration test model for the current iteration. Next, it uses the carried-on component test models of previously integrated components and the component test model of the currently integrated component to generate additional test cases. The generated integration test model and

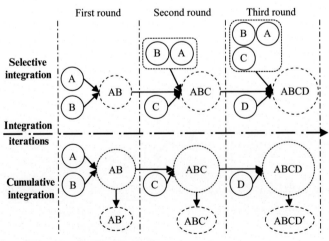

Fig. 12 Cumulative vs selective integration.

the component test model of the integrated components, including the currently integrated component, are carried to the subsequent integration iteration. In this technique, we carry on individual component test models throughout the integration-level testing.

In the cumulative technique, we build a global model by merging the given component test models. In each integration iteration, we merge the component test model of the currently integrated system component with the global model and generate the integration test model for each iteration by selecting test cases from the global model that capture interactions between the integrated components. In this technique, we have one reference to carry on throughout the integration-level testing, which is the global model. However, during our investigations, we encountered that the cumulative technique may produce invalid test behavior. Therefore, we ignored it and used only the selective technique.

4. ACCEPTANCE TEST OPTIMIZATION

The approach maps test cases of the acceptance level to test cases of the integration level. The mapping technique is based on the comparison of the involved test cases. We consider that part of these test cases target the same system functionalities since they describe the same system from different perspectives. We aim to reduce the acceptance test execution time by reducing the number of acceptance test cases. This can be achieved by eliminating acceptance test cases that have already been exercised on the system during integration-level testing. However, one needs to be careful as integration test cases are mainly applied on subsystems. Usually, they emulate some of the system components that have not yet been integrated. Hence, they cannot substitute acceptance test cases that aim at testing the whole system. There are two situations where the integration test cases are suitable to substitute acceptance and system test cases. The first situation includes test cases applied on the last iteration of the integration-level testing. These test cases are exercises during the integration of the last component to the subsystem to build a complete system. Therefore, they are applied on a complete system. The second situation includes integration test models applied on subsystems that fulfill completely the requirements of some of the system functionalities. Hence, test cases of such test models that examine these functionalities are actually applied on complete subsystems. In other words, the test cases do not emulate system components. Therefore, we need to examine the given integration test cases in order to select the ones that can be mapped to the acceptance test cases. The approach is composed of two processes: the selection process and the mapping process as shown in Fig. 13. The approach is described in terms of acceptance test models but it is applicable to system test model optimization as well.

4.1 Integration Test Case Selection

The integration test models should not contain any emulation of system components in order to qualify for comparison against the acceptance test model. We have to examine the given integration test models for the use of test stubs of system components. The test stubs may be specified in some test cases and not specified in other test cases of the same test model. Hence, our examination will be on the level of the test cases instead of the level of the test models. Test cases of the last integration test model are qualified to be mapped to the acceptance test cases. Hence, we select them directly without further examination. For the rest of the integration test models, we compare the behavior of their test stubs and test controls to the behavior of the CUTs of the subsequent integration test models as shown in Fig. 14. More

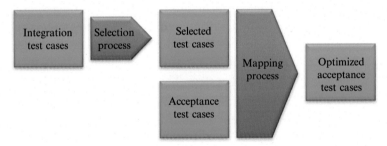

Fig. 13 Acceptance test optimization approach.

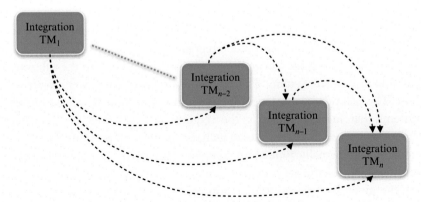

Fig. 14 The selection process for acceptance test optimization.

specifically, the approach compares the behavior of the test stubs and test controls of each test case in an integration test model to the behavior of the integrated components of each test case in the subsequent integration test models.

The selection process selects test cases that do not include test stubs of system components in their specifications. The selection criterion is given formally in Definition 9.

Definition 9. (Selection Condition)

Let $T_{kh} = (I_{kh}, E_{kh}, R_{kh})$ be the integration test case h at the integration iteration k and $T_{ij} = (I_{ij}, E_{ij}, R_{ij})$ be the integration test case j at integration iteration i, where $i > k$, then T_{kh} does not emulate the system component of T_{ij}, if and only if the following condition is satisfied:

$$Sel_{kh} = \frac{\forall (e_j, e_h) . e_j \in E_{ij}, e_h \in E_{kh} |\ (e_j \neq e_h) \vee}{((e_j = e_h) \wedge (e_j.owner.st \neq SUT))}$$

The selection process stops the comparison as soon as the condition is no longer satisfied, i.e., it returns false. Accordingly, the corresponding test case is excluded from the selection when the selection condition is evaluated to false.

The results of the selection process depend on the integration order since the usage of test stubs of system components depends on the integration order. We may not require any test stubs when we choose the right integration order. There is a lot of research being done currently on the selection of the right integration order [35–37].

4.2 Mapping Acceptance Test Cases to Integration Test Cases

The mapping process compares the acceptance test cases against the selected integration test cases. The process removes acceptance test cases from the test model if they are included in the selected integration test cases. However, the acceptance-level testing has a different perspective of the system than the integration-level testing. In the acceptance-level testing, we see the system as a block and we examine it through its external interfaces, while in the integration-level testing, we see fragments of the system, and we examine it through its external interfaces as well as through the internal interfaces of the currently integrated component. Consequently, the test cases are different with respect to the test objects described in each testing level. Acceptance test cases require at least two test objects: test control and SUT, while integration test cases require at least three test objects: test control, CUT, and subsystem.

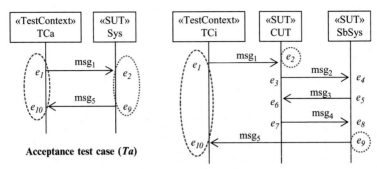

Fig. 15 Distributed Events

Furthermore, we have to take into account that the events specified on a lifeline of a test object in an acceptance test case may be distributed over several lifelines in the mapped integration test case as shown in Fig. 15. The behavior of the two test objects, TCa and Sys, in the acceptance test case is distributed over three test objects, TCi, CUT, and SbSys, in the integration test case. Moreover, integration test cases may have extra behaviors that reflect internal interactions between the integrated component and the subsystem. In other words, we should not expect the acceptance test case to be a complete fragment/block within the integration test case.

The test case inclusion as specified by Fig. 15 is used to map test cases of the same test model. It cannot be used in this process because it examines the instances. As we mentioned earlier, instances in this mapping are fundamentally different. This inclusion cannot be used to compare integration test cases from different integration iterations. We derive a new inclusion relation that does not depend on the instances of the test cases.

Definition 10. (Test Case Inclusion)

Let $T_a = \{I_a, E_a, R_a\}$ be an acceptance test case and $T_i = \{I_i, E_i, R_i\}$ be an integration test case, then the acceptance test is included in the integration test case if and only if the following conditions are satisfied:

1. $E_a \subseteq E_i$
2. $R_a \subseteq R_i$

5. A CASE STUDY: A LIBRARY MANAGEMENT SYSTEM

To illustrate our framework and partially demonstrates its effectiveness we built a prototype tool. We ran several case studies. In this chapter, we present the library management system case study and briefly discuss the results. We considered a library management system that is composed of four components to provide users with the main library services. These services are covered by test cases designed to build component test models as well as the acceptance test model. Fig. 16 shows the system architecture and some of the test models. In this case study, we apply our generation approach on the component test models to generate integration test models. We ran the prototype tool twice using two different integration orders to demonstrate the properties of the test generation approach. Next, we use the prototype tool to map the generated integration test models to the acceptance test model to reduce the acceptance test model.

The tool integrated four component test models through three iterations. It generated three test models for both integration orders with, of course, different sets of test cases as shown in Table 3. This is similar to what we have experienced in other case studies during this research. The tool generated the same number of test cases for both integration orders: seven test cases. Furthermore, the generated test cases cover all of the specified system services. Two test cases were repeated in the second integration order since they emulated a system component in the second iteration.

The optimization approach removed all of the acceptance test cases as shown in Table 4. In both integration orders, seven integration test cases that do not emulate system components were selected. The complete acceptance test cases are removed since they matched (included in) the selected integration test cases. Therefore, there is no need to execute the given acceptance test model during the acceptance-level testing for this particular case study as they have already been exercised during integration testing.

6. RELATED WORK

To the best of our knowledge, systematic reuse of test models to generate next level test models has not been covered in MBT [38]. On the other hand, different techniques, such as test coverage [1,38,39], have been proposed to minimize the number of tests. However, the scope of such techniques is the reduction of the number of tests within the same level of testing.

The work of Le [13] is the only research work closely related to our work. The author proposes a composition approach based on UML 1.x collaboration diagrams. The test model is built manually and is composed of two roles/players: the component under test role and the tester role. The

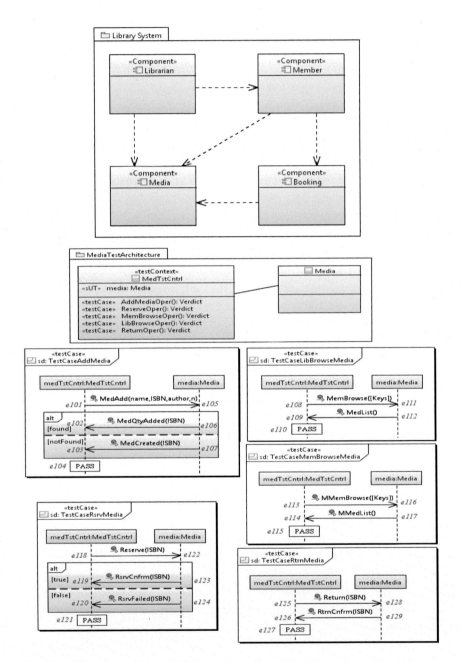

Fig. 16 A library management system.

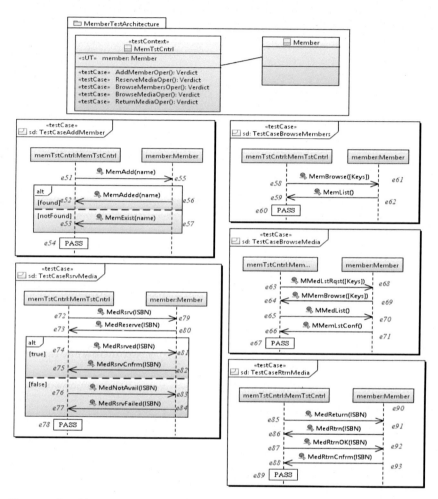

Fig. 16—Cont'd

tester role controls and performs the test-suite and simulates all necessary stubs and drivers. The author demonstrated the reusability of the tester role from component-level testing to integration-level testing through the introduction of adaptors between the component test models. This approach does not address the synchronization between events of the test behavior. The test case selection is not clear, since not all the component test cases are suitable for the integration-level testing.

There are a lot of research activities on model merging specially in the domain of version control systems (VCS) [40]. These approaches are based on the assumption that the input models have evolved from the same base

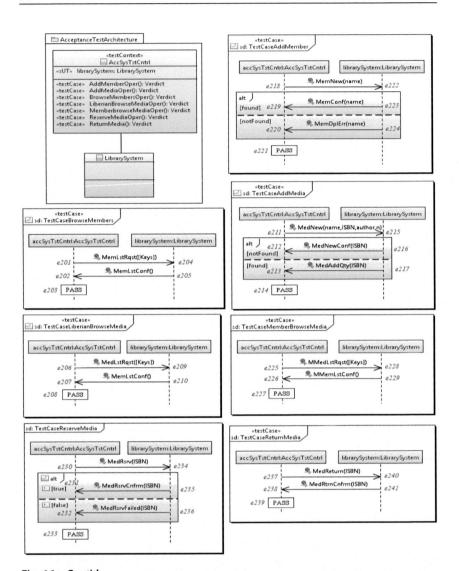

Fig. 16—Cont'd

Table 3 Integration Test Generation Results for the Library Management System

Iteration	Integrated Components	First Integration Order Generated Test Cases	Second Integration Order Generated Test Cases
1	2	2	2
2	3	3	2
3	4	2	5
Total		7	7 + 2

Table 4 Acceptance Test Optimization Results for the Library Management System

	First Integration Order # Test Cases	Second Integration Order # Test Cases
Integration test models	7	9
Selected test cases	7	7
Acceptance test model	7	7
Excluded test cases	7	7
Optimized acceptance test model	0	0

model [16,17], and some approaches even require the existence of the base model [16,17]. These approaches are not applicable in the testing domain since test models are usually built by different engineers with different views. The model comparison approaches use different calculation methods to identify similarities and differences between different models [41,42]. In our approach, we use two methods for comparing model elements, name-based matching, and feature-based matching. While not all UML model elements have names, practical studies show the effectiveness of this method [42].

Hélouët et al. [25,44] propose a merging approach for message sequence charts (MSCs) [15]. The approach merges all scenarios to build the global behavior of the system. The approach covers both basic MSCs (bMSCs) and high-level MSCs (HMSCs). These investigations focused more on the theoretical aspects and decidability-related issues. Inline operators, similar to UML combined fragments, are not covered since they can be substituted by HMSCs. We support UML combined fragments. The approach uses different composition operators, sequential, alternative, parallel, and iteration, that are specified in HMSCs. We only use the merge operator. More important, we are dealing with finite behaviors where merging and comparison can be done.

7. CONCLUSION

In this chapter, we proposed a MBT framework that relates and links different software testing levels, enables automation, reusability, and optimization. Two approaches have been concretely proposed in this framework, test generation and test optimization. Both approaches assume component test cases are well formed and cover all component interfaces and services. Test models are specified using UTP, which enables their systematic transformations into test code that can be exercised on the IUT using well-known test execution environments, such as JUnit and TTCN-3 [9]. Usage of standard notations enhances the collaboration and certainly helps bridging the gap between the development and testing activities.

The proposed framework enables reusability across the software testing levels. Test models are systematically generated from preceding test models. We discussed in details the generation of integration test models from component test models. We defined a test case merging operator to integrate component test cases that have shared behavior.

The proposed framework also enables systematic test optimization across the software testing levels. Test models are related to preceding test models to remove the ones that have already been exercised. Test optimization reduces the size of the test models, shortens test execution time, and reduces the cost of software testing. We discussed an approach that optimizes acceptance test models by relating them to the integration test models. This approach is also applicable to system test models.

We built a prototype tool and experienced with several case studies. In this chapter, we reported on the library management system case study and showed how the acceptance test model can be reduced because acceptance test cases have been covered during integration testing. However, further validation is required with larger size and industrial case studies to demonstrate the applicability and the efficiency of our framework.

MBT is a maturing field of research and practice. It is gaining in popularity in several domains including safety critical domains like avionic and automotive. MBT enables abstraction, reuse, and automation which are much needed to improve the quality of complex software systems. It alleviates the testers from routine tasks such as test cases generation, coverage evaluation, transformations, etc. However, its complete adoption by practitioners depends on the availability of industrial-strength tools, especially for the next generation cyber physical and Internet of Things based systems which will be more complex than current software systems.

ACKNOWLEDGMENTS

This work has been partially supported by the Natural Sciences and Engineering Research Council (NSERC) of Canada. We would like to thank Dr. Reinhard Gotzhein for comments and feedbacks on earlier versions of this work.

APPENDIX A: PROPERTIES OF THE INTEGRATION TEST GENERATION APPROACH

System integration may take different strategies: top-down, bottom-up, ad hoc, and big-bang, and different sequences/orders to integrate the system components. The generated test behavior for the same set of system components must be equivalent regardless of the adopted integration strategy and order. The intermediate results, at a given step, may not be equivalent since they integrate different sets of components.

Test cases are equivalent when they specify the same behavior. We define the equivalence between two test cases, t_1 and t_2 as follows.

Definition A.1. (Test Case Equivalence)

Let $t_1 = (I_1, E_1, R_1)$ and $t_2 = (I_2, E_2, R_2)$ be two test cases, then t_1 is equivalent to t_2 if and only if the following three conditions are satisfied:
1. $I_1 = I_2$
2. $E_1 = E_2$
3. $R_1 = R_2$.

The generated test cases, from different integration orders, are equivalent if and only if our approach has two properties: commutativity and associativity. The merging operation (Definition 7) uses the union operator and two special functions, $f()$ and $g()$. We need to investigate the commutativity and the associativity of our merging operation.

A.1 System Specification

Systems are composed of a set of components. Each component has internal and/or external interfaces. Internal interfaces are used to communicate among the system components. External interfaces are used to communicate with the system environment. The general system architecture can be described as shown in Fig. A.1. A system with three components is adequate to investigate the commutative and associative properties.

To simplify our investigation, we assume test cases consist of two instances only, CUT and test control. The test control represents the behavior of the test environment in addition to controlling the test execution. The test environment represents the system environment as well as system components that are not yet realized during the test execution. We also assume, for simplicity, that each component has one component test case.

The system is composed of three components, A, B, and C, and each component has one component test case: t_1, t_2, and t_3, respectively. We

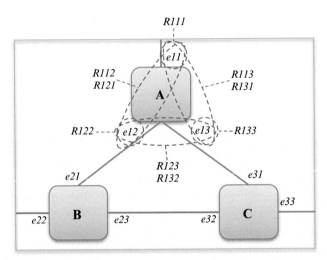

Fig. A.1 General system architecture.

assume there is an interaction between these components, and the test cases capture these interactions. The events of each component are organized into several sets to represent the corresponding component interfaces. Accordingly, sets and relations for each test case are split into several subsets to indicate such organization. The specification for each component test case is given as follows:

$t_1 = (I_1, E_1, R_1)$

$\quad I_1 = \{tc_1, a\}$

$\quad E_1 = e_{11} \cup e_{12} \cup e_{13}$, where

$\qquad e_{11}$ a set of events specified only in t_1

$\qquad e_{12}$ a set of events specified in both t_1 and t_2

$\qquad e_{13}$ a set of events specified in both t_1 and t_3

$R_1 = R_{111} \cup R_{112} \cup R_{113} \cup R_{121} \cup R_{122} \cup R_{123} \cup R_{131} \cup R_{132} \cup R_{133}$, where

$\qquad R_{111} \subseteq e_{11} \times e_{11}$

$\qquad R_{112} \subseteq e_{11} \times e_{12}$

$\qquad R_{113} \subseteq e_{11} \times e_{13}$

$\qquad R_{121} \subseteq e_{12} \times e_{11}$

$\qquad R_{122} \subseteq e_{12} \times e_{12}$

$\qquad R_{123} \subseteq e_{12} \times e_{13}$

$\qquad R_{131} \subseteq e_{13} \times e_{11}$

$\qquad R_{132} \subseteq e_{13} \times e_{12}$

$\qquad R_{133} \subseteq e_{13} \times e_{13}$

$t_2 = (I_2, E_2, R_2)$

$\quad I_2 = \{tc_2, b\}$

$\quad E_2 = e_{21} \: U \: e_{22} \: U \: e_{23}$, where

$\qquad e_{21}$ a set of events specified in both t_2 and t_1

$\qquad e_{22}$ a set of events specified only in t_2

$\qquad e_{23}$ a set of events specified in both t_2 and t_3

$R_2 = R_{211} \: U R_{212} \: U R_{213} \: U R_{221} \: U R_{222} \: U R_{223} \: U R_{231} \: U R_{232} \: U R_{233}$, where

$\qquad R_{211} \subseteq e_{21} \: x \: e_{21}$

$\qquad R_{212} \subseteq e_{21} \: x \: e_{22}$

$\qquad R_{213} \subseteq e_{21} \: x \: e_{23}$

$\qquad R_{221} \subseteq e_{22} \: x \: e_{21}$

$\qquad R_{222} \subseteq e_{22} \: x \: e_{22}$

$\qquad R_{223} \subseteq e_{22} \: x \: e_{23}$

$\qquad R_{231} \subseteq e_{23} \: x \: e_{21}$

$\qquad R_{232} \subseteq e_{23} \: x \: e_{22}$

$\qquad R_{233} \subseteq e_{23} \: x \: e_{23}$

$t_3 = (I_3, E_3, R_3)$

$\quad I_3 = \{tc_3, c\}$

$\quad E_3 = e_{31} \: U \: e_{32} \: U \: e_{33}$, where

$\qquad e_{31}$ a set of events specified in both t_3 and t_1

$\qquad e_{32}$ a set of events specified in both t_3 and t_2

$\qquad e_{33}$ a set of events specified only in t_3

$R_3 = R_{311} \: U R_{312} \: U R_{313} \: U R_{321} \: U R_{322} \: U R_{323} \: U R_{331} \: U R_{332} \: U R_{333}$, where

$\qquad R_{311} \subseteq e_{31} \: x \: e_{31}$

$\qquad R_{312} \subseteq e_{31} \: x \: e_{32}$

$\qquad R_{313} \subseteq e_{31} \: x \: e_{33}$

$\qquad R_{321} \subseteq e_{32} \: x \: e_{31}$

$\qquad R_{322} \subseteq e_{32} \: x \: e_{32}$

$\qquad R_{323} \subseteq e_{32} \: x \: e_{33}$

$\qquad R_{331} \subseteq e_{33} \: x \: e_{31}$

$\qquad R_{332} \subseteq e_{33} \: x \: e_{32}$

$\qquad R_{333} \subseteq e_{33} \: x \: e_{33}$

Notice that

$e_{12} = e_{21}$

$e_{13} = e_{31}$

$e_{23} = e_{32}$

$R_{122} = R_{211}$

$$R_{133} = R_{311}$$
$$R_{233} = R_{322}$$

Note that if there is no interaction between two components, then their corresponding variables, sets, and relations will be empty; for examples, suppose there is no interaction between A and C then

$e_{13} = \{\},$

$e_{31} = \{\},$

$R_{113} = \{\},$

$R_{123} = \{\},$

$R_{131} = \{\},$

$R_{132} = \{\},$

$R_{133} = \{\},$

$R_{311} = \{\},$

$R_{312} = \{\},$

$R_{313} = \{\},$

$R_{321} = \{\}$ *and*

$R_{331} = \{\}$

The approach creates the test control for the generated test model and builds its behavior by merging the behavior of the test controls of the given test models, which we call *tci*.

A.2 Commutativity

To demonstrate the commutativity of our approach for any two components, say A and B, we should demonstrate that the integration of their component test cases, t_1 and t_2, respectively, generates equivalent behaviors independently of the integration order: $(A + B)$ or $(B + A)$. That means

$$t_1 + t_2 = t_2 + t_1 \qquad (A.1)$$

Using Definitions 3 and 7, we get

$(g(I_1) \cup g(I_2), f(E_1) \cup f(E_2), f(R_1) \cup f(R_2)) = (g(I_2) \cup g(I_1), f(E_2) \cup f(E_1), f(R_2) \cup f(R_1))$

Hence, to validate Eq. (A.1), we need to show that

$$g(I_1) \cup g(I_2) = g(I_2) \cup g(I_1) \qquad (A.2)$$
$$f(E_1) \cup f(E_2) = f(E_2) \cup f(E_1) \qquad (A.3)$$
$$f(R_1) \cup f(R_2) = f(R_2) \cup f(R_1) \qquad (A.4)$$

Let us evaluate the left side of Eq. (A.2) first by substituting the values of I_1 and I_2 and using our definition of equivalence (Definition A.1).

$g(I_1) \cup g(I_2) = g(\{tc_1, a\}) \cup g(\{tc_2, b\})$

Then, we apply the $g()$ function:

$g(I_1) \cup g(I_2) = \{tc_i, a\} \cup \{tc_i, b\}$

Then, we apply the union operator:

$g(I_1) \cup g(I_2) = \{tc_i, a, b\}$

Next, we perform the same sequence on the right side of Eq. (A.2)

$g(I_2) \cup g(I_1) = g(\{tc_2, b\}) \cup g(\{tc_1, a\})$

$= \{tc_i, b\} \cup \{tc_i, a\}$

$= \{tc_i, b, a\}$

The two sides are equivalent. Thus, we say Eq. (A.2) is true. We take the same evaluation approach with Eq. (A.3). First, we evaluate the left side of Eq. (A.3).

$f(E_1) \cup f(E_2) = f(e_{11} \cup e_{12} \cup e_{13}) \cup f(e_{21} \cup e_{22} \cup e_{23})$

Since $e_{12} = e_{21}$, the $f()$ function replaces e_{21} with e_{12}

$f(E_1) \cup f(E_2) = e_{11} \cup e_{12} \cup e_{13} \cup \boldsymbol{e_{12}} \cup e_{22} \cup e_{23}$

$= e_{11} \cup e_{12} \cup e_{13} \cup e_{22} \cup e_{23}$

Then, we evaluate the right side of Eq. (A.3)

$f(E_2) \cup f(E_1) = f(e_{21} \cup e_{22} \cup e_{23}) \cup f(e_{11} \cup e_{12} \cup e_{13})$

Since $e_{12} = e_{21}$, the $f()$ function replaces e_{21} with e_{12}

$f(E_2) \cup f(E_1) = \boldsymbol{e_{12}} \cup e_{22} \cup e_{23} \cup e_{11} \cup e_{12} \cup e_{13}$

$= e_{12} \cup e_{22} \cup e_{23} \cup e_{11} \cup e_{13}$

Hence, the two sides are equivalent, and this is prove that Eq. (A.3) is true as well. The same evaluation approach is applied to Eq. (A.4). We take the left side of the equation first

$f(R_1) \cup f(R_2) = f(R_{111} \cup R_{112} \cup R_{113} \cup R_{121} \cup R_{122} \cup R_{123} \cup R_{131}$
$\cup R_{132} \cup R_{133}) \cup f(R_{211} \cup R_{212} \cup R_{213} \cup R_{221} \cup R_{222} \cup R_{223}$
$\cup R_{231} \cup R_{232} \cup R_{233})$

Since $R_{122} = R_{211}$, the $f()$ function replaces R_{211} with R_{122}

$f(R_1) \cup f(R_2) = R_{111} \cup R_{112} \cup R_{113} \cup R_{121} \cup R_{122} \cup R_{123} \cup R_{131}$
$\cup R_{132} \cup R_{133} \cup \boldsymbol{R_{122}} \cup R_{212} \cup R_{213} \cup R_{221} \cup R_{222} \cup R_{223}$
$\cup R_{231} \cup R_{232} \cup R_{233}$

$= R_{111} \cup R_{112} \cup R_{113} \cup R_{121} \cup R_{122} \cup R_{123} \cup R_{131} \cup R_{132} \cup R_{133}$
$\cup R_{212} \cup R_{213} \cup R_{221} \cup R_{222} \cup R_{223} \cup R_{231} \cup R_{232} \cup R_{233}$

The next step is to evaluate the right side of Eq. (A.4)

$f(R_2) \cup f(R_1) = f(R_{211} \cup R_{212} \cup R_{213} \cup R_{221} \cup R_{222} \cup R_{223} \cup R_{231}$
$\cup R_{232} \cup R_{233}) \cup f(R_{111} \cup R_{112} \cup R_{113} \cup R_{121} \cup R_{122} \cup R_{123}$
$\cup R_{131} \cup R_{132} \cup R_{133})$

$= \boldsymbol{R_{122}} \cup R_{212} \cup R_{213} \cup R_{221} \cup R_{222} \cup R_{223} \cup R_{231} \cup R_{232} \cup R_{233}$
$\cup R_{111} \cup R_{112} \cup R_{113} \cup R_{121} \cup R_{122} \cup R_{123} \cup R_{131} \cup R_{132} \cup R_{133}$

$$= R_{122} \ U \ R_{212} \ U \ R_{213} \ U \ R_{221} \ U \ R_{222} \ U \ R_{223} \ U \ R_{231} \ U \ R_{232} \ U \ R_{233}$$
$$U \ R_{111} \ U \ R_{112} \ U \ R_{113} \ U \ R_{121} \ U \ R_{123} \ U \ R_{131} \ U \ R_{132} \ U \ R_{133}$$

The results of both sides of (A.4) are equivalent. Since Eqs. (A.2), (A.3), and (A.4) are evaluated to true; then Eq. (A.1) is true too. Hence, the commutativity property of the integration approach is proven.

A.3 Associativity

To demonstrate the associativity of the integration approach for any three components, A, B, and C, we need to demonstrate that:

$$t_1 + (t_2 + t_3) = (t_1 + t_2) + t_3 \tag{A.5}$$

Using Definitions 3 and 7, we can refactor Eq. (A.5) as follows:

$$g(I_1) \ U \ (g(I_2) \ U \ g(I_3)) = (g(I_1) \ U \ g(I_2)) \ U \ g(I_3) \tag{A.6}$$
$$g(I_1) \ U \ (g(I_2) \ U \ g(I_3)) = (g(I_1) \ U \ g(I_2)) \ U \ g(I_3) \tag{A.7}$$
$$f(R_1) \ U \ (f(R_2) \ U f(R_3)) = (f(R_1) \ U f(R_2)) \ U f(R_3) \tag{A.8}$$

Hence, we have to prove that Eqs. (A.6), (A.7), and (A.8) are satisfied. Let us start by examining Eq. (A.6). First, we evaluate the left side of the equation.

$$g(I_1) \ U \ (g(I_2) \ U \ g(I_3)) = g(\{tc_1, \ a\}) \ U \ (g(\{tc_2, \ b\}) \ U \ g(\{tc_3, \ c\}))$$

Then, we apply $g()$

$$= \{tc_i, \ a\} \ U \ (\{tc_i, \ b\} \ U \ \{tc_i, \ c\})$$
$$= \{tc_i, \ a\} \ U \ \{tc_i, \ b, \ c\}$$
$$= \{tc_i, \ a, \ b, \ c\}$$

Then, we take the right side of Eq. (A.6)

$$(g(I_1) \ U \ g(I_2)) \ U \ g(I_3) = (g(\{tc_1, \ a\}) \ U \ g(\{tc_2, \ b\})) \ U \ g(\{tc_3, \ c\})$$
$$= (\{tc_i, \ a\} \ U \ \{tc_i, \ b\}) \ U \ \{tc_i, \ c\}$$
$$= \{tc_i, \ a, \ b\} \ U \ \{tc_i, \ c\}$$
$$= \{tc_i, \ a, \ b, \ c\}$$

The two sides are equal. Thus, we can say Eq. (A.6) is true. We use the same evaluation approach for Eq. (A.7). First, we evaluate the left side of Eq. (A.7).

$$f(E_1) \ U \ (f(E_2) \ U f(E_3)) = f(e_{11} \ U \ e_{12} \ U \ e_{13}) \ U \ (f(e_{21} \ U \ e_{22} \ U \ e_{23}) \ U f(e_{31} \ U \ e_{32} \ U \ e_{33}))$$

Then, we apply $f()$, which replaces the following sets

$$e_{12} = e_{21},$$
$$e_{13} = e_{31}, \ and$$
$$e_{23} = e_{32}.$$

$f(E_1) \cup (f(E_2) \cup f(E_3)) = (e_{11} \cup e_{12} \cup e_{13}) \cup ((e_{12} \cup e_{22} \cup e_{23}) \cup (e_{13} \cup e_{23} \cup e_{33}))$

$= (e_{11} \cup e_{12} \cup e_{13}) \cup (e_{12} \cup e_{22} \cup e_{23} \cup e_{13} \cup e_{33})$

$= e_{11} \cup e_{12} \cup e_{13} \cup e_{22} \cup e_{23} \cup e_{33}.$

Then, we evaluate the right side of Eq. (A.7).

$(f(E_1) \cup f(E_2)) \cup f(E_3) = (f(e_{11} \cup e_{12} \cup e_{13}) \cup f(e_{21} \cup e_{22} \cup e_{23})) \cup f(e_{31} \cup e_{32} \cup e_{33})$

$= ((e_{11} \cup e_{12} \cup e_{13}) \cup (e_{12} \cup e_{22} \cup e_{23})) \cup (e_{13} \cup e_{23} \cup e_{33})$

$= (e_{11} \cup e_{12} \cup e_{13} \cup e_{22} \cup e_{23}) \cup (e_{13} \cup e_{23} \cup e_{33})$

$= e_{11} \cup e_{12} \cup e_{13} \cup e_{22} \cup e_{23} \cup e_{33}.$

Therefore, the two sides are equal, and that proves that Eq. (A.7) is satisfied. The same evaluation approach is used for Eq. (A.8). We take the left side of the equation first.

$f(R_1) \cup (f(R_2) \cup f(R_3)) = f(R_{111} \cup R_{112} \cup R_{113} \cup R_{121} \cup R_{122} \cup R_{123} \cup R_{131} \cup R_{132} \cup R_{133}) \cup (f(R_{211} \cup R_{212} \cup R_{213} \cup R_{221} \cup R_{222} \cup R_{223} \cup R_{231} \cup R_{232} \cup R_{233}) \cup f(R_{311} \cup R_{312} \cup R_{313} \cup R_{321} \cup R_{322} \cup R_{323} \cup R_{331} \cup R_{332} \cup R_{333}))$

Then, we apply $f()$, which replaces the following relations

$R_{122} = R_{211},$

$R_{133} = R_{311},$ and

$R_{233} = R_{322}.$

$f(R_1) \cup (f(R_2) \cup f(R_3)) = (R_{111} \cup R_{112} \cup R_{113} \cup R_{121} \cup R_{122} \cup R_{123} \cup R_{131} \cup R_{132} \cup R_{133}) \cup ((R_{122} \cup R_{212} \cup R_{213} \cup R_{221} \cup R_{222} \cup R_{223} \cup R_{231} \cup R_{232} \cup R_{233}) \cup (R_{133} \cup R_{312} \cup R_{313} \cup R_{321} \cup R_{233} \cup R_{323} \cup R_{331} \cup R_{332} \cup R_{333}))$

$= (R_{111} \cup R_{112} \cup R_{113} \cup R_{121} \cup R_{122} \cup R_{123} \cup R_{131} \cup R_{132} \cup R_{133}) \cup (R_{122} \cup R_{212} \cup R_{213} \cup R_{221} \cup R_{222} \cup R_{223} \cup R_{231} \cup R_{232} \cup R_{233} \cup R_{133} \cup R_{312} \cup R_{313} \cup R_{321} \cup R_{323} \cup R_{331} \cup R_{332} \cup R_{333})$

$= R_{111} \cup R_{112} \cup R_{113} \cup R_{121} \cup R_{122} \cup R_{123} \cup R_{131} \cup R_{132} \cup R_{133} \cup R_{212} \cup R_{213} \cup R_{221} \cup R_{222} \cup R_{223} \cup R_{231} \cup R_{232} \cup R_{233} \cup R_{312} \cup R_{313} \cup R_{321} \cup R_{323} \cup R_{331} \cup R_{332} \cup R_{333}.$

The next step is to evaluate the right side of Eq. (A.8).

$(f(R_1) \cup f(R_2)) \cup f(R_3) = (f(R_{111} \cup R_{112} \cup R_{113} \cup R_{121} \cup R_{122} \cup R_{123} \cup R_{131} \cup R_{132} \cup R_{133}) \cup f(R_{211} \cup R_{212} \cup R_{213} \cup R_{221} \cup R_{222} \cup R_{223} \cup R_{231} \cup R_{232} \cup R_{233})) \cup f(R_{311} \cup R_{312} \cup R_{313} \cup R_{321} \cup R_{322} \cup R_{323} \cup R_{331} \cup R_{332} \cup R_{333}).$

Then, we apply $f()$

$$= ((R_{111} \cup R_{112} \cup R_{113} \cup R_{121} \cup R_{122} \cup R_{123} \cup R_{131} \cup R_{132} \cup R_{133})$$
$$\cup (\boldsymbol{R_{122}} \cup R_{212} \cup R_{213} \cup R_{221} \cup R_{222} \cup R_{223} \cup R_{231} \cup R_{232} \cup R_{233}))$$
$$\cup (\boldsymbol{R_{133}} \cup R_{312} \cup R_{313} \cup R_{321} \cup R_{233} \cup \boldsymbol{R_{323}} \cup R_{331} \cup R_{332} \cup R_{333})$$
$$= (R_{111} \cup R_{112} \cup R_{113} \cup R_{121} \cup R_{122} \cup R_{123} \cup R_{131} \cup R_{132} \cup R_{133}$$
$$\cup R_{212} \cup R_{213} \cup R_{221} \cup R_{222} \cup R_{223} \cup R_{231} \cup R_{232} \cup R_{233}) \cup (\boldsymbol{R_{133}}$$
$$\cup R_{312} \cup R_{313} \cup R_{321} \cup \boldsymbol{R_{233}} \cup R_{323} \cup R_{331} \cup R_{332} \cup R_{333})$$
$$= R_{111} \cup R_{112} \cup R_{113} \cup R_{121} \cup R_{122} \cup R_{123} \cup R_{131} \cup R_{132} \cup R_{133}$$
$$\cup R_{212} \cup R_{213} \cup R_{221} \cup R_{222} \cup R_{223} \cup R_{231} \cup R_{232} \cup R_{233} \cup R_{312}$$
$$\cup R_{313} \cup R_{321} \cup R_{323} \cup R_{331} \cup R_{332} \cup R_{333}.$$

The results of both sides of (A.8) are equal. Since Eqs. (A.6), (A.7), and (A.8) are satisfied; therefore, Eq. (A.5) holds. Hence, the associativity of the integration approach is proven.

REFERENCES

[1] A. Bertolino, Software testing research: achievements, challenges, dreams, in: 2007 Future of Software Engineering, IEEE Computer Society, Washington, DC, 2007, pp. 85–103.

[2] J. Grossmann, I. Fey, A. Krupp, M. Conrad, C. Wewetzer, W. Mueller, TestML-A test exchange language for model-based testing of embedded software, in: M. Broy, I. Krüger, M. Meisinger (Eds.), Model-Driven Development of Reliable Automotive Services, Springer, Berlin/Heidelberg, 2008, pp. 98–117.

[3] J. Hutchinson, J. Whittle, M. Rouncefield, S. Kristoffersen, in: Empirical assessment of MDE in industry, Proceeding of the 33rd International Conference on Software Engineering, ACM, New York, NY, 2011, pp. 471–480.

[4] M. Utting, A. Pretschner, B. Legeard, A taxonomy of model-based testing approaches, Softw. Test. Verif. Rel. 22 (2012) 297–312.

[5] A. Ulrich, in: Introducing model-based testing techniques in industrial projects, Software Engineering (Workshops), 2007. Available at http://subs.emis.de/LNI/Proceedings/Proceedings106/gi-proc-106-002.pdf. last accessed 2015.

[6] A.C. Dias-Neto, G.H. Travassos, Evaluation of {model-based} testing techniques selection approaches: an external replication, 3rd International Symposium on Empirical Software Engineering and Measurement, ESEM, 2009, pp. 269–278.

[7] OMG: Unified Modeling Language, Available at http://www.uml.org, 2014. Accessed July 2017.

[8] OMG: Object Management Group, Available at http://www.omg.org, 2014. Accessed July 2017.

[9] OMG: UML Testing Profile (UTP), Version 1.2, (Formal/2013–04–03), Available at http://www.omg.org/spec/UTP/1.2, 2013. Accessed August 2017.

[10] P. Iyenghar, E. Pulvermueller, C. Westerkamp, in: Towards model-based test automation for embedded systems using UML and UTP, 2011 IEEE 16th Conference on Emerging Technologies Factory Automation (ETFA), IEEE, 2011, pp. 1–9.

[11] B.P. Lamancha, P.R. Mateo, I.R. de Guzmán, M.P. Usaola, M.P. Velthius, in: Automated model-based testing using the UML testing profile and QVT, Proceedings of the 6th International Workshop on Model-Driven Engineering, Verification and Validation, ACM, New York, NY, 2009, pp. 6:1–6:10.

[12] P. Krishnan, P. Pari-Salas, Model-based testing and the UML testing profile, in: Semantics and Algebraic Specification, Springer, Berlin/Heidelberg, 2009, pp. 315–328.

[13] H. Le, A collaboration-based testing model for composite components, in: 2011 IEEE 2nd International Conference on Software Engineering and Service Science (ICSESS), Institute of Electrical and Electronics Engineers (IEEE), Beijing, China, 2011, pp. 610–613.

[14] D. Liang, K. Xu, Test-driven component integration with UML 2.0 testing and monitoring profile, 7th International Conference on Quality Software, QSIC 2007, IEEE Computer Society, Washington, DC, 2007, pp. 32–39.

[15] W. Chen, Q. Ying, Y. Xue, C. Zhao, Software testing process automation based on UTP—a case study, in: M. Li, B. Boehm, L. Osterweil (Eds.), Unifying the Software Process Spectrum, Springer, Berlin/Heidelberg, 2006, pp. 222–234.

[16] P. Iyenghar, in: Test framework generation For model-based testing in embedded systems, 2011 37th EUROMICRO Conference on Software Engineering and Advanced Applications (SEAA), IEEE, 2011, pp. 267–274.

[17] P. Baker, C. Jervis, Testing UML2.0 models using TTCN-3 and the UML2.0 testing profile, in: E. Gaudin, E. Najm, R. Reed (Eds.), SDL 2007: Design for Dependable Systems, Springer, Berlin/Heidelberg, 2007, pp. 86–100.

[18] M. Busch, R. Chaparadza, Z.R. Dai, A. Hoffmann, L. Lacmene, T. Ngwangwen, G.C. Ndem, H. Ogawa, D. Serbanescu, I. Schieferdecker, J. Zander-Nowicka, Model transformers for test generation from system models, Proceedings of Conquest 2006, 10th International Conference on Quality Engineering in Software Technology, September, Hanser Verlag, 2006, pp. 1–16.

[19] M. Mussa, F. Khendek, Towards a model based approach for integration testing, in: I. Ober, I. Ober (Eds.), SDL 2011: Integrating System and Software Modeling, LNCS, vol. 7083, Springer, Berlin/Heidelberg, 2012, pp. 106–121.

[20] M. Mussa, F. Khendek, Identification and selection of interaction test scenarios for integration testing, in: Ø. Haugen, R. Reed, R. Gotzhein (Eds.), SAM2012: System Analysis and Modeling: Theory and Practice, LNCS, vol. 7744, Springer, Berlin/Heidelberg, 2013, pp. 16–33.

[21] M. Mussa, F. Khendek, in: Merging test models, 18th International Conference on Engineering of Complex Computer Systems (ICECCS), IEEE, 2013, pp. 167–170.

[22] M. Mussa, F. Khendek, Acceptance test optimization, in: D. Amyot, P. Fonseca, i. Casas, G. Mussbacher (Eds.), System Analysis and Modeling: Models and Reusability, SAM2014, LNCS, vol. 8769, Springer, 2014, pp. 158–173.

[23] S. Fortsch, B. Westfechtel, in: Differencing and merging of software diagrams: state of the art and challenges, ICSOFT 2007—International Conference on Software and Data Technologies, INSTICC Press, 2007, pp. 90–99.

[24] S. Nejati, M. Sabetzadeh, M. Chechik, S. Easterbrook, P. Zave, in: Matching and merging of statecharts specifications, 29th International Conference on Software Engineering, ICSE 2007, IEEE Computer Society, 2007, pp. 54–64.

[25] L. Hélouët, T. Hénin, C. Chevrier, Automating scenario merging, in: R. Gotzhein, R. Reed (Eds.), System Analysis and Modeling: Language Profiles, LNCS, vol. 4320, Springer, Berlin/Heidelberg, 2006, pp. 64–81.

[26] T. Mens, A state-of-the-art survey on software merging, IEEE Trans. Softw. Eng. 28 (2002) 449–462. IEEE.

[27] F. Khendek, G.V. Bochmann, Merging behavior specifications, J. Formal Methods Syst. Des. 6 (1995) 259–293.

[28] M. Lund, K. Stølen, J. Misra, T. Nipkow, E. Sekerinski (Eds.), A fully general operational semantics for UMLÂ 2.0 sequence diagrams with potential and mandatory choice. in: Proceedings of the Australian Software Engineering Conference, 2004, Springer, Berlin/Heidelberg, 2006, pp. 380–395.

[29] X. Li, Z. Liu, H. Jifeng, in: A formal semantics of UML sequence diagram, Proceedings of Software Engineering Conference, Australian, 2004, pp. 168–177.

[30] ITU-T Recommendation: Z.120, Message Sequence Charts (MSC), Geneva, Switzerland, 1999.

[31] B.K. Aichernig, F. Lorber, S. Tiran, in: Integrating model-based testing and analysis tools via test case exchange, 2012 Sixth International Symposium on Theoretical Aspects of Software Engineering (TASE), IEEE, 2012, pp. 119–126.

[32] T. Mens, A state-of-the-art survey on software merging, IEEE Trans. Softw. Eng. 28 (2002) 449–462.

[33] M. Stephan, J.R. Cordy, A survey of model comparison approaches and applications, 1st International Conference on Model-Driven Engineering and Software Development, MODELSWARD, 2013, pp. 265–277.

[34] S. Fortsch, B. Westfechtel, Differencing and merging of software diagrams: state of the art and challenges, ICSOFT 2007—International Conference on Software and Data Technologies, 2007, pp. 90–99.

[35] Z. Wang, B. Li, L. Wang, Q. Li, A brief survey on automatic integration test order generation, In SEKE 2011—Proceedings of the 23rd International Conference on Software Engineering and Knowledge Engineering, July 7, 2011–July 9, Knowledge Systems Institute Graduate School, Miami, FL, 2011, pp. 254–257.

[36] A. Abdurazik, J. Offutt, Using coupling-based weights for the class integration and test order problem, Comput. J. 52 (2009) 557–570.

[37] L.C. Briand, Y. Labiche, Y. Wang, An investigation of graph-based class integration test order strategies, IEEE Trans. Softw. Eng. 29 (2003) 594–607.

[38] P. Ammann, J. Offutt, Introduction to Software Testing, Cambridge University Press, New York, 2008.

[39] M. Shirole, R. Kumar, UML behavioral model based test case generation: a survey, SIGSOFT Softw. Eng. Notes 38 (2013) 1–13.

[40] L. Jingyue, O.P.N. Slyngstad, M. Torchiano, M. Morisio, C. Bunse, A state-of-the-practice survey of risk management in development with off-the-shelf software components, IEEE Trans. Softw. Eng. 34 (2008) 271–286.

[41] N. Budhija, S.P. Ahuja, Review of software reusability, International Conference on Computer Science and Information Technology (ICCSIT'2011), Pattaya, 2011, pp. 113–115.

[42] G.N.K.S. Babu, D.S.K. Srivatsa, Analysis and measures of software reusability, Int. J. Rev. Comput. 1 (2009) 41–46. Available at http://www.ijric.org/volumes/Vol1/5Vol1.pdf. last accessed 2014.

[43] T.J. Biggerstaff, A.J. Perlis, Software Reusability, ACM Press, New York, NY, 1989.

[44] J. Klein, B. Caillaud, L. Hélouët, Merging scenarios, Proceedings of the Ninth International Workshop on Formal Methods for Industrial Critical Systems (FMICS 2004), June 25–June 27. Electr. Notes Theor. Comput. Sci. 133, Elsevier, Amsterdam, The Netherlands, 2005, pp. 193–215.

ABOUT THE AUTHORS

Mohamed Mussa received a PhD in Electrical and Computer Engineering from Concordia University in 2015. He worked on a model-based framework for test cases reuse and optimization. He obtained a Master's degree from the same university in 2000. Mohamed has worked as a software designer/developer for several years with several institutions. Mohamed is interested in model-based software engineering and testing.

Ferhat Khendek received his PhD from University of Montreal, Canada. He is a full professor in the Department of Electrical and Computer Engineering of Concordia University where he also holds since 2011 the NSERC/Ericsson Senior Industrial Research Chair in Model Based Management, a major collaboration between Ericsson and Concordia University. Ferhat Khendek has published more than 200 conference/journal papers. He is a co-inventor of six granted patents and of ten patents currently under review. Ferhat Khendek's research interests are in model-based software engineering and management, formal methods, validation and testing, and service engineering and architectures.

Three Open Problems in the Context of E2E Web Testing and a Vision: NEONATE

Filippo Ricca*, Maurizio Leotta*, Andrea Stocco†
*DIBRIS, Università di Genova, Genova, Italy
†University of British Columbia, Vancouver, BC, Canada

Contents

1. Introduction	90
2. The Three Open Problems in the Context of E2E Web Testing	92
2.1 The Fragility Problem	93
2.2 The Strong Coupling and Low Cohesion Problem	95
2.3 The Incompleteness Problem	97
3. State of the Art on the Three Open Problems	97
3.1 State of the Art on the Fragility Problem	98
3.2 State of the Art on the Strong Coupling and Low Cohesion Problem	101
3.3 State of the Art on the Incompleteness Problem	103
4. Overcoming the Three Open Problems: The NEONATE Vision	104
4.1 The Stuck Situation	104
4.2 The Vision	105
4.3 Existing Integrated Testing Environments	106
5. Architecture of the NEONATE Integrated Testing Environment	107
5.1 Robust Web Element Locators With ROBULA+	108
5.2 Automatic Generation of Page Objects With APOGEN	110
5.3 Generating Visual Test Suites With PESTO	111
5.4 Separating Test Specification from Test Implementation With APORES	113
5.5 Suggesting and Executing Repairs for Broken Code With AUTOREPAIR	114
5.6 Extending Existing Test Suites With Ts-EXT	116
5.7 Supporting the Tester During Maintenance/Development With ASSISTANT	119
6. NEONATE's Examples of Use	119
6.1 Automated Test Suite Development (Scenario 1)	120
6.2 Automated Test Suite Refactoring (Scenario 2)	122
7. NEONATE's Long-Term Impact	124
7.1 Scientific	124
7.2 Practical	125
7.3 Industrial	125

Advances in Computers, Volume 113
ISSN 0065-2458
https://doi.org/10.1016/bs.adcom.2018.10.005

8. Conclusions   126
References   128
About the Authors   132

Abstract

Web applications are critical assets of our society and thus assuring their quality is of undeniable importance. Despite the advances in software testing, the ever-increasing technological complexity of these applications makes it difficult to prevent errors.

In this work, we provide a thorough description of the three open problems hindering web test automation: fragility problem, strong coupling and low cohesion problem, and incompleteness problem. We conjecture that a major breakthrough in test automation is needed, because the problems are closely correlated, and hence need to be attacked together rather than separately. To this aim, we describe NEONATE, a novel integrated testing environment specifically designed to empower the web tester.

Our utmost purpose is to make the research community aware of the existence of the three problems and their correlation, so that more research effort can be directed in providing solutions and tools to advance the state of the art of web test automation.

1. INTRODUCTION

Web applications pervade the lives of billions of individuals and have a significant impact on all aspects of our society, being crucial for a multitude of economic, social, and educational activities. Indeed, a considerable slice of modern software consists of web applications executed in the browser, running both on desktop and on smartphone devices [1]. Associations, enterprises, governmental organizations, companies, scientific groups use the web as a powerful and convenient way to promote activities/products and to carry out their core business. People daily use online services as source of information, means of communication, source of entertainment, and venue for commerce [2].

Hence, the quality and correctness of web applications are of undeniable importance. However, developing and maintaining a complex web system has become challenging, because the advances of web technologies in the last decade have also changed the way in which software running on the web is developed and maintained [3]. Unfortunately, the evolution of the tools for the analysis and testing of such complex software systems is not proceeding at the same pace.

Functional testing is one of the main approaches for assuring the quality of web applications. The goal of functional testing is to exercise the web application under test to detect failures, where a failure can be considered as a deviation from the system's intended behavior. In many software projects, functional testing is neglected because of time or cost constraints [4]. However, the impact of failures in a web application may range from simple inconveniences (e.g., a malfunction that causes users dissatisfaction), up to huge economic problems (e.g., interruption of business).

Today, web developers mostly test their applications manually, i.e., they manually interact with the web applications to check if they behave as expected. Unfortunately, this practice is prone to errors, time-consuming, and ultimately not very effective. For these reasons, most teams try to automate manual testing activities by means of test automation tools. This process involves a manual step which consists of implementing the *test code* able to instrument the web application and run predefined test-oriented user scenarios. Test code provides input data, operates on GUI components, and retrieves information to be compared with oracles (e.g., using assertions).

In the web domain, test automation tools usually operate at user interface (GUI) level, interacting with the web elements that are displayed on the web page, as seen by the end users. This kind of testing is called end-to-end (E2E), because the application is tested as a whole, in its entirety, and from the perspective of the end user. Different categories of E2E test automation tools are available on the marketplace. For example, DOM-based tools use objects in the Document Object Model (DOM), the hierarchical structure underlying a HTML page, to locate and interact with web elements. Such tools have reached a high level of maturity and popularity as, for example, the case of Selenium [5]. A competing category of visual tools has also appeared in the last years based on the use of image recognition techniques to identify the web elements (e.g., JAutomate [6] and Sikuli [7]).

The economic convenience of test automation is strongly correlated to the maintenance cost of the test suites [8]. We have identified two major problems that contribute to increase such a cost. First, all the existing test automation technologies suffer—in different proportions—from the *fragility problem*. When a web application evolves to accommodate requirements changes or functionality extensions, existing automated test code can easily break, and testers must correct it. This task is expensive, because it is usually performed manually. Often breakages occur in response to minor changes, e.g., a change of the page layout only. In these cases the test code is named fragile.

Second, test code is usually strongly coupled with the application under test, and full of technical details that limit greatly its readability and an easy maintenance. This is the *strong coupling and low cohesion problem*.

Moreover, automated test suites are inherently incomplete, because they cover only a subset of the input and functionalities space of the application. This third limitation, which we called *incompleteness problem*, is an open testing problem. As we shall see in this work, the presence of these three problems makes the activity of testing adequately the web applications challenging.

Existing test automation tools offer little to no help to overcome the three aforementioned problems. A huge amount of resources is still required by the testers to cope with the creation and maintenance of a web test suite. Thus, we believe a paradigm shift is necessary to advance in the state of the art of web test automation. To this aim, we envision NEONATE (Novel algorithms/ techniques for building and maintaining Web Test Code), an integrated testing environment able to empower the web tester limiting the three open problems. We have envisioned NEONATE out of the knowledge gained during industrial collaborations and the desire to solve concrete problems [9, 10].

The contributions of this work are as follows:

- a detailed description of three big open problems in the context of web test automation, namely, fragility problem, strong coupling and low cohesion problem, and incompleteness problem;
- a comprehensive analysis of the state-of-the-art concerning existing tools and solutions related to the three open problems;
- our vision of how the three problems hindering web test automation can be mitigated. Specifically, this concerns the development and adoption of a novel integrated testing environment called NEONATE;
- the overall description of the NEONATE integrated testing environment prototype, together with some usage scenarios.

The chapter is organized as follows: Section 2 introduces the three open problems in the context of web test automation, and Section 3 describes the pertinent literature. Section 4 illustrates our vision to overcome the three open problems, while Section 5 details the NEONATE integrated testing environment. At last, Section 6 reports two simulated usage examples of NEONATE pertaining to the creation and refactoring of a web test suite, and Section 7 describes the possible impact of NEONATE from both the perspectives of the researchers and practitioners. Section 8 concludes the chapter and outlines the future work.

 ## 2. THE THREE OPEN PROBLEMS IN THE CONTEXT OF E2E WEB TESTING

E2E testing is a type of black box testing based on the concept of *test scenario* (or test logic), a sequence of steps/actions performed on the web application under test. One or more *test cases* can be derived from a test scenario by specifying the actual input data to use in each step and the expected outcomes. Test cases can be manually executed by a human on the browser or they can be implemented into *test code* (also known as test scripts). To this extent, testers use a high-level programming language (e.g., Java, Python, Ruby) to develop scripts that consist of commands simulating the user's actions on the GUI and retrieving information to be used in assertions verifying the expected outcomes.

The main benefits of adopting test automation are the possibility of (1) executing the test cases more often, for instance overnight, so that to increase the probability of bugs detection, (2) finding bugs on the early stages of development, and (3) reusing test code across successive releases of the web app under test (i.e., to catch regressions[a]).

However, despite these benefits, test automation is not "a silver bullet" and has limitations well known by testers. A thorough list can be found in [11], where the authors overview benefits and challenges of test automation based on empirical studies and experience reports in the industry. Among the outlined limitations, two are particularly severe when the web domain is considered: the *difficulty in maintenance of test automation artefacts* and the fact that *automation cannot replace manual testing*.

The first limitation can be better defined in two concrete problems namely, the *fragility problem* and the *strong coupling and low cohesion problem*.

The second limitation has deep roots and can be also associated with the *incompleteness problem*. In fact, not all testing tasks can be—or are worth to be—automated [8], thus developers automate only the test cases that they judge to be more crucial in their setting. As a consequence, an automated test suite covers only a small suboptimal portion of the application.

The authors of this work have matured similar opinions as Rafi's and colleagues by both working within industrial projects [9, 10] and performing academic empirical studies [12–14]. Other researchers have acknowledged the existence of these issues and have proposed initial solutions (see Section 3 for further details).

2.1 The Fragility Problem

The maintenance of test code during software evolution is the chief problem of web test automation, because the cost is expected to grow with: (1) the application size, (2) the number of test cases, and (3) the inevitable

---

[a] Regression testing aims at verifying that software previously developed and tested still behaves correctly even after the occurrence of an evolution or maintenance activity.

application evolution. Research works have singled out the manual maintenance of *locators* to be particularly problematic and expensive [13, 15].
Locators are specific commands used by test automation tools to identify
the web elements on the GUI, before to perform actions on them. Examples
of actions consist of clicking on a link, or filling in a text field in a form.
Locators are used, for instance, to identify and fill the input portions of a
web page (e.g., the form fields), to execute navigations (e.g., by locating
and clicking on links) or to verify the correctness of the output (e.g., by
locating the web page elements showing the result of a computation). It
has been shown that even slight modifications of the application under test
have a massive impact on locators [13]. For instance, changes as simple as
renaming a page element or altering the choices in a dropdown list can cause
locators to break (e.g., they become unable to select the desired element in
the web page). Thus, the specific characteristics of the web applications
make the test cases more *fragile*, rendering their maintenance extremely difficult and expensive as compared to maintaining test cases for a desktop
application.

Let us clarify this problem by considering a simplified web application
composed of two web pages—**insertInfo.php** and **showInfo.php**—that allow
to insert and visualize personal information of the users. The test code for
Version 1 of this web application (**Test** in Fig. 1) opens the **insertInfo.php**
page, fills the form (shown in Fig. 1, top), submits the information and verifies that the inserted data are correctly displayed in the **showInfo** page (not
shown for brevity). In this way, it is possible to test the correct insertion of
the information in the application. To implement this test, it is necessary to
locate some web page elements as, for instance, the field of the form for
inserting the mobile phone number (i.e., the highlighted target element
1234 shown in Fig. 1). To this aim, a developer can use tools for the automatic generation of locators as FirePath,[b] which generates XPath expressions
for the elements in a web page that can be used as locators. In our example,
suppose that FirePath would produce the following XPath locator for the
"Mobile" text field: //*[@id="userInfo"]/tr[3]/td[2].

We now consider a new version of the web application (Version 2). In
this evolved version, the user is required to type also the gender information,
and the corresponding text field has been inserted between the "Surname"
and the "Mobile" text fields (see Fig. 1, bottom). In this scenario, Test is no
longer able to select correctly the mobile phone number because its locator

---

[b] https://addons.mozilla.org/firefox/addon/firepath/.

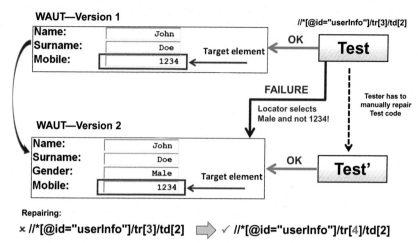

Repairing:

✗ //*[@id="userInfo"]/tr[3]/td[2] ⟹ ✓ //*[@id="userInfo"]/tr[4]/td[2]

Fig. 1 Fragility problem explained by means of an example.

"points to" another element, i.e., the "Gender" text field. As a consequence, let us assume that the test would break, because the input validation of the application would not consider a phone number as a valid gender (e.g., the application could stop the insertion and visualize an error message). Hence, Test must be manually repaired.

In particular, the locator could be modified from //*[@id="userInfo"]/tr [3]/td[2] to //*[@id="userInfo"]/tr[4]/td[2]) in order to select the correct web element.

2.2 The Strong Coupling and Low Cohesion Problem

Test code produced by means of test automation tools often tends to merge two different notions: the test scenario (i.e., *what* to test) with test implementation (i.e., *how* to test). As a result, the test code is full of implementation details, that do not pertain to the test scenario. Thus, tests become difficult to read and understand, and costly to maintain and evolve. This is a common pitfall that leads to the strong coupling and low cohesion problem, i.e., the produced test code is *strongly coupled* with the web page structure and merges two different concerns. As an example, let us consider the fragment of code in Listing 1, which is a possible partial implementation of the Test of Fig. 1, developed with Selenium WebDriver [5]. The test code implements a portion of a simple test scenario which requires to enter the form data (name, surname, mobile number), click on the Enter button, and then (not shown) verify that the user has been correctly saved. As evident from our example, the test code is strongly coupled with the web pages composing the application under test. The highlighted portions of the test are indeed related to technicalities such as locators used for retrieving the web elements to interact with (e.g., //*[@id="userInfo"]/tr[3]/td[2]), or command calls to the browser-specific APIs of the test automation tool (e.g., driver.findElement(...)) that fall outside the test scenario.

Listing 1 An excerpt of strongly coupled and low cohesive test code. Implementation details are highlighted in red.

```
@Test
public void testInsertInfo(){
  WebDriver driver = new FirefoxDriver();
  driver.get("localhost/webappundertest/insertInfo.php");
  driver.findElement(By.xpath("//*[@id='userInfo']/tr[1]/td[2]")).
      sendKeys("John");
  driver.findElement(By.xpath("//*[@id='userInfo']/tr[2]/td[2]")).
      sendKeys("Doe");
  driver.findElement(By.xpath("//*[@id='userInfo']/tr[3]/td[2]")).
      sendKeys("1234");
  driver.findElement(By.xpath("//*[@value='Enter']")).click();
  ...
}
```

It is important to highlight that the strong coupling and low cohesion problem emphasizes the fragility problem discussed in Section 2.1. In fact, if the test suite is strongly coupled with the web application under test, several locators are inevitably repeated in the test code. If such locators become fragile after an evolution step, multiple repair activities would need to be carried out [14].

Fortunately, there is a solution to mitigate the strong coupling and low cohesion problem in the test code. The test scenario can be well separated from its technical implementation by using the Page Object (PO) design pattern[c] [16]. Page objects serve as an interface of the web app: they represent the GUIs as a series of object-oriented classes that encapsulate the features offered by each page into methods. Thanks to its adoption, the implementation details are moved into the page objects, a bridge between web pages and test cases, with the latter only containing the test logics.

---

[c] https://github.com/SeleniumHQ/selenium/wiki/PageObjects.

2.3 The Incompleteness Problem

Even if techniques for automatically generating test cases for web applications have been proposed [17–19], typically test suites are still developed manually. Web testers study requirements documents and create test cases that cover the requirements of the web application under test. Then, they implement the test cases in test code. Being manual, these activities are time consuming, expensive, and not very effective. The direct consequence is the inability of the test suites to cover the input space of the application thoroughly, because only specific paths of the web application under test are exercised, leading to a low coverage of the functionalities and thus to a poorly tested web application.

We have observed this problem repeatedly during our industrial collaborations. For example, during a project with a company we found out that only a small portion of their tests was automated [10].

At last, we want to highlight that the fragility and strong coupling problems contribute to have incomplete test suites. Indeed, testers spend most of their time correcting/maintaining existing fragile/strong coupled code, rather than actually developing new test cases. In short, limiting the first two problems (fragility and strong coupling) implies to (indirectly) mitigate also the third one (incompleteness).

In conclusion, the three problems are correlated and, in order to find an effective solution, they need to be addressed *together*. We believe that by limiting the first two problems (fragility and strong coupling), also the incompleteness problem would be (indirectly) mitigated, leading to better tested web applications.

3. STATE OF THE ART ON THE THREE OPEN PROBLEMS

In the last 15 years, the research community has been particularly active in proposing new approaches and tools to advance the state of the art and practice in web test automation. Fig. 2 overviews the most relevant contributions and their mapping with the three open problems affecting web test automation and described in Section 2. In this section, we provide a detailed description of the investigations carried out by researchers and their findings. We, by no means, claim that our list represents all the relevant and noteworthy research performed in the area of web testing. However, we present the research solutions that, according to the experience and personal opinion of the authors, are mostly correlated with the three open problems.

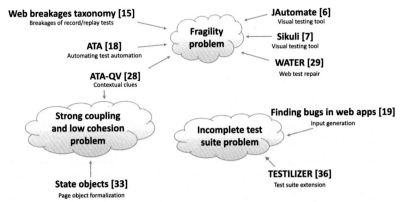

Fig. 2 Existing research proposals and tools addressing the three open problems individually.

3.1 State of the Art on the Fragility Problem

The brittleness of web test cases developed with test automation tools is a well-known problem among practitioners, and it has also been acknowledged and studied by many researchers [20–23]. However, a first study on the causes behind test breakages was by performed only recently, by Hammoudi and colleagues [15]. In this paper, they developed test suites for eight popular web applications and simulated a regression testing scenario, through the manual evolution of those test suites on 453 releases. Then, they characterized and collected the reasons for which tests broke. As an outcome, they developed a taxonomy of breakages for record/replay tests. As a confirmation of the findings of our previous research work [14, 24], also Hammoudi and colleagues have singled out *locators* as constructs that are particularly fragile in the face of software evolution, accounting for the three-quarters of the total amount of breakages.

The fragility of web test code can be mitigated in two different yet complementary ways. First, one can *prevent* the occurrence of test breakages, by creating test code which is designed to be *robust* to minor application changes. A second mitigation rule would be to use *repair* techniques to fix the broken test code automatically.

3.1.1 Robust Data Extraction

In the context of information retrieval and web data mining, researchers have proposed techniques for the robust extraction of information from evolving structured documents (e.g., XML documents). As an example, Dalvi and colleagues [25] propose to generate high-level XPaths that are resilient to minor page changes and thus can be used to retrieve the same information in evolving versions of an HTML document. In another work [26] two models were used to study robustness: the adversarial model, which includes the worst-case robustness of wrappers, and the probabilistic model, which is based on the expected robustness of wrappers, as web pages evolve. By using both models, robust wrapper can be constructed. An evaluation on real websites demonstrated that such algorithms are highly effective in coping up with changes in websites and reduce the wrapper breakage by up to 500% over existing techniques.

The downside of such techniques is that they require learning a probabilistic model from a corpus of documents. For this reason, they cannot be directly adopted in a typical software engineering scenario, where the testing phase is usually characterized by strict timing constraints. In order to be useful, wrapper generation techniques must be adapted for generating robust locators—rather than wrappers—to be used in web testing environments, where there is a similar need by the tests to target the same web element, across different releases of the same web application.

3.1.2 Breakage Prevention

To this extent, the ROBULA+ algorithm [24] uses heuristics adapted from the information retrieval field in order to generate robust XPaths that can be used as locators in web test cases. A robust locator is likely to work correctly also if the web application undergoes minor GUI changes on new releases. The intuition behind ROBULA+ is to carefully combine XPath predicates in order to maintain the locators as short and compact as possible, while retaining a low level of fragility. Empirical results have demonstrated such an intuition: ROBULA+ locators exhibited a lower fragility than the state-of-the-practice/art locator generator algorithms such as FirePath or Selenium IDE.

LED [27] is a programming-by-example tool that automatically synthesizes web element locators based on positive and negative examples of DOM elements provided as input by the developer. LED casts the problem of finding a locator to solving a constraint satisfaction problem over the group of valid DOM states in a web application. Results show that LED can synthesize DOM element locators with a 98% recall and 92% precision, with as low as five positive and five negative relevant examples.

Other notable solutions are ATA [20] and its successor ATA-QV [28]. For brevity, we describe only the enhanced latest version ATA-QV. The tool uses visual landmarks (named contextual clues in the paper) to identify objects on the GUI. For each element with which the test interacts, ATA-QV retrieves a list of potential visual "clues" to be used as reliable landmarks to identify such elements uniquely. Let us consider the form of Fig. 1. ATA-QV would automatically retrieve the labels that are associated to each text field (i.e., that are positioned above, or besides them), so that, for example, the first text field would be associated to its descriptive label "Name", the second text field with the label "Surname", and so on. By means of this mapping, ATA-QV then converts manual test steps into more abstract commands, which should be more resilient to the evolution of the applications, if the set of GUI changes remains limited. Even if very promising, ATA and ATA-QV suffer two major limitations: (1) if a descriptive label cannot be found, the tool will use the typical DOM-based locators such as the XPath of the element, making the script susceptible of the fragility problem, (2) the tool does not guarantee that the labels will be persistent across versions (for instance, if the tagname of the element gets modified—from input to button—then the tool would fail in targeting the correct element), and (3) such techniques are embedded into commercial tools, and are not available "as-is" to practitioners.

3.1.3 Web Test Repair

In the context of Web, the state-of-the-art test repair algorithm is WATER [29]. WATER is based on differential testing and compares the execution of the tests on two different releases: one in which the tests run properly, and another in which the tests break. By gathering data about these executions, WATER uses heuristics to find a set of potential fixes for the broken tests. While the repair algorithm of WATER has a straightforward design and can manage a good number of cases (such as locators or assertions), it has limitations that derive from its DOM-related narrowness. First, the algorithms can produce a great number of false positives, as recognized by the authors of the paper [29]. Second, only relying on the DOM may be insufficient to find candidate repairs. Third, the algorithm only triggers repairs at the point in which the test stops, which makes it impossible to handle propagated breakages [15], i.e., cases in which a breakage appears in a later point of the test execution. In many of such cases, it is imperative for the tester to inspect the GUI or replay the tests to find the root cause behind a breakage, because existing repair algorithms are inapplicable.

Recently, an enhanced repair algorithm has been proposed, named WATERFALL [30]. WATERFALL is built on top of WATER and suggests repairs for broken Selenium IDE test scripts. The algorithm is also based on differential testing, but it does take into account all the intermediate minor versions that occur between two major releases of a web application and uses the WATER heuristics to trigger the repairs. The results of the empirical study on seven web applications show that WATERFALL is more effective than WATER (with a 209% improvement on the number of correct repairs suggested). This is because the number of code changes between two major releases is typically larger than the number of changes between any pair of successive releases or commits, and are also be intertwined in manners that render repair heuristics less effective. Applying a repair technique iteratively to intermediate versions or commits leaves the technique with fewer, less intertwined changes to consider per application, increasing its chances of success [30].

3.2 State of the Art on the Strong Coupling and Low Cohesion Problem

As far as maintenance is concerned, Page Object has proven successful in web test automation, where it has emerged as the leading pattern for enhancing test maintenance, reducing code duplication and lowering the coupling between test cases and web applications. Page objects provide an unique location for maintenance activities. Since no duplications are present, a single code fragment needs to be corrected, and the modification is propagated on the entire test suite.

Martin Fowler first described the page object pattern under the name Window Driver [16]. Fowler illustrates basic rules of thumb for page object creation, as that "*it should provide an easy programmable interface to the program, hiding the underlying widgetry in the window.*" Furthermore, Fowler advocates assertions-free page objects, meaning that "*although they are commonly used for testing, they should not make assertions themselves. Their responsibility is to provide access to the state of the underlying page, leaving the testers to carry out the assertion logic*" [16]. However, the term "page object" has been popularized by the Selenium web testing framework, which has become the generally used name. Selenium's wiki strongly encourages the use of page objects as a best practice and provides advices on how they should be implemented [31].

Berner and colleagues [8] report on observation and experiences made in a dozen of projects in the area of test automation. Both works push on the concepts of reuse and single responsibility, of which the page object pattern is an important candidate for the implementation of such best practices within a web test suite. In another empirical study [32], studies on the prevalence of Selenium-based tests among open-source applications are presented. The most frequent subjects to change are quantitatively and qualitatively assessed, e.g., it has been found that 75% of web tests are no longer maintained after two/three releases due to impracticality to coevolve them together with the software under test. In their future work, the authors suggest the use of design pattern, such as the Page Object, to maintain the traceability between unit tests and web application portions.

The benefits associated with the adoption of the Page Object pattern in the maintenance of web test suites have been empirically measured, both within an industrial environment and in academia. A first empirical work shows that test suites developed with a programmable approach (e.g., Selenium WebDriver) exhibit higher benefits than those developed with a capture and replay method (e.g., Selenium IDE) [12]. In addition, maintaining a programmable test suite required less effort, thanks to the introduction of the Page Object design pattern, able to decouple the test logics from that of the application. This insight is confirmed by another investigation within a real world industrial setting [10]. Thanks to the Page Object, the maintenance effort is drastically reduced when the software evolves, thus justifying their initial implementation effort. Unfortunately, in case of big web applications, the creation of page objects can be a laborious task.

Concerning how page objects should be implemented, a state-based generalization, based on UML state charts, is proposed in [33]. For Van Deursen, a behavioral state machine of the web application can effectively guide a tester toward the generation of page objects, following the typical user scenarios. In his proposal, each page object corresponds to a state in the state machine, and hence becomes a *state object*. A state object has a set of well-defined responsibilities, defined by its methods. There are two kinds of behavioral methods. The *inspection methods* return the textual value of web elements displayed on the browser, when it is in a given state (typically, they can be used in test scenarios to verify that the browser displays the expected values, e.g., the user name of the current logged user must be present on the home page). The *trigger methods*, on the other hand, correspond to functions that make the browser change state. Inspection methods can also be useful as a self-check, to verify that a series of constraints hold when the application is in a particular state (for example, in the authenticating state, one would expect to find input fields for the insertion of user name and password).

In a recent work, Yu and colleagues [34] propose an automatic test generation technique for dynamic web applications. The technique decouples test code from web pages by automatically generating page objects. Based on the page objects, tests are created performing an iterative feedback-directed random test generation. On top of the retrieved methods, sequences of calls are generated with Randoop [35] for the automatic construction of test cases.

3.3 State of the Art on the Incompleteness Problem

Achieving total coverage in testing is definitely impossible, due to the limited amount of resources or the combinatorial explosion of the inputs in case of complex applications. However, research efforts have been directed to improve the state of the art with test augmentation and test generation techniques.

To the best of our knowledge, TESTILIZER is the first work aimed at *extending* an existing web application test suite by leveraging existing test cases [36]. This approach reuses knowledge in existing human-written test cases and uses a web crawler to extend the navigation paths. In particular, TESTILIZER exploits input values in existing tests to explore alternative paths and mined oracles for regenerating assertions for such alternative paths. Results show that, on average, TESTILIZER can generate test suites with improvements of up to 150% on the fault detection rate and up to 30% on the code coverage, compared to the original test suite.

Current web testing techniques simplify the test oracle problem in the generated test cases by using "soft" oracles, such as invalid HTML and runtime exceptions [36]. However, "soft" oracles are limited and not able to find bugs pertaining to the functional requirements of the application. To be really useful, automatically generated test cases should contain "strong" oracles (i.e., assertions) to determine whether the application under test works as requested. Indeed, it has been empirically evaluated that assertions are strongly correlated with the test suite effectiveness [37]. Code coverage is certainly a desirable characteristic for a test suite. However, that paper shows that there is a very strong correlation between the effectiveness of a test suite and its size, with the quantity and quality of assertions it contains.

In the context of web test generation, ARTEMIS is a feedback-directed automated test generation for JavaScript in which execution is monitored to collect information that directs the test generator toward inputs that yield increased coverage [38]. The generated tests can be used for instance to detect HTML validity problems and other programming errors. Mesbah and colleagues [17] propose a new testing technique that features the power of automated exploration (by means of a crawler) with invariant-based testing. In this technique, the user interface is checked against different constraints, expressed as invariants, which can act as oracles to automatically conduct sanity checks at a DOM level. For example, generic invariants include the absence of broken links, or the validity of the HTML. Concerning JavaScript applications, JSEFT is a framework that targets test generation for JavaScript applications. The approach employs a combination of function coverage maximization and function state abstraction algorithms to efficiently generate unit test cases with automatically generated mutation-based oracles [39]. ATRINA [40] infers test oracles from existing UI-level test cases to generate JavaScript unit tests. SUBWEB [41] is a new search-based web test generation technique, in which page objects and genetic operators are used to drive the generation of both test inputs and feasible navigation paths. On a first case study, SUBWEB was able to achieve higher coverage of the navigation model than a typical crawling-based approach.

4. OVERCOMING THE THREE OPEN PROBLEMS: THE NEONATE VISION

After the insights gained in years devoted to research in the web testing field [12, 23, 24, 42–46] and our analysis of the state of the art and practice related to this field, we have matured a vision of how to overcome the existing open problems affecting web test automation.

4.1 The Stuck Situation

First, we want to summarize the current problematic situation in three points as follows.

− First, web applications are complex, heterogeneous, distributed systems that are highly dynamic with unpredictable control flows. Specific requirements characterize them, such as huge time-to-market pressure, distributed infrastructure, high asynchronicity, and constant shifts in user requirements. Thus, in those years, developing high-quality web applications and testing web systems have become one of the most challenging goals among software engineers, demanding for novel approaches/techniques.

- Second, web testers are often not able to develop robust and maintainable test code. One might argue that this is due to the testers' inexperience or the lack of skills. While this is likely possible, we believe the increasing complexity of web applications and the three aforementioned problems (see Section 2) to have a predominant role.
- Third, when test code is produced, it often exhibits characteristics as fragility, strong coupling with the web application under test, and incompleteness. Thus, such automated test suites are underused or quickly abandoned, despite their potential value to catch errors and regressions. As a result of this tangled situation, we currently live with "buggy" web applications.

Thus, what we really need is a major breakthrough in web application testing, i.e., a new way of doing test automation.

4.2 The Vision

The vision behind the NEONATE project is to *empower the Web tester* with an *integrated testing environment* (ITE) composed by techniques and tools that are specifically designed to deal with (and hopefully overcome) the three big problems affecting web test automation.

In our vision, NEONATE will offer the tester automatic support in the: (1) creation of robust test code, (2) repair of broken test code, (3) separation of the test logic from the implementation details, (4) extension of existing test suites, and (5) migration to novel visual testing technologies.

We believe that NEONATE will facilitate the development and maintenance of web test suites. As a result, web applications will be tested more intensively, and hopefully this will benefit their correctness.

The need of integrating various testing tools in a ITE has been pointed out by several researchers in the last 20 years. For instance, one of the first works on this topic is the one of Gao et al. [47]. They developed a web-based system for test information sharing, control and management. Moreover, they report that big companies have strong demand for integrated testing environments providing: (1) test information bank (supporting the test information sharing among engineers in different phases), (2) facility tools for test information tracking, analysis and reporting, and (3) a systematic transparent interface to plug-in various test tools. Subsequently, Williams et al. in [48] describe an architecture for integrating both new and existing testing tools into solutions adopted in testing organizations across IBM. More in detail, this integrated testing environment supports the control and data integration across tools and repositories, provides an unique graphical user interface, and can be extended to support new tools.

Agreed with the authors of these works, we designed the architecture of NEONATE on a plug-in infrastructure. In this way the functionalities offered by our ITE can be easily extended to include the most novel state-of-the-art solutions that will be eventually proposed in the future. Moreover, NEONATE is equipped with a GUI-based module that first analyses the test code looking for code deficiencies, and then presents—in a coherent, unified way—several wizards conceived to help the tester in executing the suggested improvement tasks (see additional details in Section 5).

4.3 Existing Integrated Testing Environments

To the best of our knowledge, no integrated testing environments (ITE) have been proposed to face simultaneously the three open problems in the context of web testing.

A notable ITE is FITTEST [49], whose purpose is to automatically generate test cases for web applications. Such test cases are generated using a combination of different techniques: model-based testing, combinatorial testing, mutation, and search-based testing techniques. NEONATE, on the other hand, encompasses a set of prototype tools for improving, maintaining, and extending a web test suite. Thus, the goals of the two projects are orthogonal and can be eventually utilized together (however, the engineering integration cost between the artifacts produced by two different ITEs might be nonnegligible).

FITE [50] is a static and dynamic analysis ITE, not specific for web applications. We share with the authors of FITE the same basic idea: developing an integrated testing environment able to continuously analyze the increments and produce recommendations to the tester. As in our proposal, the role of the user is essential, and the purpose of the testing environment is to empower the human tester. One of the most interesting aspects of FITE is its pluggable view-based approach, i.e., the tester can select a kind of analysis (e.g., performance analysis) and the tool produces only recommendations related to that selection. For example, for the "performance" selection, the tool might show code paths with the highest execution times. The main difference with respect to NEONATE is in the level of testing: FITE is able to assist the developer with unit tests whereas NEONATE manages E2E test code. Moreover, the considered quality factors are different: FITE targets security and performance aspects, while NEONATE focuses on the quality and correctness of the test source code.

One of the first works emphasizing the importance of having an ITE in the web application context is by Margaria and colleagues [51] where the authors present a methodology with the aim of granting coverage of all the functionalities of a web application. The core module is the Test Coordinator that drives the generation, execution, evaluation, and management of the system-level tests. Differently from NEONATE, in [51] a model checker and the concept of property are at the base of the proposal, as well as a graphical test cases design facility that is not present in NEONATE.

5. ARCHITECTURE OF THE NEONATE INTEGRATED TESTING ENVIRONMENT

In this section, we present the architecture of the NEONATE integrated testing environment. In our vision, NEONATE will be a full-fledged toolset composed by a set of joinable integrated prototypes (hereafter referred also as *modules*), each of which addresses a particular development/maintenance task on web test code. NEONATE will be built on top of Selenium WebDriver, the flagship open-source test automation tool for web applications. We have opted for the WebDriver framework because it is a remarkable open-source solution, widely adopted both in the academia and in the industry [32]. The maturity of WebDriver on the worldwide testing scenario has been also recognized by the W3C consortium, of which it has become a standard.[d] NEONATE will be realized by relying on the Eclipse platform and its plug-in-oriented architecture. Fig. 3 gives a high-level overview of NEONATE: each module will communicate by means of a lightweight protocol, will share a common repository containing the test code (repository software architecture), and will eventually be combined in a tool-chain. On top of the plug-in architecture, a GUI will allow the web tester to use the tools in stand-alone modality or combine them. NEONATE will support modules for the development of test code (i.e., those labeled with "D") and the maintenance of existing test code (i.e., those labeled with "M"). NEONATE will feature the ASSISTANT module, i.e., an orchestrator mechanism based on wizards able to suggest the testers which plug-in, or set of plug-ins, needs to be executed for a specific purpose. For instance, if the tester wants to refactor an existing Selenium DOM based web test suite, to enhance its robustness and structural quality, the ASSISTANT will suggest to apply in series the modules ROBULA+ and APORES. In the

---

[d] http://www.w3.org/TR/webdriver/.

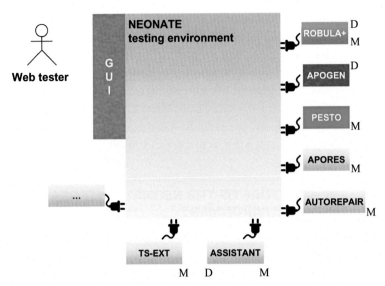

Fig. 3 High-level architecture of the NEONATE integrated testing environment.

following sections, we describe in turn each module of NEONATE. We underscore that some prototypes have already been implemented as part of our research (the colored modules in Fig. 3) while some others are part of our ongoing work (the gray modules).

5.1 Robust Web Element Locators With ROBULA+

Usually, the first aspect on which it is important to intervene is the test suite robustness. As already said, locators play a major role in the fragility problem. Thus, the first module of NEONATE integrates ROBULA+ (ROBUst Locator Algorithm+) [24], a state-of-the-art algorithm for automatically generating robust web element locators. We define a robust locator as a selection command that continues to point to the target web element, even if the web page has changed because of a new release of the web application. ROBULA+ is based on the XPath query language. The intuition behind ROBULA+ is to carefully combine XPath predicates in order to maintain the locators as short and smart as possible, while retaining a low level of fragility. In short, the algorithm starts with the most generic XPath locator that selects all nodes in the DOM tree (//*). It then iteratively refines the locator until only the element of interest is selected. In such iterative refinement, ROBULA+ applies seven refinement transformations, according to a set of heuristic XPath specialization steps, prioritization, and black listing techniques. The prioritization is used to rank candidate XPath expressions in terms of expected robustness, while the black list excludes attributes that are intrinsically considered fragile. For further technical details, we refer the reader to [24].

We have demonstrated the effectiveness of ROBULA+ in producing reliable locators and limiting the fragility of test cases. In brief, we have compared the robustness of the locators generated by state-of-the-art/practice tools and algorithms (i.e., FirePath absolute and ID-relative, Selenium IDE, and Montoto [52]) with the ones generated by ROBULA+. Empirical results (see Fig. 4) indicate that the locators generated by ROBULA+ are significantly better in terms of robustness than all the other kinds of locators (63%–90% fragility reduction) which is expected to be associated with a corresponding reduction of the maintenance effort required to repair the test cases. Moreover, the time required by ROBULA+ for generating the XPath locators is acceptable for a human web tester (only 0.16 s per locator on average).

We have implemented ROBULA+ as a freely available Java program, able to generate (in batch mode) XPath locators for hundreds of web elements (useful, for instance, when the locators of an entire test suite have to be changed with the ones generated by ROBULA+). Moreover, ROBULA+ is also available as a Firefox add-on that can be used by web testers to generate locators during the development of test suites. We are also considering the idea of including in NEONATE the capability of combining the locators produced by a set of different algorithms (including ROBULA+) into a single, consolidated *multilocator* [23] based on a voting mechanism that assigns different voting weights to different locator generation algorithms. The two implementations of ROBULA+ are available on the tool's web site: http://sepl.dibris.unige.it/ROBULA.php.

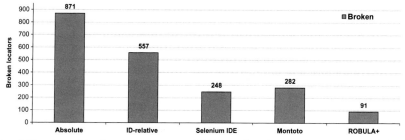

Fig. 4 Number of broken locators out of 1110 analyzed web elements in eight web applications (see [24]).

5.2 Automatic Generation of Page Objects With APOGEN

While developing a web test suite, it is of great importance to separate the test logic from its technical low-level implementation. With this aim in mind, the second module of NEONATE integrates APOGEN (Automatic Page Object GENerator) [53], a prototype tool for the *automatic generation of page objects* to support E2E web testing. Automatically generated page objects can alleviate the manual work of web testers so as to reduce the development costs.

APOGEN consists of five main modules (see Fig. 5): a Crawler, a Clusterer, a Cluster Visual Editor, a Static Analyser, and a Code Generator. The input of APOGEN is any web application, together with the input data necessary for the login and forms navigation. The output is a set of Java page objects as supported by the Selenium WebDriver framework. In short, APOGEN infers a model of the target web application by reverse engineering it by means of an event-based crawler (we chose Crawljax, a state-of-the-art tool for the automatic crawling of interactive web applications [17, 54]). Then, similar web pages are clustered into syntactically and semantically meaningful groups. The event-based model and the additional information (e.g., DOMs and clusters) are statically analyzed to generate a state object-based model. At last, this model is transformed into Java page objects, via model to text transformations. Since our ultimate goal is the automatic generation of meaningful page objects we share and tried to incorporate in the development of APOGEN the guidelines and best practices reported in [33].

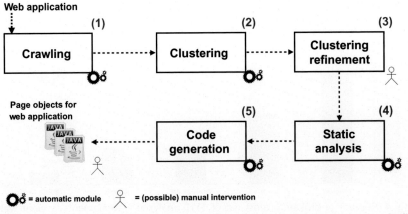

Fig. 5 High-level overview of APOGEN's approach for web page objects creation (see [53]).

Experimental results on six existing web applications indicate that APOGEN is effective to group semantically related web pages [55]. Furthermore, the page objects obtained from the output of clustering are very similar to the page objects that a developer would create manually. Indeed, 75% of the code generated by APOGEN can be used "as-is" by a Web tester, breaking down the manual effort for page object creation. Moreover, a big portion of the page object methods (84%) created to support assertion definition corresponds to meaningful and useful behavioral abstractions.

For further technical details, we refer the reader to [53]. All details about APOGEN, demo video, and experimental data can be found on the tool web site: http://sepl.dibris.unige.it/APOGEN.php.

5.3 Generating Visual Test Suites With PESTO

Depending on the characteristics of the web application under test, changing the approach used for localizing the web elements to interact with could change the fragility of the locators and thus the associated maintenance effort. More in detail, nowadays there are two main approaches to web element localization [13, 14, 56]. (1) the DOM-based approach (supported by, e.g., Selenium WebDriver[e]), where test cases access the web page Document Object Model (DOM) to locate the web elements (e.g., anchors, buttons) by accessing their properties (e.g., identifier or text), or by navigating the DOM tree by means of XPath queries. On the other hand, (2) the visual approach (adopted by, e.g., JAutomate [6] and Sikuli [7]) relies on image recognition techniques to identify and control GUI components.

PESTO (PagE object tranSformation TOol) [57, 58] is a tool able to transform a Selenium WebDriver test suite into a Sikuli visual test suite automatically, while retaining the same coverage and all the assertions.

PESTO can be useful when the web application under test is evolved to adopt modern visual widgets such as Google Maps. In these cases, DOM-based tools are not adequate, because the DOM of these visual components is complicated to retrieve (if not impossible), and such tools do not support visual testing (e.g., they cannot assert that an image is *visually* present on the screen). Moreover, with PESTO, companies can evaluate the benefits of third generation tools at minimum cost, without taking the risk of a substantial investment, necessary for the manual migration of existing test suites. Automatically migrated test cases can be smoothly introduced in the existing

---

[e] http://www.seleniumhq.org/projects/webdriver/.

testing process, so as to evaluate their effectiveness and robustness in comparison with the existing DOM-based suites.

To the best of our knowledge no other solution, both in the academia and in the industry, yet exists to carry out such migration task for Selenium-web based test cases. A proposal in the context of desktop applications is by Alégroth and colleagues [56], which allows to migrate existing automated component-based GUI test cases (GUITAR [59]) to the visual approach (VGT GUITAR).

PESTO relies on aspect-oriented programming, computer-vision, and code-transformations. PESTO executes the transformation by means of two main modules (see Fig. 6). The Visual Locators Generator generates a visual locator for each web element used by the DOM-based test suite (i.e., a unique image representing that web element on the web application GUI). Specifically, PESTO automatically retrieves the bounding rectangle of the web elements, and an image is automatically captured for each web element surrounded by such rectangles. This is technically realized by capturing the command calls to the web elements with aspect-oriented programming (AOP). Based on the captured images, the original test suites

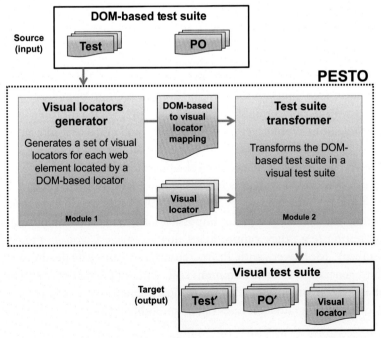

Fig. 6 High-level logical architecture of PESTO.

are automatically rewritten as visual test suites by the Visual Test Suite Transformer. This module converts the source code of the DOM-based test suite in order to adopt the visual approach; the majority of the changes are concentrated in the page objects code, since page objects are responsible for the interaction with the web pages.

The effectiveness of PESTO has been evaluated on a set of DOM-based validation test suites, developed by an independent, professional web tester for different open-source web applications [57]. PESTO generated the corresponding visual test suites for the DOM-based validation test suites requiring only a small amount of alignment work. The visual test suites automatically generated by PESTO were then executed and the test cases exhibited the correct, expected behavior. In our study, PESTO was able to migrate 100% of the command calls used in the existing DOM-based validation test suites. Moreover, by analyzing other existing DOM-based test suites, we found that PESTO can handle more than 95% of the employed command calls, when an abstraction as the page object is used. The visual locators automatically generated by PESTO were checked for readability by the professional tester involved in our experiments and they were judged easy to understand.

For the interested reader, a demo video of PESTO and the source code can be found on the tool web site: http://sepl.dibris.unige.it/PESTO.php.

5.4 Separating Test Specification from Test Implementation With APORES

The goal of the APORES (Automatic Page Objects REStructurer) prototype is separating test specification from test implementation in test code suffering the strong coupling and low cohesion problem. More concretely, APORES (which is one of the modules still to be implemented) will attack the challenging problem of restructuring a test suite built without the page objects into an equivalent one adopting them (see Fig. 7). While APOGEN creates page objects from scratch (thus supporting development), APORES will infer page objects from existing test code. Further, the two prototype tools can be used together by the web tester to alleviate the incompleteness problem: APORES will be used to transform a legacy test suite and APOGEN to generate fresh page objects that can be used to extend the initial test suite.

We are aware that the task is challenging to be executed in an automatic way. However, we are confident that a big portion of the transformation can be conducted automatically, with techniques similar to those of the already implemented prototypes of NEONATE. From the technical point of view,

Fig. 7 Input and output of APORES: it is evident the presence of the novel created page objects.

APORES could re-use part of the AOP infrastructure fine-tuned for PESTO and the crawler used for APOGEN. Further, automatic code transformation techniques (e.g., using the TXL language [60]) will be employed to effectively execute the migration.

5.5 Suggesting and Executing Repairs for Broken Code With AUTOREPAIR

The goal of the AUTOREPAIR prototype is suggesting and eventually executing repairs for broken test code due to web application evolution. Automatic repairing of test suites is an alternative way to face the fragility problem. ROBULA+ would *prevent* and limit the amount of broken locators; on the other hand, AUTOREPAIR would intervene in all cases in which a *repair* is needed. In our vision, AUTOREPAIR would manage also breakages that go beyond locator problems, for instance, a breakage in the test workflow due to a missing statement. Self-repairing test suites is a big open research problem that goes beyond the web context (e.g., techniques could be adapted to work also in the mobile environment). Indeed, several research groups are currently working on the test suite self-repairing problem [23, 29, 30, 61, 62].

As a first step toward the automatic repair of web test suites, we intend to use differential testing to compare the test executions of two successive releases of a web application: a first in which the test suite runs properly, and a second in which breakages occur (as done in WATER [29]). Then, by means of static and dynamic analyses techniques, we aim at (1) automatically determine the root cause behind the breakage, (2) automatically determine repairs for the broken statements, (3) rank the candidate repairs according to some heuristics, (4) suggest the ordered list of repairs to the developer for inspection, and (5) automatically fix the test with the chosen repair.

This approach is simple but effective because it is based on the assumption that the evolution of the web application under test is limited. In the comparison, we will use information contained in error stack traces and output messages produced by previously instrumented test code (as done for example in the ReAssert tool [61]). AUTOREPAIR would manage programmable WebDriver test code instead of capture and replay (C&R) test code (in the case of WATER specifically Selenium IDE test code) and we intend to use effective similarity measures [63] and prioritize the given suggestions to reduce the number of false positives repair suggestions.

In our vision, AUTOREPAIR will consider the test suites self-repairing problem in its entirety, also managing changes at level of steps of the test scenario (due to business logic changes) and input data (e.g., the format of the data is changed from DD/MM/YYYY to MM/DD/YYYY). In particular, repairing automatically the steps of a test scenario is extremely challenging. Indeed, repair operations can range from simple insertion and deletion of steps (e.g., deleting a confirmation action) to repair of multiple steps that require to execute multiple operations in series. Considering input data, one factor limiting the usage of web application test automation techniques is the cost of finding appropriate input values. To mitigate this problem, Elbaum and colleagues proposed a family of techniques based on user-session data [64]. In general, user-session-based techniques: (1) collect user interactions when users use a web application, (2) store the clients requests in the form of URLs and name-value pairs, and then (3) apply strategies to generate test cases. Several mechanisms can be used to capture and store user-session data (e.g., one of the simplest is configuring the web server to log all the received requests). An example of access log for an e-commerce application could be the following:

```
(2018-01-01 12:00:00) Client1 Home
(2018-01-01 12:00:01) Client1 SearchProd, prodCat=notebook
(2018-01-01 12:00:02) Client2 Home
(2018-01-01 12:00:03) Client2 FindShop, place=London,
country=UK
(2018-01-01 12:00:03) Client1 FindShop, place=Rome,
country=IT
(2018-01-01 12:00:04) Client3 Home
(2018-01-01 12:00:06) Client3 AdvSearchProd, pMin=200,
pMax=800
(2018-01-01 12:00:07) Client3 AdvSearchProd, date=01-01-
2018
```

In particular, time stamps are assumed to be associated with each entry of the log. Then a column contains the name of the host requesting a web page (e.g., Client1). The next column contains the name of the requested page followed by the name-value pairs provided to the web server (e.g., via GET requests). When requests coming from the same host are found within a proper time interval (which depends on the specific implementation of the user-session data storage mechanism), it is assumed that navigation from a previously accessed page to a new one is taking place. Otherwise, a direct request of a page is considered to occur. Finally, when a request from a host is not followed by any other request from the same host, this is interpreted as the termination of the navigation session. From this simple example it is possible to extract useful input data such as geographical locations (e.g., London, UK and Rome, IT), dates (e.g., 01-01-2018), ranges of price (e.g., from 200 to 800), and name of products (e.g., notebook). These user-session data, collected in previous releases can be then reused as input data in the current release. However, also user-session data suffer the evolution problem: session data may become invalid due to changes in the application. Input data repairing techniques [65] are able to alleviate this problem; in this way input data are able to survive the application evolution.

Finally, meaningful input generation is known to be a challenging task, due to the difficulty to find meaningful values for forms (interesting proposals exist [66–68]), and the correlation they may have (e.g., the zip code value and the country to which it refers).

5.6 Extending Existing Test Suites With Ts-EXT

Human-written test suites can be a gold mine, i.e., a valuable source of domain knowledge related to which interactions are more important to cover, which data have been used as inputs, and what elements on the page need to be asserted and how.

The goal of the Ts-Ext (Test Suite EXTender) prototype is to improve the effectiveness of the target test suite by extending it. With improving the effectiveness of the target test suite we mean: augmenting the coverage and fault finding capability of the test suite.

To implement Ts-Ext a web crawler is necessary, i.e., a program that automatically retrieves web pages of a target application and builds a web graph model where, in the simplest case, nodes are web pages and edges are hyperlinks between pages. There are three specific problems to address: (1) automated input generation, (2) paths selection (among the unbounded number of behaviors) and (3) assertions generation. Similarly to Fard et al., who proposed the TESTILIZER tool [36], we believe that these problems can be faced by combining the manual approach with an automated one.

The idea is the following: by starting from an existing limited test suite, Ts-EXT will extend it automatically adding more test cases so as to cover unexplored paths in the web graph model. More in detail, Ts-EXT will: (1) mine existing test code to infer existing input data and assertions, (2) execute the test code to find already explored web application paths, and (3) expand the initial human-written test suite with other novel test cases. To produce these novel test cases a Web crawler will be employed to discover unexplored paths while input data and assertions will be computed by means of an input generator, leveraging available user-session data [64] (i.e., data gathered when users exercise a Web application, as seen in Section 5.5), and using a tool for generating assertions.

More in detail, assertions can be created by levering on dynamic detection of likely invariants techniques similar to the ones implemented in Daikon [69]. An invariant is a property that can be relied upon to be true during the execution of a program or a portion of it. Invariants can be used in assert statements, documentation, and formal specifications. Examples include: being constant ($x == a$), nonzero ($x\ != 0$), being in a range ($a <= x <= b$), linear relationships ($y = ax + b$), ordering ($x <= y$), functions from a library ($x = f(y)$), containment (x belongs y), sortedness (x is sorted). The assertion generator of Ts-EXT can benefit from a Daikon-like automatic invariant detector. As an example of possible additional assertion, consider a simple online shop web application. As a first step, it is required to trace a large number of user sessions. The resulting log will include information concerning interesting values shown on the web pages (e.g., the number of products in the carts during the users' navigations) and the info about every action performed by the users (e.g., adding or removing products). By analyzing the execution traces, the invariant detector is able, for instance, to find the following invariant: the number of products in the cart of a user must be always equal to the number of products she/he added in—or removed from—the cart (e.g., if five products have been added and then two have been removed, an automatically generated invariant will check that the cart contains three products). This kind of invariants can be added *automatically* as assertions in different points of the test scripts (for instance every time a product is added or removed from the cart).

Fig. 8 exemplifies our idea. Tc0 is the test code written by the Web tester. Its execution produces the path composed by blue nodes on the web graph

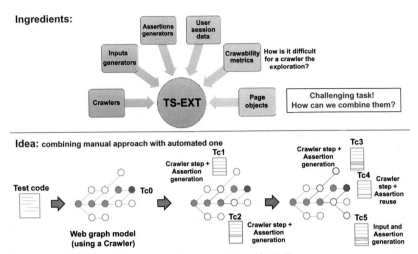

Fig. 8 Improving the effectiveness of a test suite with Ts-Ext.

Testilizer assertion

1	AssertTrue(lastRow.getParent().getTag() == "table"
2	AND lastRow.getParent().getAttributes() == "list of table's attributes"
3	AND lastRow.getTag()=="tr"
4	AND lastRow.getAttributes()=="list of tr's attributes"
5	AND for each TD (X=1..n):
6	lastRow.getChild(X).getTag()=="td"
7	AND lastRow.getChild(X).getAttributes=="list of td X's attributes"
8	)

"Human Like" assertion

1	AssertTrue(table.getName=="John" AND table.getSurname=="Doe")

Fig. 9 An example of a Testilizer (*top*) and a *Human-like* (*bottom*) assertion.

model (see Fig. 8). On the contrary, red nodes (Fig. 8, center and right) represent further steps carried out by or via the aforementioned "ingredients" (e.g., input generator). Tc1 ...Tc5 represent the code of test cases diversifying Tc0. For example, for generating Tc1 it is sufficient to make two steps similarly to Tc0 (i.e., the first two blue steps), a Crawler step and finally generating a fresh assertion by means of the generation assertion tool.

An ambitious goal would be to outperform Testilizer both in terms of effectiveness of the extended test suite (i.e., more coverage and more faults finding capability) and in terms of quality of the produced assertions. Indeed, the assertions produced by Testilizer are quite effective but difficult to understand and maintain because based on the structure of the page (see Fig. 9 top for an example). On the contrary, we intend to automatically produce "human-like" assertions, i.e., assertions simple to understand and more similar to those that would have produced a human (see Fig. 9 bottom for an example).

5.7 Supporting the Tester During Maintenance/Development With ASSISTANT

The ASSISTANT module will support the web tester in the use of NEONATE and its modules. The goal of the ASSISTANT is suggesting to the web tester the actions/transformations to apply to the test code that are most appropriate, in order to improve its quality, or to extend an initial test suite.

In our vision, the ASSISTANT is a recommendation system that will analyze the test code looking for code smells. A repository of *test code smells* containing potential problems identified in the test code during our analyses will be maintained and continuously updated, as new potential threats manifest. The test code manually written by the tester will be continuously analyzed by means of a static analyzer (*diagnosis phase*), starting from the earliest stages of development. The ASSISTANT will consult the test code smells repository and provide the tester with suggestions/recommendations. The suggestion can be: (1) *on demand*, i.e., explicitly requested by the tester or (2) *in automatic modality*, when the ASSISTANT will automatically warn the tester about a potential threat in the code, and suggest the NEONATE module that can be adopted to mitigate/solve it (e.g., ROBULA+, APORES, and TS-EXT). Thus, suggestions are displayed based on the task that the tester is performing on the test suite. For instance, if the tester is editing a test code using fragile absolute XPath locators in the editor, the ASSISTANT will automatically highlight the identified smell suggesting her/him the usage of the ROBULA+ module.

A dialog-based GUI will guide the tester through the steps necessary to execute the selected refactoring. Depending on the complexity of the refactoring (in some cases, multiple modules have to be called in series), a wizard or a simple dialog will be used to gather the necessary information. Similarly to the Eclipse refactoring wizard, the ASSISTANT will be equipped with a preview module, able to show the changes that the refactoring will perform. This will open a view containing a list of changes to be made to the current test code along with a link that allow to execute the refactoring on the test code.

6. NEONATE'S EXAMPLES OF USE

This section aims at showing how NEONATE can be useful in simplifying typical test code tasks. Two scenarios will be considered: (1) creation from scratch of an automated test suite for a web application and (2) refactoring of an existing automated test suite. Besides these two examples, NEONATE can be utilized for many other maintenance activities

such as: generating robust locators for new or modified web elements with
ROBULA+, generating page objects for new portions of the web application
with APOGEN, realigning broken test code with the updated web application
with AUTOREPAIR.

6.1 Automated Test Suite Development (Scenario 1)

Let us assume that the quality assurance team (QAT) currently verifies the
correct behavior of the web application employing several web testers that
manually follow the steps of a predefined set of test cases. This is actually a
typical solution adopted in the industrial practice, as observed in our experience. However, as discussed before, this practice is error prone, time consuming, and ultimately not very effective.

For this reason, the QAT wishes to improve the level of automation of
the testing activities by means of automated testing tools. The desired goal is
to bring all the main benefits of automation in the current inefficient testing
setting, e.g., a fast and unattended execution of the test cases after every
change made to the web application under test (i.e., for regression purposes).

Without a ITE like NEONATE, the QAT has to manually implement the
test code able to instrument the web application and verify its behavior.
QAT can follow two strategies [14]:
- adopt a capture and replay (C&R) tool (e.g., Selenium IDE). In this former case, the web testers perform actions on the web application GUI and
  execute the test scenario. The tool records such actions and generates
  replayable test scripts that, in a second time, can redeliver the same actions
  to the browser automatically.
- rely on a programmable test automation framework (e.g., Selenium
  WebDriver). In this latter case, testers have to implement the test code
  using a programming language (such as Java or Ruby) and APIs providing
  commands to control the browser and interact with the web elements
  composing the pages (e.g., click a button, fill in a field, or submit a form).
Both strategies have shown nonnegligible development/maintenance effort
[12] that often represents a barrier to the adoption of test automation. More
in detail, C&R test suites requires less development effort as compared to
the programmable ones. On the other hand, programmable tools show, if
the test code is well-engineered, a lower overall cost (i.e., considering both
development and maintenance costs) since the adoption of specific design
patterns. As already discussed, the page object pattern can drastically reduce
the maintenance effort of a test suite during the web application evolution.

By adopting NEONATE, instead, the QAT can get all the benefits of high-quality test automation code at a cost comparable with the one of a C&R solution. In fact, the ASSISTANT contains a wizard procedure that suggests the steps to follow during the creation of a new test suite. The first step consists in developing high-quality page objects representing the web pages of the application. This step is fully automated in NEONATE. Indeed, the ASSISTANT asks only a few information about the target web application such as its URL and the data to insert in the forms that cannot be filled with autogenerated inputs (e.g., the login credentials). At this point, APOGEN is executed and a set of page objects encapsulating the web pages functionalities is created. Thanks to the integration with ROBULA+, the locators used in the page objects for locating the web page elements to interact with are as robust as the best locators that a human expert would create. Fig. 10 shows a fragment of a sample web application which is composed by various pages such as Login, Index, Messages, Settings, etc. For each of them, APOGEN generates a page object encapsulating their functionalities (reported as gray boxes in Fig. 10). For instance, the Login page contains a form allowing the users to

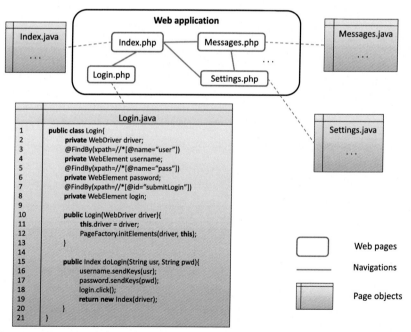

Fig. 10 Fragment of the web application and associated page objects created by APOGEN.

```
1    @Test
2    public void testLogin(){
3        Login lp = new Login(driver);
4        Index ip = lp.doLogin("admin", "secret");
5        assertEquals(ip.getLoggedUser(), "(admin)");
6    }
```

Fig. 11 Example of page object-based programmable automated test.

log in the application. Correspondingly, a Login page object is created containing a method that allows to perform the authentication (see lines 15–20).

In the next step, the QAT has to manually implement the test cases into test code by simply calling the various methods exposed by the page objects and following the tests specifications (before driving the manual testing activities). Fig. 11 reports an example of test code for verifying the correct behavior of the login functionality relying on the page objects created by APOGEN.

At this point, the automated test suite creation is completed and the test suite executes the same test cases that were previously executed in a manual fashion. In addition, the ASSISTANT suggests to improve the coverage of the test suite by using the TS-EXT module. In this way several additional test cases can be added to the original test suite. For instance, in our example, the test suite contains only a positive login test (i.e., log in using correct credentials). Thus, TS-EXT is able to find the different navigation paths that appear when inserting incorrect or incomplete credentials and leading to one or more "error" pages displaying specific messages to the user (e.g., invalid credentials, empty password). Hence, new test cases are created, increasing the coverage of the initial test suite.

Finally, whether the QAT has decided that it is better to adopt the visual localization strategy for the web elements to interact with (instead of the DOM-based one that is the default option in NEONATE), because for example a refactoring toward complex visual components to offer increased user-friendliness and responsiveness is forthcoming, it is possible to automatically migrate (a portion of) the freshly created DOM-based test suite to the visual approach by mean of the PESTO module.

6.2 Automated Test Suite Refactoring (Scenario 2)

In this second example of use, let us assume that the QAT has already adopted Selenium IDE as a reference tool for web testing. Fig. 12 shows a Selenium IDE test script used to verify the correct behavior of the login

#	Action	Locator	Value
1	open	localhost/webapp/login.php	
2	type	//*[@id='LoginForm']/input[1]	"admin"
3	type	//*[@id='LoginForm']/input[2]	"secret"
4	click	//input[@id='submitLogin']	
5	assertText	//*[@id='loggedUser']	"(admin)"

Fig. 12 Example of C&R automated test.

```
1    @Test
2    public void testLogin(){
3        WebDriver driver = new FirefoxDriver();
4        driver.get("localhost/webapp/login.php");
5        driver.findElement(By.xpath("//*[@id='LoginForm']/input[1]")).sendKeys("admin");
6        driver.findElement(By.xpath("//*[@id='LoginForm']/input[2]")).sendKeys("secret");
7        driver.findElement(By.xpath("//*[@id='submitLogin']")).click();
8        assertEquals (driver.findElement(By.xpath("//*[@id='loggedUser']")).getText(), "(admin)");
9        driver.quit();
10   }
```

Fig. 13 Example of programmable automated test not adopting page object and factory patterns.

functionality. However, QAT quickly discovered several limitations in its usage. For instance, a C&R tool does not provide natively some useful features, such as conditional statements, loops, logging functionality, exception handling, and parameterized (also known as data-driven) test cases. Moreover, as mentioned in the previous section, C&R solutions have been shown to be very expensive during maintenance. For this reason, QAT could plan to migrate the test suite to Selenium WebDriver, i.e., adopting a programmable approach.

Without NEONATE, the effort of such migration is, in practice, similar to developing a new programmable test suite from scratch. Indeed, the only support provided by Selenium IDE is to export the test cases in Java, one of the languages supported by Selenium WebDriver. However, in order to take full advantage of the benefits of the programmable approach it is required to provide the test code with a completely different organization with respect to the one created by the simple one-to-one translation from the Selenium IDE C&R test code. Indeed, the adoption of the Page Object and Factory patterns drastically affects the test suite structure. Fig. 13 shows the test code (written using the Selenium WebDriver framework), implementing the same test case, that does not adopt any design pattern (the same result could be obtained using the export functionality of Selenium IDE). This test code, even if small, is largely suboptimal considering the two factors: robustness of the locators and structural quality. On the other hand, the test code reported in Fig. 11 is by far simpler and more understandable since all the technical details are moved to the page objects.

With NEONATE such onerous refactoring activities are executed automatically. Indeed, the ASSISTANT, analyzing the Java code exported by Selenium IDE, suggests the use of the APORES module for restructuring the test suite built without the page objects into an equivalent one adopting them. As reported in the previous section for APOGEN, also APORES takes advantage of ROBULA+ for generating robust web element locators. At this point, the programmable automated test suite is completed and executes the same test cases that were previously executed with Selenium IDE. As described in the previous section, following the ASSISTANT suggestions, it is then possible, for instance, to improve the test suite coverage by means of the TS-EXT module or migrate it to the visual approach with the PESTO module.

7. NEONATE'S LONG-TERM IMPACT

We believe that NEONATE will represent a major breakthrough in the web testing domain. Moreover, since web apps are similar to mobile apps many of our proposals and prototype tools can also be applied to the mobile context increasing the impact of our work. More in detail, we believe that the impact of NEONATE will be on three lines: scientific, practical, and industrial.

7.1 Scientific

From the scientific viewpoint, NEONATE will boost the web test automation research in novel directions such as generating robust web test code, suggesting improvements to test code, migrating existing state-of-the-practice test suites to novel approaches. The research of this project is strategic, because it aims at raising the quality of all daily used apps and is original, because only few researchers in the world are facing these problems and with solutions complementary to ours. Currently, there is scientific evidence that test automation is hindered by the fragility problem, but its exact nature is unknown and unaddressed. Moreover, it is not clear whether specific ad hoc remedies and shortcuts are adopted in the industrial practice. Finally, possible relationships between the problem and the available testing tools are totally unknown. An industrial survey will be useful to answer all these questions.

7.1.1 Benchmarks

The possibility to compare the results of our work with those of other competitors is of vital importance. In general, well-defined *benchmarks* allow researchers to empirically validate their proposals and compare them against the state-of-the-art solutions. For instance, examples of benchmarks used in testing research involve SIR [70], Defect4J [71], and SF100 [72]. A benchmark of this kind is absent in web testing, thus another follow up of our research concerns building a benchmark of web applications, along with test suites, bug reports, and bug fixes. Such a benchmark could be a reference for the web testing community once made publicly available. We intend to build benchmarks on which we will validate our proposals and compare them with state-of-the art algorithms and tools. These benchmarks will be a reference for the community and we will make them publicly available. Note that, currently, although really useful, benchmarks are not available in our community.

7.2 Practical

From a practical viewpoint, the causal implication of our project is the following: we aim at nullifying the impact of the fragility problem. This will lead to a breakthrough in the practical usage of test automation. Extending the usage of test automation will increase the coverage of tested apps (deeply tested apps) with the indirect consequence of improving the quality of all web apps. This will have a remarkable economic relevance and direct impact on the daily lives of the users. It is worth noting that although our prototype tools are dependent on the chosen testing platform (e.g., Selenium), they will be almost independent from the technology used for developing the web app under test (e.g., Ajax, Angular, JQuery) because E2E testing tools consider the app as a black box and focus only on GUI interactions. Moreover, our solutions can be easily adapted/generalized to any E2E testing tool and framework.

7.3 Industrial

Even if it is outside the scope of the project to conduct technology transfer from academia to industry, the postproject outcome may include the development of full-fledged industrial tools and the creation of high-tech start-ups. Big companies and nonprofit organizations could integrate our solutions (e.g., the algorithms for generating locators currently embedded in the browsers produce locators that are extremely fragile) into their browsers as plug-ins or into specialized IDE for web developers and testers.

8. CONCLUSIONS

Nowadays web applications are critical assets of our society. Billions of people use them every day as source of information, means of communication and venue for commerce. For these reasons, checking their quality and correctness is of undeniable importance. Despite the advances in software testing, i.e., one of the possible ways to improve web applications quality, the ever-increasing technological complexity of modern web applications makes current testing techniques not adequate.

To face this issue, test automation tools are adopted by software engineers to automate the creation and execution of the test cases. However, such tools in industrial practice often leads to create test code which is (1) fragile to minor application changes, (2) strongly coupled with the web application under test and thus poorly maintainable, and (3) incomplete, i.e., having a low coverage of the web application input and functionalities. As a consequence, automated test suites are often abandoned, in spite of their potential value to catch errors and regressions.

In this work, we have discussed the three open problems that, according to our experience in web testing [12, 23, 24, 42–46] and our analysis of the state of the art and practice of this field, are among the major causes that limit the adoption of test automation in the web domain: fragility problem, strong coupling and low cohesion problem and incompleteness problem. We have then analyzed the existing research solutions to these problems, and we have presented our vision to overcome them, which concerns the development and usage of a new integrated testing environment called NEONATE. Finally, we have sketched our ongoing work on the NEONATE project which has the utmost purpose of representing a major breakthrough in the web testing domain.

As seen in the chapter, NEONATE, when completed, will be a powerful and flexible toolset focused on the development/maintenance/evolution of web test code. Currently three out of seven expected modules (i.e., ROBULA+, PESTO, and APOGEN) have already been implemented and the corresponding empirical evaluation shown that they outperform the state-of-the-art proposals. Future work will be devoted to the design, implementation, and validation of the remaining prototype tools and to empirically show the effectiveness of NEONATE as a whole testing framework.

In conclusion, we strongly believe that the NEONATE long-term project will demonstrate to the research community and practitioners that it is possible to create robust test code, to repair broken test code, to increase the maintainability of test code, and, in the end, to improve the effectiveness of existing test suites.

In our opinion, web technologies will further increase their relevance in the next years. That is, everyday we use a web interface to access and use our applications and data. For this reason, we hope that this work could help the testing community to be more aware about the problems hindering web test automation and foster more researchers to find solutions. In fact, in the web context, new technologies emerge continuously and, each of them could require specific solutions to solve the three problems. For example, if a new web technology comes into the scene, the impact of the fragility problem might be more or less severe depending on the intrinsic characteristics of technology itself and that of the tool utilized for testing it (e.g., DOM-based or visual). NEONATE has been specifically designed to empower the web tester and limiting the three open problems. Thus, it has a human-centric vision aimed at supporting the tester activities. On other hand, novel computer-centric testing solutions based on artificial intelligence and machine learning are emerging with the aim of automatically creating and maintaining the test code. Their goal is to replace, as much as possible, the humans. We believe that the two approaches, human and computer centric, could (and probably should) coexist and as to reinforce each other with the overall goal to improve the final quality of the future web applications.

Moreover, also mobile applications are today very important for a great variety of activities and businesses; and this will be even more true in the next decade. NEONATE can influence closely related contexts such as mobile and IoT testing research. Especially for the former, we expect mobile apps testers to face similar problems as in the web context. Thus, we believe that our testing framework can be adapted also to solve issues affecting the mobile testing environment and novel solutions drew on the research presented in this chapter. As an example, the Android applications have XML-based interfaces that are conceptually similar to the HTML of the web applications. For this reason, similar challenges/research problems may concern robust localization of GUI elements, mobile page objects creation, visual image recognition, or hybrid test suites. Thus, it is realistic to hypothesize that the NEONATE ITE could be ported or implemented also to support mobile applications (for instance, by leveraging existing mobile testing frameworks and tools like Selendroid[f] or Appium[g]).

---

[f] http://selendroid.io/.

[g] http://appium.io/.

REFERENCES

[1] I.V. Yakovlev, Web 2.0: is it evolutionary or revolutionary? IT Prof. 9 (6) (2007) 43–45. ISSN: 1520-9202. https://doi.org/10.1109/MITP.2007.123.

[2] T. O'Reilly, What is web 2.0? Design patterns and business models for the next generation of software, 2005. http://oreillynet.com/pub/a/oreilly/tim/news/2005/09/30/what-is-web-20.html.

[3] A. Mesbah, Software analysis for the web: achievements and prospects, in: Proceedings of 23rd International Conference on Software Analysis, Evolution, and Reengineering, SANER 2016, IEEE, 2016, pp. 91–103. https://doi.org/10.1109/SANER.2016.109.

[4] H.S. Chaini, S.K. Pradhan, Test script execution and effective result analysis in hybrid test automation framework, in: Proceedings of International Conference on Advances in Computer Engineering and Applications, IEEE, 2015, pp. 214–217. https://doi.org/10.1109/ICACEA.2015.7164698.

[5] SeleniumHQ Web Browser Automation, 2017. http://www.seleniumhq.org/.

[6] E. Alégroth, M. Nass, H.H. Olsson, JAutomate: a tool for system- and acceptance-test automation, in: Proceedings of Sixth International Conference on Software Testing, Verification and Validation, ICST 2013, ISBN: 978-0-7695-4968-2, pp. 439–446.

[7] T.-H. Chang, T. Yeh, R.C. Miller, GUI testing using computer vision, in: Proceedings of 28th Conference on Human Factors in Computing Systems, CHI 2010, ACM, 2010, pp. 1535–1544.

[8] S. Berner, R. Weber, R.K. Keller, Observations and lessons learned from automated testing, in: Proceedings of 27th International Conference on Software Engineering, ICSE 2005, IEEE, 2005, pp. 571–579.

[9] M. Leotta, D. Clerissi, F. Ricca, C. Spadaro, Comparing the maintainability of Selenium WebDriver test suites employing different locators: a case study, in: Proceedings of First International Workshop on Joining AcadeMiA and Industry Contributions to Testing Automation, JAMAICA 2013, ACM, 2013, ISBN: 978-1-4503-2161-7, pp. 53–58. https://doi.org/10.1145/2489280.2489284.

[10] M. Leotta, D. Clerissi, F. Ricca, C. Spadaro, Improving test suites maintainability with the page object pattern: an industrial case study, in: Proceedings of Sixth International Conference on Software Testing, Verification and Validation Workshop, ICSTW 2013, IEEE, 2013, ISBN: 978-1-4799-1324-4, pp. 108–113. https://doi.org/10.1109/ICSTW.2013.19.

[11] D.M. Rafi, K.R.K. Moses, K. Petersen, M.V. Mäntylä, Benefits and limitations of automated software testing: systematic literature review and practitioner survey, in: Proceedings of Seventh International Workshop on Automation of Software Test, AST 2012, IEEE, 2012, ISBN: 978-1-4673-1822-8, pp. 36–42. http://dl.acm.org/citation.cfm?id=2663608.2663616.

[12] M. Leotta, D. Clerissi, F. Ricca, P. Tonella, Capture-replay vs programmable web testing: an empirical assessment during test case evolution, in: Proceedings of 20th Working Conference on Reverse Engineering, WCRE 2013, IEEE, 2013, ISBN: 978-1-4799-2931-3, pp. 272–281. https://doi.org/10.1109/WCRE.2013.6671302.

[13] M. Leotta, D. Clerissi, F. Ricca, P. Tonella, Visual vs DOM-based web locators: an empirical study, in: S. Casteleyn, G. Rossi, M. Winckler (Eds.), LNCS, Proceedings of 14th International Conference on Web Engineering (ICWE 2014), vol. 8541, Springer, 2014, pp. 322–340. https://doi.org/10.1007/978-3-319-08245-5_19.

[14] M. Leotta, D. Clerissi, F. Ricca, P. Tonella, Approaches and tools for automated end-to-end web testing. Adv. Comput. 101 (2016) 193–237. ISSN: 0065-2458. https://doi.org/10.1016/bs.adcom.2015.11.007.

[15] M. Hammoudi, G. Rothermel, P. Tonella, Why do record/replay tests of web applications break? in: Proceedings of Ninth International Conference on Software Testing, Verification and Validation, ICST 2016, IEEE, 2016, pp. 180–190.

[16] M. Fowler, Page Object, 2013. http://martinfowler.com/bliki/PageObject.html.
[17] A. Mesbah, A. van Deursen, D. Roest, Invariant-based automatic testing of modern web applications, IEEE Trans. Softw. Eng. 38 (1) (2012) 35–53. http://doi. ieeecomputersociety.org/10.1109/TSE.2011.28.
[18] S. Thummalapenta, K.V. Lakshmi, S. Sinha, N. Sinha, S. Chandra, Guided test generation for web applications, in: Proceedings of 35th International Conference on Software Engineering, ICSE 2013, IEEE, 2013, ISBN: 978-1-4673-3076-3. pp. 162–171. http://dl.acm.org/citation.cfm?id=2486788.2486810.
[19] S. Artzi, A. Kiezun, J. Dolby, F. Tip, D. Dig, A. Paradkar, M.D. Ernst, Finding bugs in dynamic web applications, in: Proceedings of International Symposium on Software Testing and Analysis, ISSTA 2008, ACM, 2008, ISBN: 978-1-60558-050-0, pp. 261–272. https://doi.org/10.1145/1390630.1390662.
[20] S. Thummalapenta, S. Sinha, N. Singhania, S. Chandra, Automating test automation, in: Proceedings of 34th International Conference on Software Engineering, ICSE 2012, IEEE, 2012, pp. 881–891.
[21] F. Ricca, M. Leotta, A. Stocco, D. Clerissi, P. Tonella, Web testware evolution, in: Proceedings of 15th International Symposium on Web Systems Evolution, WSE 2013, IEEE, 2013, ISBN: 978-1-4799-1608-5, pp. 39–44. https://doi.org/10.1109/WSE.2013.6642415.
[22] M. Leotta, A. Stocco, F. Ricca, P. Tonella, Reducing web test cases aging by means of robust XPath locators. in: Proceedings of 25th International Symposium on Software Reliability Engineering Workshops, ISSREW, IEEE, 2014, pp. 449–454. https://doi.org/10.1109/ISSREW.2014.17.
[23] M. Leotta, A. Stocco, F. Ricca, P. Tonella, Using multi-locators to increase the robustness of web test cases, in: Proceedings of Eighth IEEE International Conference on Software Testing, Verification and Validation, ICST 2015, IEEE, 2015, ISBN: 978-1-4799-7125-1, pp. 1–10. https://doi.org/10.1109/ICST.2015.7102611.
[24] M. Leotta, A. Stocco, F. Ricca, P. Tonella, ROBULA+: an algorithm for generating robust XPath locators for web testing, J. Softw. Evol. Process 28 (3) (2016) 177–204. ISSN: 2047-7481. https://doi.org/10.1002/smr.1771.
[25] N. Dalvi, P. Bohannon, F. Sha, Robust web extraction: an approach based on a probabilistic tree-edit model, in: Proceedings of International Conference on Management of Data, SIGMOD 2009, ACM, 2009, ISBN: 978-1-60558-551-2, pp. 335–348. https://doi.org/10.1145/1559845.1559882.
[26] A. Parameswaran, N. Dalvi, H. Garcia-Molina, R. Rastogi, Optimal Schemes for Robust Web Extraction, vol. 4, VLDB Endowment, 2011.
[27] K. Bajaj, K. Pattabiraman, A. Mesbah, Synthesizing web element locators, in: Proceedings of 30th International Conference on Automated Software Engineering, ASE 2011, IEEE, 2015, pp. 331–341. https://doi.org/10.1109/ASE.2015.23.
[28] R. Yandrapally, S. Thummalapenta, S. Sinha, S. Chandra, Robust test automation using contextual clues, in: Proceedings of 25th International Symposium on Software Testing and Analysis, ISSTA 2014, ACM, 2014, ISBN: 978-1-4503-2645-2, pp. 304–314. https://doi.org/10.1145/2610384.2610390.
[29] S.R. Choudhary, D. Zhao, H. Versee, A. Orso, WATER: web application test repair, in: Proceedings of First International Workshop on End-to-End Test Script Engineering, ETSE 2011, ACM, 2011, ISBN: 978-1-4503-0808-3, pp. 24–29.
[30] M. Hammoudi, G. Rothermel, A. Stocco, WATERFALL: an incremental approach for repairing record-replay tests of web applications, in: Proceedings of 24th International Symposium on the Foundations of Software Engineering, FSE 2016, ACM, 2016, pp. 751–762.
[31] S. Stewart, Page Objects–Selenium wiki, 2017 https://github.com/SeleniumHQ/selenium/wiki/PageObjects.

[32] L. Christophe, R. Stevens, C.D. Roover, W.D. Meuter, Prevalence and maintenance of automated functional tests for web applications, in: Proceedings of 30th International Conference on Software Maintenance and Evolution, ICSME 2014, IEEE, 2014.

[33] A. van Deursen, Testing web applications with state objects. Commun. ACM 58 (8) (2015) 36–43. ISSN: 0001-0782. https://doi.org/10.1145/2755501.

[34] B. Yu, L. Ma, C. Zhang, Incremental web application testing using page object, in: Proceedings of Third Workshop on Hot Topics in Web Systems and Technologies, HOTWEB 2015, 2015, ISBN: 978-1-4673-9688-2, pp. 1–6.

[35] C. Pacheco, S.K. Lahiri, M.D. Ernst, T. Ball, Feedback-directed random test generation, in: Proceedings of 29th International Conference on Software Engineering, ICSE 2007, IEEE, 2007, pp. 75–84.

[36] A. Milani Fard, M. Mirzaaghaei, A. Mesbah, Leveraging existing tests in automated test generation for web applications, in: Proceedings of 29th International Conference on Automated Software Engineering, ASE 2014, ACM, 2014, ISBN: 978-1-4503-3013-8, pp. 67–78. https://doi.org/10.1145/2642937.2642991.

[37] Y. Zhang, A. Mesbah, Assertions are strongly correlated with test suite effectiveness, in: Proceedings of 10th Joint Meeting on Foundations of Software Engineering, ESEC/FSE 2015, ACM, 2015, ISBN: 978-1-4503-3675-8, pp. 214–224. https://doi.org/10.1145/2786805.2786858.

[38] S. Artzi, J. Dolby, S.H. Jensen, A. Møller, F. Tip, A framework for automated testing of javascript web applications, in: Proceedings of 33rd International Conference on Software Engineering, ICSE 2011, ACM, 2011, ISBN: 978-1-4503-0445-0, pp. 571–580. https://doi.org/10.1145/1985793.1985871.

[39] S. Mirshokraie, A. Mesbah, K. Pattabiraman, PYTHIA: Generating test cases with oracles for JavaScript applications, in: Proceedings of 28th International Conference on Automated Software Engineering, ASE 2013, IEEE, 2013, pp. 610–615. https://doi.org/10.1109/ASE.2013.6693121.

[40] S. Mirshokraie, A. Mesbah, K. Pattabiraman, Atrina: inferring unit oracles from GUI test cases, in: Proceedings of International Conference on Software Testing, Verification, and Validation, ICST 2016, IEEE, 2016, pp. 330–340.

[41] M. Biagiola, F. Ricca, P. Tonella, Search based path and input data generation for web application testing, in: Proceedings of Ninth International Symposium on Search Based Software Engineering, SSBSE 2017, Springer, 2017, pp. 18–32. https://doi.org/10.1007/978-3-319-66299-2_2.

[42] F. Ricca, P. Tonella, Analysis and testing of web applications, in: Proceedings of 23rd International Conference on Software Engineering, ICSE 2001, IEEE, 2001, pp. 25–34. https://doi.org/10.1109/ICSE.2001.919078.

[43] P. Tonella, F. Ricca, Statistical testing of web applications, J. Softw. Maint. 16 (1–2) (2004) 103–127.

[44] F. Ricca, P. Tonella, Detecting anomaly and failure in web applications, IEEE MultiMedia 13 (2) (2006) 44–51.

[45] A. Marchetto, F. Ricca, P. Tonella, A case study-based comparison of web testing techniques applied to AJAX web applications. Int. J. Softw. Tools Technol. Transfer 10 (6) (2008) 477–492. ISSN: 1433-2787. https://doi.org/10.1007/s10009-008-0086-x.

[46] A. Marchetto, P. Tonella, F. Ricca, State-based testing of Ajax web applications, in: Proceedings of First International Conference on Software Testing, Verification and Validation, ICST 2008, IEEE, 2008, pp. 121–130.

[47] J. Gao, C. Chen, Y. Toyoshima, D.K. Leung, Developing an integrated testing environment using the world wide web technology, in: Proceedings of the 21st International Computer Software and Applications Conference, COMPSAC 1997, IEEE, Washington, DC, USA, 1997, ISBN: 0-8186-8105-5, pp. 594–601. http://dl.acm.org/citation.cfm?id=645979.675991.

[48] C. Williams, H. Sluiman, D. Pitcher, M. Slavescu, J. Spratley, M. Brodhun, J. McLean, C. Rankin, K. Rosengren, The STCL test tools architecture, IBM Syst. J. 41 (1) (2002) 74–88. ISSN: 0018-8670.

[49] T.E.J. Vos, P. Tonella, J. Wegener, M. Harman, W. Prasetya, E. Puoskari, Y. Nir-Buchbinder, Future internet testing with FITTEST, in: Proceedings of 15th European Conference on Software Maintenance and Reengineering, CSMR 2011, IEEE, ISSN 1534-5351, 2011, pp. 355–358. https://doi.org/10.1109/CSMR.2011.51.

[50] M.W. Whalen, P. Godefroid, L. Mariani, A. Polini, N. Tillmann, W. Visser, FITE: future integrated testing environment, in: Proceedings of Workshop on Future of Software Engineering Research, FoSER 2010, ACM, 2010, ISBN: 978-1-4503-0427-6, pp. 401–406. https://doi.org/10.1145/1882362.1882444.

[51] T. Margaria, O. Niese, B. Steffen, Demonstration of an automated integrated test environment for web-based applications, in: Proceedings of Ninth International SPIN Workshop, LNCS, vol. 2318, Springer, 2002, pp. 250–253.

[52] P. Montoto, A. Pan, J. Raposo, F. Bellas, J. Lopez, Automated browsing in AJAX websites, Data Knowl. Eng. 70 (3) (2011) 269–283. ISSN: 0169-023X. https://doi.org/10.1016/j.datak.2010.12.001.

[53] A. Stocco, M. Leotta, F. Ricca, P. Tonella, APOGEN: automatic page object generator for web testing, Softw. Qual. J. 25 (3) (2017) 1007–1039. ISSN: 1573-1367. https://doi.org/10.1007/s11219-016-9331-9.

[54] A. Mesbah, A. van Deursen, S. Lenselink, Crawling Ajax-based web applications through dynamic analysis of user interface state changes, ACM Trans. Web 6 (1) (2012) 3:1–3:30.

[55] A. Stocco, M. Leotta, F. Ricca, P. Tonella, Clustering-aided page object generation for web testing, in: Proceedings of 16th International Conference on Web Engineering (ICWE 2016), LNCS, vol. 9671, Springer, 2016, ISBN: 978-3-319-38790-1, pp. 132–151. https://doi.org/10.1007/978-3-319-38791-8_8.

[56] E. Alégroth, Z. Gao, R. Oliveira, A. Memon, Conceptualization and evaluation of component-based testing unified with visual GUI testing: an empirical study, in: Proceedings of Eighth International Conference on Software Testing, Verification and Validation, ICST 2015, 2015, pp. 1–10. https://doi.org/10.1109/ICST.2015.7102584.

[57] M. Leotta, A. Stocco, F. Ricca, P. Tonella, PESTO: automated migration of DOM-based web tests towards the visual approach. J. Softw. Test. Verification Reliab 28 (4) (2018) e1665. ISSN: 1099-1689. https://doi.org/10.1002/stvr.1665 (John Wiley & Sons).

[58] M. Leotta, A. Stocco, F. Ricca, P. Tonella, Automated generation of visual web tests from DOM-based web tests, in: Proceedings of 30th Symposium on Applied Computing, SAC 2015, 2015, ISBN: 978-1-4503-3196-8, pp. 775–782.

[59] B.N. Nguyen, B. Robbins, I. Banerjee, A. Memon, GUITAR: an innovative tool for automated testing of GUI-driven software, Autom. Softw. Eng. 21 (1) (2014) 65–105. ISSN: 1573-7535. https://doi.org/10.1007/s10515-013-0128-9.

[60] J.R. Cordy, TXL—a language for programming language tools and applications, in: Proceedings of Fourth Workshop on Language Descriptions, Tools, and Applications (LDTA 2004), Electronic notes in theoretical computer science, ISSN 1571-0661, vol. 110, Elsevier, 2004, pp. 3–31. https://doi.org/10.1016/j.entcs.2004.11.006.

[61] B. Daniel, D. Dig, T. Gvero, V. Jagannath, J. Jiaa, D. Mitchell, J. Nogiec, S.H. Tan, D. Marinov, ReAssert: a tool for repairing broken unit tests, in: Proceedings of 33rd International Conference on Software Engineering, ICSE 2011, IEEE, ISSN 0270-5257, 2011, pp. 1010–1012. https://doi.org/10.1145/1985793.1985978.

[62] M. Mirzaaghaei, F. Pastore, M. Pezzè, Automatic test case evolution, J. Softw. Test. Verification and Reliab. 24 (5) (2014) 386–411. ISSN: 0960-0833.

[63] S. Barman, S. Chasins, R. Bodik, S. Gulwani, Ringer: web automation by demonstration, in: Proceedings of International Conference on Object-Oriented Programming, Systems, Languages, and Applications, OOPSLA 2016, ACM, 2016, ISBN: 978-1-4503-4444-9. pp. 748–764. https://doi.org/10.1145/2983990.2984020.

[64] S. Elbaum, G. Rothermel, S. Karre, M. Fisher II, Leveraging user-session data to support web application testing, IEEE Trans. Softw. Eng. 31 (3) (2005) 187–202. ISSN: 0098-5589.

[65] M. Harman, N. Alshahwan, Automated session data repair for web application regression testing, in: Proceedings of the International Conference on Software Testing, Verification, and Validation, ICST 2008, 2008, ISBN: 978-0-7695-3127-4, pp. 298–307.

[66] G. Wassermann, D. Yu, A. Chander, D. Dhurjati, H. Inamura, Z. Su, Dynamic test input generation for web applications, in: Proceedings of International Symposium on Software Testing and Analysis, ISSTA 2008, ACM, 2008, ISBN: 978-1-60558-050-0, pp. 249–260. https://doi.org/10.1145/1390630.1390661.

[67] L. Mariani, M. Pezzè, O. Riganelli, M. Santoro, Link: exploiting the web of data to generate test inputs, in: Proceedings of International Symposium on Software Testing and Analysis, ISSTA 2014, ACM, 2014, ISBN: 978-1-4503-2645-2, pp. 373–384. https://doi.org/10.1145/2610384.2610397.

[68] J. Lin, F. Wang, Using semantic similarity in crawling-based web application testing, in: Proceedings of 10th International Conference on Software Testing, Verification and Validation, ICST 2017, 2017.

[69] M.D. Ernst, J.H. Perkins, P.J. Guo, S. McCamant, C. Pacheco, M.S. Tschantz, C. Xiao, The Daikon system for dynamic detection of likely invariants, Sci. Comput. Program. 69 (1–3) (2007) 35–45, ISSN: 0167-6423.

[70] H. Do, S. Elbaum, G. Rothermel, Supporting controlled experimentation with testing techniques: an infrastructure and its potential impact, Empir. Softw. Eng. 10 (4) (2005) 405–435. ISSN: 1573-7616. https://doi.org/10.1007/s10664-005-3861-2.

[71] R. Just, D. Jalali, M.D. Ernst, Defects4J: a database of existing faults to enable controlled testing studies for java programs, in: Proceedings of 2014 International Symposium on Software Testing and Analysis, ISSTA 2014, ACM, 2014, ISBN: 978-1-4503-2645-2, pp. 437–440. https://doi.org/10.1145/2610384.2628055.

[72] G. Fraser, A. Arcuri, Sound empirical evidence in software testing, in: Proceedings of 34th International Conference on Software Engineering, ICSE 2012, IEEE, 2012, ISBN: 978-1-4673-1067-3, pp. 178–188.

ABOUT THE AUTHORS

Filippo Ricca is an associate professor at the University of Genova, Italy. He received his PhD degree in Computer Science from the same University, in 2003, with the thesis "Analysis, Testing and Re-structuring of Web Applications". In 2011 he was awarded the ICSE 2001 MIP (Most Influential Paper) award, for his paper: "Analysis and Testing of Web Applications". He is author or coauthor of more than 100 research papers published in international journals and conferences/workshops. Filippo Ricca was Program

Chair of CSMR/WCRE 2014, CSMR 2013, ICPC 2011, and WSE 2008. His current research interests include: Software modeling, Reverse engineering, Empirical studies in Software Engineering, Web applications and Software Testing.

Maurizio Leotta is a researcher at the University of Genova, Italy. He received his PhD degree in Computer Science from the same University, in 2015, with the thesis "Automated Web Testing: Analysis and Maintenance Effort Reduction". He is author or coauthor of more than 60 research papers published in international journals and conferences/workshops. His current research interests are in software engineering, with a particular focus on the following themes: Web\Mobile\IoT application testing, functional test automation, empirical software engineering, business process modelling and model-driven software engineering.

Andrea Stocco is a postdoctoral fellow at the department of Electrical and Computer Engineering (ECE) of the University of British Columbia, Canada. He received his PhD in Computer Science at the University of Genova, Italy, in 2017, with the thesis "Automatic page object generation to support E2E testing of web applications". He is the recipient of the Best Student Paper Award at the 16th International Conference on Web Engineering (ICWE 2016). His research interests include web testing and empirical software engineering, with particular emphasis on test breakage detection and automatic repair, robustness and maintainability of test suites for web applications.

CHAPTER FOUR

Advances in Using Agile and Lean Processes for Software Development

Pilar Rodríguez, Mika Mäntylä, Markku Oivo, Lucy Ellen Lwakatare, Pertti Seppänen, Pasi Kuvaja

Faculty of Information Technology and Electrical Engineering, University of Oulu, Finland

Contents

1. Introduction   136
   1.1 Evolution of Software Development Processes   137
   1.2 The Emergence of Agile Software Development   139
   1.3 Lean Thinking and Continuous Deployment (CD)   141
2. Trends on Agile, Lean, and Rapid Software Development   142
   2.1 Overview of Agile Literature   143
   2.2 Trends in Agile   145
   2.3 Summary   148
3. A Walk Through the Roots of Agile and Lean Thinking   148
   3.1 Lean Thinking   149
   3.2 Lean Manufacturing Toolkit   151
   3.3 Agility   155
   3.4 Combining Lean and Agile in Manufacturing   157
4. Agile and Lean in Software Development   158
   4.1 Agile Software Development   159
   4.2 Lean Software Development   171
5. Beyond Agile and Lean: Toward Rapid Software Development, Continuous Delivery, and CD   182
   5.1 Need for Speed   183
   5.2 Continuous Delivery and Continuous Deployment   185
   5.3 Key Elements to Get Speed in Your Software Development Process   186
6. DevOps   190
   6.1 The Origins of DevOps   190
   6.2 DevOps Practices   193
   6.3 The Benefits and Challenges of DevOps   196

Advances in Computers, Volume 113
ISSN 0065-2458
https://doi.org/10.1016/bs.adcom.2018.03.014

7. The Lean Startup Movement   197
   7.1 Origin and Key Concepts   197
   7.2 Derivatives and Further Developments   200
   7.3 Lean Startup and Other Startup Research   201
   7.4 Summary of the Lean Startup Movement   201
8. Miscellany   202
   8.1 Metrics in Agile and Lean Software Development   202
   8.2 Technical Debt in Agile and Lean Software Development   205
9. Conclusions and Future Directions   209
References   212
About the Authors   221

Abstract

Software development processes have evolved according to market needs. Fast changing conditions that characterize current software markets have favored methods advocating speed and flexibility. Agile and Lean software development are in the forefront of these methods. This chapter presents a unified view of Agile software development, Lean software development, and most recent advances toward rapid releases. First, we introduce the area and explain the reasons why the software development industry begun to move into this direction in the late 1990s. Section 2 characterizes the research trends on Agile software development. This section helps understand the relevance of Agile software development in the research literature. Section 3 provides a walk through the roots of Agile and Lean thinking, as they originally emerged in manufacturing. Section 4 develops into Agile and Lean for software development. Main characteristics and most popular methods and practices of Agile and Lean software development are developed in this section. Section 5 centers on rapid releases, continuous delivery, and continuous deployment, the latest advances in the area to get speed. The concepts of DevOps, as a means to take full (end-to-end) advantage of Agile and Lean, and Lean start-up, as an approach to foster innovation, are the focus of the two following Sections 6 and 7. Finally, Section 8 focuses on two important aspects of Agile and Lean software development: (1) metrics to guide decision making and (2) technical debt as a mechanism to gain business advantage. To wrap up the chapter, we peer into future directions in the area.

1. INTRODUCTION

This chapter will provide you with advanced knowledge on Agile software development, Lean software development, and the latest progress toward Rapid software development. You may wonder *why you should care about software processes, and, in particular, about Agile and Lean software development.* One may legitimately question the importance of software development processes as many successful software innovations have not been *initially* held by well-defined software processes. Take, for example, the case of Facebook. Facebook, which emerged as a student's project at Harvard University in 2004, is currently one of the world's most popular social networks [1]. However, as companies grow, business and technical risks also increase, more customers become dependent on the developed products, projects become larger, the development involves more people and aspects such as work coordination become essential [2]. Using the example of Facebook, what was initially composed of a group of students, rapidly became one of the Fortune 500 companies, with a revenue of US$ 27.638 billion in 2016. Today, Facebook has around 18,770 employees and over 1.94 billion monthly active users. Obviously, nowadays, Facebook needs to coordinate its work through a more sophisticated development process, which is, indeed, based on Agile and Lean software development.[a]

You could also wonder, *then, from the wide umbrella of software development processes that are available in the literature, which is the best one I should apply? Why should I care about Agile and Lean?* The answer to this question is not straightforward, and it will certainly depend on your company's business context and organizational needs. The traditional phase-gate process, commonly applied in other engineering fields, has been widely used in software development for decades (e.g., Waterfall model [3]). However, nowadays, software has acquired an essential role in our lives, which has completely changed the way in which the software development business is run these days. In a software-driven world, innovation, speed, and ability to adapt to business changes have become critical in most software business domains. If this is your case, you can greatly benefit from Agile and Lean software development processes, like companies such as Facebook, Microsoft, IBM, Google, Adobe, Spotify, Netflix, …, just to mention a few, already did. Next, we briefly describe how software processes have evolved over time to better understand the relevance of Agile software development and, also, the reasons why it caused some initial controversy in the software community.

1.1 Evolution of Software Development Processes

The software engineering literature has stressed the importance of software development processes and their influence on product quality for decades [2,4]. Consequently, efforts for improving software development processes

---

[a] Scaling Facebook to 500 million users and beyond, Facebook engineering webpage, http://www. facebook.com/note.php?note_id=409881258919 (last accessed, 15.06.2017).

have made different contributions over time. Among the landscape of contributions to software processes, we can find process models such as the Waterfall model [3] or the Spiral model [5], process standards, which are widely used in the industry, such as the ISO/IEC 12207:2008 [6], ISO/IEC 90003:2004(E) [7], model-based improvement frameworks such as CMMI [8], SPICE (ISO/IEC 15504:1998) [9], and Bootstrap [10], and, more recently, adaptive approaches for software development such as Agile and Lean methods [11]. Fig. 1 depicts a timeline with some of these important contributions to software development processes.[b] It bases on the review of process models conducted by Münch et al. [2].

Software development processes have gradually evolved from heavy-weight, stage-based processes, which built on the assumption that require-ments are relatively stable and split the development process on distinct sequential phases based on those requirements (such as the Waterfall model [3]), to lighter/more flexible processes that try to cope with current business turbulence. As illustrated in Fig. 1, process-centric models, which were considered deterministic and repeatable [18], have given way to flexible approaches, which emphasize continuous evolution of software functionality, flexibility, speed, face-to-face communication, customer involvement, and a minimal number of lightweight artifacts.

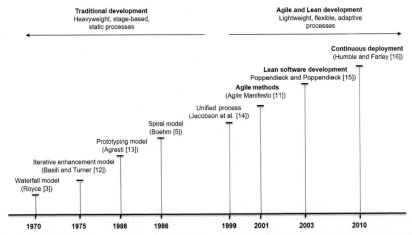

Fig. 1 Contributions to software development processes over last 4 decades [12–16].

---

[b] The purpose of Fig. 1 is to show trends in software development processes. Therefore, it includes representative contributions with illustrative purposes. Classical software engineering books such as [17] and guides such as the *Software Engineering Body of Knowledge* (SWEBOK) provide a more complete inventory on software engineering processes and related topics.

1.2 The Emergence of Agile Software Development

The main driver behind the evolution of software processes toward Agile and Lean software development is the environment in which the software development industry works today, which is fundamentally different from the environment in which it used to work 20 years ago. On the one hand, the prevalence of software products has shaped a software-driven society. Software is all around us, from the (electronic) toothbrush that we use in the morning, to the alarm that we set-up before going to sleep. Thus, the business around software products is open to countless opportunities. On the other hand, the *Age of Information*,[c] which is strongly based on the Internet, has made digital resources constantly available for everyone. Information flows are accelerated, global communication and networking are faster and individuals can explore their personal needs easier. Overall, software relevance, globalization, and market dynamics, characterized by rapid and unpredictable change, have shifted the software development land-scape. These market features put pressure on software and software-intensive organizations to develop what Eisenhardt and Martin [20] called *dynamic capabilities*. Software development organizations need to be crea-tive, innovative, and flexible, while working with incomplete information and pursuing economic efficiency. This situation is particularly relevant for organizations in domains such as web or mobile applications and services. In these contexts, applications are frequently developed in a matter of weeks or months, rather than years, as it was common when using tradi-tional processes [1]. Yet, the emergence of trends like internet of things (IoT) will shape software development processes of more traditional indus-tries in the near future as well.

To cope with this situation, a community of practitioners formulated the Agile Manifesto in 2001 [11]. In a 2 days meeting in the Wasatch moun-tains of Utah in February 2001,[d] 14 independent software practitioners proposed the Agile Manifesto as an alternative to software processes that were considered too inflexible and slow to adapt to business changes. It pro-voked a change in the way in which the software engineering community was addressing software development processes. In essence, Agile recognizes the limitations of anticipating external business changes and provides means

---

[c] *Information Age* refers to the age that began in the 1970s in which society started to work in networks through a constant flow of information based on technology. Since then, company competitiveness is very dependent on its knowledge on technology, information, and network access [19].

[d] http://agilemanifesto.org/history.html

oriented to implement minimum viable products (MVPs) quickly as a key competitive advantage [21]. The idea is that MVPs are incrementally developed in collaboration with customers, as more information about the product is known. Contrary to traditional phase-based software development processes, Agile appraises iterative development. Traditional software development phases (e.g., requirements, development, and testing) are run in small cycles to achieve continuous delivery and feedback and, therefore, improve the company's capability to adapt to market and customer fluctuations [21]. Furthermore, Agile emphasizes the human side of the sociotechnical software development activity as a primary driver of project success [22,23]. Agile methods advocate that the key role that people plays in software development makes the development process less deterministic and repeatable than previously considered by traditional prescriptive methods [18]. Kruchten [24] explained it using the party metaphor: *"Consider you organize a party this Saturday at your home. You invite a group of people, order the food, move the furniture, and you record carefully: who comes, when, who eats what, drinks what, etc., and everyone leaves having had a good time. The next Saturday, to have another great time, you decide to have the very same people, the very same food and drinks, you ask them to dress the same way, arrive at the same time, you introduce them to the same people, bring them the same glasses with the same drinks, make the same jokes,... will you be guaranteed to have a nice party again...?"* In the same line, Boehm stressed that *"one of the most significant contributions of the agile methods community has been to put to rest the mistaken belief that there could be a one-size-fits-all software process by which all software systems could be developed"* [2]. The bad news is that, unfortunately, there is no silver bullet for software development processes. That is, there is no definitive answer to the question: *what is the best software development process I should apply?* However, the good news is that Agile and Lean software development offer means to cope with current software business dynamics, which can be adapted to different contexts.

Regarding to the adoption level of Agile and Lean software processes in the software development industry, what was initially considered as a fad has been progressively consolidated. Nowadays, a great extent of the software industry follows these methods. For example, the latest State of Agile Survey by VersionOne [25] found that 43% of its 3880 respondents worked in software organizations with more than 50% of their teams using Agile. Only 4% of respondents indicated that in their organization no team was using Agile. Similarly, a survey study conducted in Finland in 2012 concluded that the Finnish software industry has moved toward these methods, being the 58% of the survey's respondents working in organizational units using Agile software development [26]. Nevertheless, it is also worth mentioning that the adoption of Agile has not always been a smooth process. Indeed, Agile software development caused some initial controversy among those that saw it as an attempt to undermine the software engineering discipline [27]. Particularly, there were some initial concerns about using it as an easy excuse to avoid good engineering practices and just code up whatever comes next [27–30]. However, as we will see in the rest of the chapter, and particularly in Sections 4 and 5, Agile software development is a much more disciplined approach than it may seem at first sight.

1.3 Lean Thinking and Continuous Deployment (CD)

A few years after the formulation of the Agile Manifesto, Lean thinking (or simply Lean) started to become also the center of attention of the software development industry [15]. However, Lean did not appear as a software development approach of its own until 2003, when Poppendieck and Poppendieck introduced it in their book *Lean Software Development: An Agile Toolkit* [15]. One of the reasons that captured the attention of the software community upon Lean thinking was the focus of Lean on the organization as a whole [15,31–33]. Agile software development was mainly conceived to work for small software development teams [34]. However, scaling Agile methods, such as Scrum [35] or eXtreme programming (XP) [36], to operate a whole organization was not straightforward; Agile does not explicitly provide support in that sense [32,37–39]. Thus, Lean thinking was discerned as a promising solution to be combined with Agile methods. Nowadays, Lean thinking is not only applied as a way of scaling Agile, but also as well as a way of improving software development processes from a wider perspective [40].

The interest on Lean thinking, which originally emerged in the automotive industry [41], is not exclusive to the software development industry. Lean has been widely adopted in different domains, from the healthcare domain to the clothing industry. Table 1 shows some examples of benefits that are attributed to Lean in diverse domains, in terms of profitability, productivity, time-to-market, and product quality.

During last years, a vogue for rapid and continuous software engineering is emerging [46]. Rapid and continuous software engineering bases on many ideas from Agile and Lean software processes. Concretely, it refers to the organizational capability to develop, release, and learn from software in rapid parallel cycles, often as short as hours or days [46–48]. *Speed* is the key aspect

Table 1 Examples of Benefits Attributed to Lean Thinking in Companies From Different Domains

Dimension	Benefit
Profitability and productivity	Toyota, which is well known as an icon of Lean thinking, cut $2.6 billion out of its $113 billion manufacturing costs, without closing a single plant in 2002 by using Lean principles [42]. Moreover, although Toyota has suffered challenging crisis in its history [43], the company is one of the leaders at global automobile sales
Time to market	In the retail clothing industry, Zara's business model is based on Lean thinking. Zara has reduced its lead time by collecting and sharing input from customers daily and using Lean inventories. Zara delivers new items to stores twice a week (as much as 12 times faster than its competitors) and bring in almost 30,000 designs each year, as opposed to the 2000–4000 new items introduced by its competitors [44] In the healthcare domain, Lean thinking has provided significant improvements in reducing patient waiting lists, floor space utilization, and lead time in laboratorial tests [45]
Product quality	Toyota is well known as an icon in quality control [43]. For example, the Toyota Lexus CT200h has received the maximum rating under the Japanese overall safety assessment in 2011

in this trend. The aim is to minimize the step between development and deployment so that new code is deployed to production environment for customers to use as soon as it is ready. CD is the term used to refer to this phenomenon [16,49–52]. Although the concept of deploying software to customers as soon as new code is developed is not new and it is based on Agile software development principles, CD extends Agile software development by moving from cyclic to continuous value delivery. This evolution requires not only Agile processes at the team level but also integration of the complete R&D organization, parallelization, and automation of processes that are sequential in the value chain and constant customer feedback [46]. Rapid releases, continuous delivery, and CD will be widely discussed in Section 5.

2. TRENDS ON AGILE, LEAN, AND RAPID SOFTWARE DEVELOPMENT

Given the importance of Agile, this section provides a more detailed account of the history and topics of Agile in the scientific literature. The analysis helps the reader understand how a movement initially promoted by practitioner's proponents of Agile methodologies has become a well-established research discipline. Particularly, we analyze the various concepts that are investigated inside the scientific literature of Agile through the lens of automated text mining of scientific studies of Agile development. To characterize the research on Agile software development, we collected and analyzed a large set of studies by searching abstracts in the Scopus database, starting from the year 2000. Using search words "agile" and "software" resulted in over 7000 published articles that we used in our trend mining.

2.1 Overview of Agile Literature

First, we explore the research topics in Agile software development over the past 2 decades. We divided the articles into four roughly equally large groups based on the year of publication and plotted a word comparison cloud of their titles. Fig. 2 shows the word comparison cloud. Early years (2000–07) are in black, second period (2008–11) is in gray, third period (2012–14) is in red,

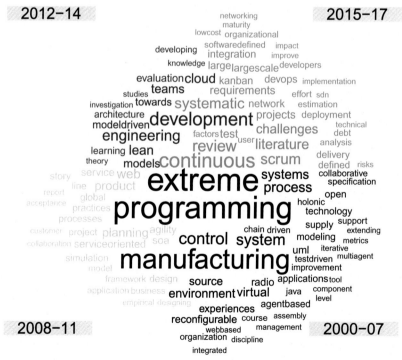

Fig. 2 Word comparison cloud of research on Agile software development from 2000 to till date.

and the most recent era (2015–17) is in blue. The words in each quadrat are more common in their given periods than in other periods, i.e., the size of each word is determined by its deviation from the overall occurrence average. The figure was plotted with R Package "wordcloud."

Although the early years contain several articles about Agile software development in general, e.g., Refs. [28,53,54], the Agile software development method that was the most investigated early was XP [36]. XP is an Agile method with the most doctrinaire approach by setting detailed development practices such as test-driven development and pair programming. The early success of XP as well as declining interest toward it later, may well be explained by its mandating approach to a set of development practices. Adopting it at first can be straightforward as it gives clear rules such that all the development code is done by pair programming. However, later people may realize that such ultimatums might not make sense in their context and, thus, they would turn to more flexible Agile methods. For example, for pair programming research shows that having two junior programmers can form a cost-effective pair while the same does not hold for two expert programmers [55]. During the early years, many papers about Agile manufacturing appeared [56] which addressed the manufacturing of physical products. Work in this area also influenced Agile software development as we shall explain in the next section.

In the second period (2008–11) research about Agile software development in the technological contexts of the web and service-oriented architectures increased in numbers. Thus, the scientists working on technical advances also connected their works to software process advances in Agile software development, e.g., Ref. [57]. During that time, researchers connected Agile software development to work on global and distributed software development, which addresses issues of developing the same software in multiple sites with diverse locations around the globe, e.g., Ref. [58]. You should also notice that papers connecting Agile to global and distributed software development were the first research steps toward scaling Agile beyond one colocated team. Research efforts in Agile software development increased also in the areas of process simulations, e.g., Ref. [59], linking business to Agile, e.g., Ref. [60], and in customer and other type of collaborations, e.g., Ref. [61].

In the third period (2012–14), the technical advances of cloud computing [62] and model-driven development [63] were linked with the process advances of Agile. The term "Lean" gained popularity as a way to advance Agile even further [64]. Learning and teaching Agile was studied in universities [65] but articles about learning Agile also include papers about self-reflection in Agile teams [66].

In the most recent period (2015–17), we can see the rise of word continuous, which increases in popularity due to the practices of continuous integration, CD, and continuous delivery [46,67]. These practices highlight the importance of running working software, which is integrated and delivered to clients continuously, rather than providing major updates biannually, for example. We also notice increase in popularity of Kanban that is a method for managing work and tasks in Agile software development. Systematic literature reviews (SLRs) about Agile increased in popularity in the most recent period as well. SLRs offer an overview of particular areas of Agile software development like using software metrics in Agile [68] or technical debt [69].

2.2 Trends in Agile

The previous section offered a general overview on Agile software development over the past 2 decades. For this section, we did individual searches on topics presented in previous section but also on engineering topics we knew that had been introduced with Agile software development such as refactoring. Thus, we searched for the keywords shown in Fig. 3 combined with the word "software". Adding word software was necessary to exclude, for example, papers discussing Scrum in context of Rugby instead of Agile software development. Research in software engineering in general has increased in past 2 decades [70], so we normalized all search results by the number of papers found with the general terms of "software development" or "software engineering". We also normalized all trend graphs to have values between 0 and 1 where max is the year with the highest normalized amount of papers.

The concepts of XP, Scrum, Lean, Kanban, DevOps, and continuous integration/deployment/delivery are shown in Fig. 3A–F. We can see that XP has been declining in research interest since 2004, Fig. 3A. Scrum, Lean, and Kanban have had good popularity since 2009 but they still appear to be gaining popularity, although the growth trends is not as steep as it has been in the past. DevOps and continuous integration/deployment/delivery seem to be gaining popularity fast and can be characterized as the hot research topics in Agile software development. These topics will be developed in Sections 5 and 6.

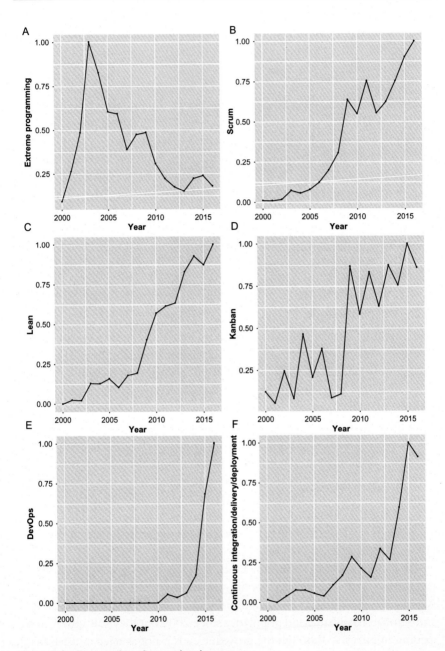

Fig. 3 Trends on Agile software development.

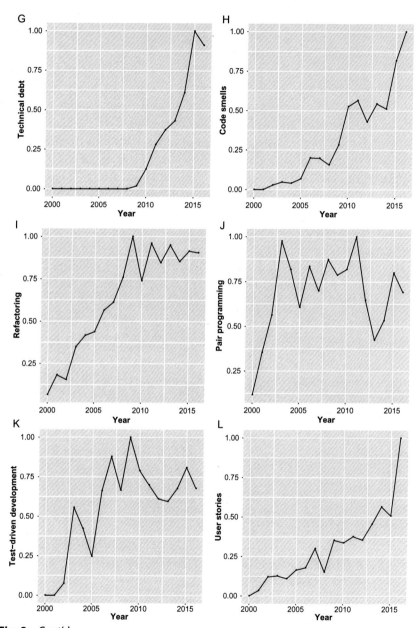

Fig. 3—Cont'd

Fig. 3G–I shows the popularity of the concepts of technical debt and code smells that are used to describe suboptimal software engineering solutions which hinder the velocity of Agile teams. We can see that the research in both areas is still increasing. The concept of technical debt appears to be more recent than code smells. The concept of technical debt is also larger than code smells. Code smells refer to code problems only, while technical debt can refer to problems with using an old version of the build server, for example. Refactoring is about code modifications to fix code smells or technical debt or due to the need for other behavior persevering changes. The popularity of refactoring research has been stable since 2009. The concept of technical debt in the context of Agile will be developed in Section 8.

Finally, Fig. 3J–L shows the popularity of three Agile software development practices that were popularized by XP that are pair programming, test-driven development, and user stories. We can see that although the popularity of XP has declined in research literature, the popularity of pair programming has remained relatively stable since 2003, and the research interest in TDD has similarly remained stable since 2007. Research on user stories that could also be titled as Agile requirements engineering is still experiencing fast growth. So, even when XP might become extinct, some of its practices are still sparking wide interest.

2.3 Summary

This section has demonstrated popularity of many Agile software development concepts and practices in the research literature. Next sections will go deeper to the foundations of Agile and help the reader in putting these concepts together as all of these concepts are essentially solutions of the business and engineering problems companies are having while developing software-based products, services, and solutions.

3. A WALK THROUGH THE ROOTS OF AGILE AND LEAN THINKING

This section describes Lean thinking, agility, and its combination (known as Leagility) as they originally arose in manufacturing (more concretely, in the automotive industry). Understanding the principles of Agile and Lean in the manufacturing domain helps better understand the essence of these paradigms. Thus, this section provides background of Agile and Lean that we believe may be interesting for the reader, although, its reading is not mandatory to understand the rest of the chapter.

3.1 Lean Thinking

Lean was born as part of the industrial renaissance in Japan after the *Second World War* in the 1940s. Traditional mass production systems, which are based on producing large batches of products through continuous assembly lines of interchangeable parts, were unsuitable in a Japanese market lashed by the war and characterized by low volumes. Thus, Japanese manufacturers found themselves surrounded by obstacles that forced them to develop new methods. Japanese new production methods led to what we know today as Lean or Lean thinking. However, it is interesting to notice that the term Lean was not used by Japanese manufacturers. Lean was first used by researchers from the Massachusetts Institute of Technology (MIT) when they began research under the International Motor Vehicle Program (IMVP) in 1986.[e] With the aim of investigating the differences across worldwide automotive industries, IMVP researchers discovered that the Japanese auto industry was far ahead when compared to the automotive industry in America. By carefully studying Japanese methods, particularly those of Toyota under its Toyota production system (TPS), they conceived an entirely different production system, which they called *Lean manufacturing* [41]. Thus, Lean or Lean thinking is the English term that western researchers used to describe Japanese original production methods.

According to MIT's researchers, Lean is about *"doing more with less"* by ideally producing *"the right things, at the right time, and in the right place"* [41]. Based on fundamental industrial engineering principles, Lean thinking, as interpreted by MIT's researchers, steeps in a philosophy of maximizing value and minimizing waste. Five principles guide this philosophy [71]: value, value stream, flow, pull, and perfection.

- *Value* is everything that a customer is willing to pay for. Thus, value is understood as customer value. Defining and understanding value from the perspective of the customer is the central focus of Lean thinking. Lean thinking stresses that every activity inside a company should contribute to create customer value. The goal is to make organizations deliver as much customer value as possible. Its counterpart, waste, or *muda* in Japanese, is everything that consumes resources but creates no customer value.
- *Value stream* is the optimized end-to-end collection of actions required to bring a product/service from customer order to customer care.

---

[e] International motor vehicle program (IMVP, http://www.imvpnet.org/) is an international network that focuses on analyzing the challenges faced by the global automotive industry. IMVP, founded at the Massachusetts Institute of Technology (MIT) in 1979, has mapped Lean methodologies and established benchmarking standards.

The activities in the value stream are classified in three groups: (1) activities that create value and, therefore, are legitimately part of the value stream; (2) activities that create no value for customers but are currently needed to manufacture the product. These activities remain in the value stream as, otherwise, the product cannot be manufactured, but the aim is to improve the production process so that they can be removed at some point; and (3) activities that clearly create no value for customers and should be immediately removed from the value stream.

- *Flow* means that production activities should be organized as a continuous flow, eliminating discontinuities. Flow requires that unnecessary steps and inventories are eliminated (waste). Opposite to mass production, in which interchangeable parts are continuously produced to assemble final products, in Lean thinking, products are made using single piece flows. In a single piece flow, operations are efficient in a way that batch sizes and inventories are replaced by working on one product at a time.

- *Pull* focuses on producing products only when they are really needed. Accordingly, in a pull system, an upstream process produces only when a downstream process is ready and "pulls" some more work from the upstream process. This minimizes inventories. In traditional push systems, work is produced in response to a preset schedule, regardless of whether the next process needs it at that time. Accordingly, the process is transformed from batch-and-queue to just-in-time (JIT) production.

- *Perfection*, or Kaizen in Japanese, refers to the enterprise-level improvement and learning cycle. Lean aims to achieve zero waste and defects based on the concept that there is no end in the strive for perfection and learning.

MIT's five principles of Lean are probably the most popular interpretation of Lean thinking. Some voices have questioned the validity of IMPV's studies by pointing out to a certain biased interpretation of the original Japanese methods [72]. For example, Toyota and its TPS, which had a great influence on Lean thinking as interpreted by MIT's researchers, was originally called the *Respect for Humanity System*. The respect for humanity system was based on two main pillars: continuous improvement and respect for people [73]. Thus, value and waste were not originally the main focuses of the paradigm. When later Lean was interpreted from a western point of view, influenced by the existing western's social–economical system, value, and waste became key components of Lean thinking. Nevertheless, biased or not, it is indisputable that MIT's studies, and the five principles outcome of these studies, have vastly influenced the shape of Lean thinking as it is understood nowadays. The book *The Machine That Changed the World*[f] [41] provides a more detailed description of the story of Lean.

Besides MIT's researchers, it is important to know that other authors have also provided slightly different interpretations of Lean thinking. Table 2 presents most popular interpretations of Lean thinking. It can be said that Lean has incrementally evolved over time [41,71,74–76]. As presented in Table 2, Onho [74] described the TPS mainly from a perspective of reducing wastes such as overproduction, waiting, transportation, overprocessing, inventory, movement, and defects. Although Onho emphasized the concepts of JIT and autonomation, many authors have taken waste as the key element of Lean thinking. For example, Cusumano referred to Lean as: *"The authors [referring to the authors of The Machine That Changed the World] used the term 'Lean' to describe any efficient management practice that minimizes waste"* [77]. In a similar vein, Shah and Ward defined Lean production as *"an integrated socio-technical system whose main objective is to eliminate waste by concurrently reducing or minimizing supplier, customer, and internal variability"* [78]. The excessive focus on waste has frequently created certain unfair reputation to Lean thinking among those who have seen it as an easy excuse to get rid of employees and resources. However, although the concept of waste is undoubtedly important in Lean, the 5, 14, and 13 principles of Lean thinking by Womack and Jones [71], Liker [75], and Morgan and Liker [76] respectively (see Table 2), clearly show that Lean thinking is more than only removing waste. As we will see in Section 4, the interpretation of Lean thinking in software development has been mainly taken from an Agile perspective, as Lean software development was largely built upon the values and principles formulated in the Agile Manifesto. Thus, although waste is an important concept in a software development context as well, Lean software development goes way beyond waste reduction.

3.2 Lean Manufacturing Toolkit

Lean thinking has no formal practices, but builds on any practice that supports its principles. Still, a toolkit of recommended practices, tools, and techniques, from which to choose from, have emerged over time to implement the fundamentals in practice. Table 3 summarizes those that are relevant

---

[f] As an indicator *"The Machine That Changed the World"* [41] is one of the most widely cited references in operations management (referenced by 16,226 sources in Google Scholar, accessed 19.01.2018).

Table 2 Main Sources of Lean Thinking in a Manufacturing Context

Source	Description
Ohno [74]	Ohno described the Toyota production system (TPS) as follows: • The basis of the TPS is the absolute elimination of waste. Ohno described seven classes of waste: (1) *overproduction* (producing more quantity of product or product's components than needed), (2) *waiting* (waiting for a previous upstream activity to be done), (3) *transportation* (unnecessary moving of materials), (4) *overprocessing* (unnecessarily processing by a downstream step in the manufacturing process), (5) *inventory* (storing of materials that are not needed for the present), (6) *movement* (unnecessary motion of people), and (7) *defects* (rework because of quality defects) • According to Ohno, two pillars support this system: (1) *Just-in-time* (JIT), which refers to producing the right items at the right time in the right amounts and (2) *autonomation*, which refers to a quality control principle that provides machines and operators the ability to stop the work immediately whenever a problem or defect is detected
Liker [75]	Liker considers 14 principles in his interpretation of Lean thinking as follows: 1. Base your management decisions on a long-term philosophy, even at the expense of short-term financial goals 2. Create a continuous process flow to bring problems to the surface 3. Use "pull" systems to avoid overproduction 4. Level out the workload (heijunka) 5. Build a culture of stopping to fix problems, to get quality right the first time 6. Standardized tasks and processes are the foundation for continuous improvement and employee empowerment 7. Use visual control so no problems are hidden 8. Use only reliable, thoroughly tested technology that serves your people and process 9. Grow leaders who thoroughly understand the work, live the philosophy, and teach it to others 10. Develop exceptional people and teams who follow your company's philosophy 11. Respect your extended network of partners and suppliers by challenging them and helping them improve 12. Go and see for yourself to thoroughly understand the situation (genchi genbutsu) 13. Make decisions slowly by consensus, thoroughly considering all options; implement decisions rapidly (nemawashi) 14. Become a learning organization through relentless reflection (hansei) and continuous improvement (kaizen)

Morgan and Liker [76]	Morgan and Liker focused on the Toyota product development process. It can be described in 3 categories and 13 principles:
	Process
	1. Establish customer-defined value to separate value added from waste
	2. Front load the product development process to explore thoroughly alternative solutions while there is maximum design space
	3. Create a leveled product development process flow
	4. Utilize rigorous standardization to reduce variation and create flexibility and predictable outcomes
	Skilled people
	5. Develop a chief engineer system to integrate development from start to finish
	6. Organize to balance functional expertise and crossfunctional integration
	7. Develop towering technical competence in all engineers
	8. Fully integrate suppliers into the product development system
	9. Build in learning and continuous improvement
	10. Build a culture to support excellence and relentless improvement
	Tools and Technology
	11. Adapt technology to fit your people and process
	12. Align your organization through simple, visual communication
	13. Use powerful tools for standardization and organizational learning

Table 3 Selection of Recommended Practices/Tools/Techniques From Lean Thinking

Practice/Tool/ Technique	Description
Autonomation (Jidoka)	Quality control principle that aims to provide machines and operators the ability to stop the work immediately whenever a problem or defect is detected. This leads to improvements in production processes by eliminating root causes of defects. Related to the concept of "Stop-the-line"
Cell	Location of processing steps immediately adjacent to each other so that products can be processed in very nearly continuous flow
Chief engineer (Shusa)	Manager with total responsibility for the development of a product. The chief engineer is the person who has responsibility to integrate the development team's work around a coherent product vision
Just-in-time (JIT)	System for producing the right items at the right time in the right amounts
Kaikaku	Process improvement through a radical change
Kaizen	Continuous process improvement through small and frequent changes
Kanban	Method based on signals for implementing the principle of pull by signaling upstream production
Lead time	Metric defining the total time that a customer has to wait to receive a product after placing the order
Mistake proofing (Poka-yoke)	Method that helps operators prevent quality errors in production by, for example, choosing wrong parts. The method ensures that an operator can only choose the right action as selecting the wrong action is impossible. For example, product designs have physical shapes that make impossible to wrongly install parts in wrong orientations
Big room (Obeya)	Project leaders room containing visual charts to enhance effective and timely communications, and to shorten the Plan–Do–Check–Act cycle
Set-based concurrent engineering	Approach to designing products and services by considering sets of ideas rather than a single idea
Self-reflection (Hansei)	Continuous improvement practice of looking back to think how a process can be improved
Six-Sigma	Management system focused on improving quality by using mathematical and statistical tools to minimize variability

Table 3 Selection of Recommended Practices/Tools/Techniques From Lean Thinking—cont'd

Practice/Tool/Technique	Description
Standardization	Precise procedures for each activity, including working on a sequence of tasks, minimum inventory, cycle time (the time required to complete an operation), and talk time (available production time divided by customer demand)
Usable knowledge	Learning process focused on capturing knowledge to be applied in other projects
Value stream mapping (VSM)	Tool for analyzing value and waste in the production process. VSM provides a standardized language for identifying all specific activities that occur along the value stream
Visual control	Tools to show the status of the production process so that everyone involved in the production process can understand it at a glance
Work-in-progress (WIP)	Items of work between processing steps
5-Whys	Method for analyzing/finding the root cause of the problems consisting in asking "why" five times whenever a problem is discovered

from a software development perspective, as they have been mentioned in the context of Lean software development (and will appear later on in the next sections). For a more thorough listing of Lean terms, with examples and illustrations, the reader is referred to The Lean Lexicon created by The Lean Enterprise Institute [79].

3.3 Agility

Similar to Lean thinking, the agility movement does not has its origins in software development but in a manufacturing context. However, agility in manufacturing emerged well after Lean thinking was already in use. Concretely, it appeared in manufacturing at the beginning of the 1990s, with the publication of a report by the Iacocca Institute [80]. This report defined agility as a solution to adjust and respond to change, and satisfy fluctuant demand through flexible production. Specifically, Nagel and Dove [80] defined agility as "*a manufacturing system with extraordinary capability to meet the rapidly changing needs of the marketplace, a system that can shift quickly among product modes or between product lines, ideally in real-time response to customer demand.*" Christopher and Towill [81] describes how "*the origins of agility as a business concept lie in flexible manufacturing systems, characterizing it as a business-wide capability that embraces organizational structures, information systems, logistics, processes, and in particular mind set.*"

Similar as well to the concept of Lean, agility seems to be "*highly poly-morphous and not amenable to simple definition*" [82]. However, two aspects are usually stressed in the literature: adaptation and flexibility. The concept of adaptation is based on the contingency theories, which are classes of the behavioral theory [83]. These theories state that there is no universal way of managing an organization. The organizing style should be adapted to the situational environmental constraints in which the organization operates. Thus, organizations are seen as open systems that must interact with their environment to be successful. Successful organizations do not keep themselves in isolation but develop their adaptation capabilities. A close related term is organizational flexibility. Organizational flexibility is considered as "*the organization's ability to adjust its internal structures and processes in response to changes in the environment*" [84]. The aim of agility is to achieve an adaptive and flexible organization able to respond to changes and exploit them to take advantage.

The literature highlights also four capabilities that are particularly important to achieve an Agile organization [85]:

- *Responsiveness*, as the ability to identify and respond fast to changes. It does not only mean identifying changes when they happen (being reactive to changes) but also proactively sensing, perceiving, and anticipating changes to get a competitive advantage.
- *Competency*, which involves all capabilities needed to achieve productivity, efficiency, and effectiveness. Competency comprehends each area in an organization, from strategic vision, appropriate technology and knowledge, product quality, cost effectiveness, operations efficiency, internal and external cooperation, etc.
- *Flexibility*, which refers to the organizational capability to make changes in products and to achieve different objectives using the same facilities. Flexibility involves from product characteristics and volumes to organizational and people flexibility.
- *Speed* or quickness, which relates to making tasks and activities in the shortest possible time in order to reduce time-to-market.

To achieve these capabilities, Agile organizations are usually flat organizations with few levels of hierarchy, and informal and changing authority. Regulations with respect to job description, work schedules, and overall organizational policies are also loose, and trust is given to teams to perform their tasks. Teams are self-organized and job descriptions are redefined based on need [84]. A more comprehensible description of the main attributes of agility as it originally emerged in the manufacturing domain can be found in Ref. [84].

3.4 Combining Lean and Agile in Manufacturing

As in software development, the combination of Lean and Agile also occurred in manufacturing [86]. However, in a manufacturing context, the migration occurred from Lean and functional systems to Agile and customized production [81]. That is, Lean thinking was already in place when many organizations started to look for agility. Accordingly, the shift was characterized by a strong influence of Lean thinking, mainly understood as efficiency and waste reduction. In this context, Lean and Agile seemed to have some intertwined ideas [86]. Agile's focus on flexibility and capacity to embrace change could challenge Lean's focus on overall economic contribution. For example, achieving flexibility in a manufacturing context may request important investments in machines and storages. Marson-Jones et al.'s matrix [87] was created to guide the combination of Lean and Agile in a manufacturing context. It is based on the idea of market winners and market qualifiers. The combination is guided according to the primary company's aim as follows:

- If the primary company's aim is on reducing cost (i.e., cost is the market winner and, therefore, dominates the paradigm as primer requirement), then Lean should be the primary strategy. In this case, aspects such as quality, lead time, and service level are highly desirable but are not strictly requested (market qualifiers).
- However, if the goal of the company is to improve its service level, Agile should take priority. In this case, cost would be a market qualifier but not a market winner.

In a similar vein, the theory of Leagility in a manufacturing domain [86,88] says that while the combination of Agile and Lean might work well, since Lean capabilities can contribute to Agile performance, the combination has to be carefully planned to prevent risks that Agile's demands may cause on

Lean capabilities [86]. Specifically, time and space restrictions are considered in the manufacturing domain, limiting the combination of Agile and Lean to three scenarios:

1. Different value streams, that is, different products.
2. The same product but at different points in time.
3. Using both paradigms at different points in the value stream by using decoupling strategies (strategies to decouple Agile and Lean).

In other words, in manufacturing, there is a tendency to suggest that although Agile and Lean can be combined, they cannot be used at the same time in the same space. Fortunately, the software domain does not have the same restrictions. Indeed, although building upon similar principles, the own characteristics of software development have made the combination of Agile and Lean for software development quite different from its combination in manufacturing. Next section develops this combination in detail.

4. AGILE AND LEAN IN SOFTWARE DEVELOPMENT

Section 3 has described the way in which Agile and Lean emerged in manufacturing. When it comes to the software development domain, we need to consider the fundamental differences between domains [89]. For example, differently from manufactured products, software products are of an intangible nature, which makes it impossible to specify physical measures, affecting the entire concept of quality [90]. The concepts of value and waste are impacted as well. Software products are malleable and its value is not limited to a single time-bound effort. Defining value is not a straightforward task in software development [91], having a whole research field on its own, value-based software engineering [92]. Similarly, waste does not have to follow necessarily the path of the original seven forms of waste as identified by Ohno [74]. Most waste in manufacturing can be directly detected by observing the physical material flow and machine/worker activities. However, many times, intangible work items in software development challenge waste recognition and process improvement. Principles such as those related to flow or visual communication are also impacted due to this intangible nature. Another key difference is on the role of people. While human presence in a manufacturing environment is mainly required to operate automated machines, in software development, people creativity, and knowledge are essential.

Not everything is more challenging, though. Compared to manufactured products, software is quickly duplicated and easily changed throughout releases. This offers innumerable more possibilities from a flexibility and agility perspective than in the case of manufactured products. The main message is that a direct copy/paste from manufacturing to software development is not possible. However, if Agile and Lean are thought of as a set of principles rather than practices (as presented in Section 3), applying these principles in a software development domain makes more sense and can lead to process improvements [40,75,77]. This section develops on how, based on the principles presented in Section 3, Agile and Lean have been interpreted for software development.

4.1 Agile Software Development

The Agile Manifesto [11] started the Agile software development movement. Similar to Lean thinking principles, the Agile Manifesto only establishes the fundamentals of ASD. Besides, a variety of methods have emerged to implement Agile's values and principles in practice. XP [36], Scrum [35], dynamic systems development method [93], crystal [94], or feature-driven development [95] are among the popular Agile methods. In this section, we will describe Scrum and XP as they are the most widely used in industry [26], having captured also a great attention in research (see Section 2). As we will see next, while XP provides concrete software development practices, Scrum could be mostly considered as a software management approach.

4.1.1 Agile Values and Principles: The Agile Manifesto
Under the slogan "*We are uncovering better ways of developing software by doing it and helping others do it*," the Agile Manifesto supposed a reaction against traditional methods that were considered unable to meet market dynamics [11]. Based on the original ideas of Agility, and taking Lean thinking as one of its inspiring sources [96], the Agile Manifesto was formulated from a software development angle.

The Agile Manifesto highlights four values:
1. *individuals and interactions* over processes and tools,
2. *working software* over comprehensive documentation,
3. *customer collaboration* over contract negotiation, and
4. *responding to change* over following a plan.

What the authors of the Manifesto wanted to mean through these values is that while there is value in the items on the right (i.e., processes and tools, comprehensive documentation, contract negotiation, and following a plan), they value the items on the left more (i.e., individuals and interactions, working software, customer collaboration, and responding to change). In other words, processes and tools are useful as far as they serve people and their interactions; documentation that serves working software is fine, but documentation that is not supporting the development process and never gets updated becomes a waste of time and resources; contracts with customers are fair to run the software business, however customer collaboration is more important in order to develop the right product that will serve real customer's needs; and being able to respond to business changes is more important than following a strict plan that might lead to create the wrong product.

In addition to these four values, the Agile Manifesto establishes 12 principles as follows [11]:

1. Our highest priority is to satisfy the customer through early and continuous delivery of valuable software.
2. Welcome changing requirements, even late in development. Agile processes harness change for the customer's competitive advantage.
3. Deliver working software frequently, from a couple of weeks to a couple of months, with a preference to the shorter timescale.
4. Business people and developers must work together daily throughout the project.
5. Build projects around motivated individuals. Give them the environment and support they need, and trust them to get the job done.
6. The most efficient and effective method of conveying information to and within a development team is face-to-face conversation.
7. Working software is the primary measure of progress.
8. Agile processes promote sustainable development. The sponsors, developers, and users should be able to maintain a constant pace indefinitely.
9. Continuous attention to technical excellence and good design enhances agility.
10. Simplicity—the art of maximizing the amount of work not done—is essential.
11. The best architectures, requirements, and designs emerge from self-organizing teams.
12. At regular intervals, the team reflects on how to become more effective, then tunes and adjusts its behavior accordingly.

These values and principles do not say how the software development process should be operationalized, but provide the basis to guide it. The idea is to provide *responsiveness*, *flexibility*, *capability*, and *speed* to the software development process (see Agile capabilities presented in Section 3.3). The Agile Manifesto caused some initial controversy among the software engineering community, especially in those who perceived Agile as an easy excuse for irresponsibility with no regard for the engineering side of the software discipline [27]. However, many others understood that agility and discipline were not so opposed [28,29,97]. Indeed, the 17 proponents of the Manifesto stated that: *"the Agile movement is not anti-methodology, in fact, many of us want to restore credibility to the word methodology. We want to restore a balance. We embrace modelling, but not in order to file some diagram in a dusty corporate repository. We embrace documentation, but not hundreds of pages of never-maintained and rarely used tomes. We plan, but recognize the limits of planning in a turbulent environment."*[g]

4.1.2 XP

eXtreme Programming [36], simply known as XP, was the most popular Agile method at the beginning of the movement. Although other Agile methods, such as Scrum [35], have acquired more popularity recently, many XP practices are still widely used in industry. Indeed, XP offers a very concrete set of practices to implement Agile. Overall, these practices aim to provide continuous feedback from short development cycles and incremental planning, and flexible implementation schedule to respond to changing business needs. XP relies on evolutionary design, automated tests to monitor the development process, a balance between short-term programmer's instincts and long-term business interest and oral communication.

As presented in Table 4, XP has its own set of values and principles, which are well aligned with the values and principles of the Agile Manifesto [36]. XP's values are the basis for XP's principles (i.e., each principle embodies all values).

However, what most probably took the attention of software practitioners on XP was not this set of values and principles, but its very concrete set of practices for software development. When creating the method, the authors tried to come back to what they argued to be the basic activities of software development: *coding* (the one artifact that development absolutely cannot live without), *testing* (and particularly automated testing, to get confidence that the system works as it should work), *listening* (so that things that have to be communicated get communicated, e.g., listening to what the customer says about the business problem), and *designing*

---

[g] http://agilemanifesto.org/history.html (last accessed 22.01.2018).

Table 4 XP Own Set of Values and Principles

XP's values	1. *Communication*. Keeping a constant flow of *right* communication 2. *Simplicity*. Finding the simplest solution that could possible work. XP argues that it is better to do it simple today and pay a little more tomorrow to change it if needed, than to do something very complicated today that may never be used in the future 3. *Feedback*. Keeping a constant feedback about the current state of the system. Feedback works from a scale of minutes (e.g., feedback from unit testing) to a scale of days or weeks (e.g., customer feedback about the most valuable requirements—stories in XP's terminology) 4. *Courage*. Going at absolutely top speed. Tossing solutions and starting over a most promising design if needed. Courage opens opportunities for more high-risk and high-reward experiments
XP's principles	1. *Rapid feedback*. Speed is important from a learning perspective because the time between an action and its feedback is critical to learn. XP aims to learn in seconds or minutes instead of days or weeks at development level, and in a matter of days or weeks at business level instead of months or years 2. *Assume simplicity*. Each problem is faced as if it could be solved using a very simple solution. The focus is on solving today's problem and add complexity in future, if needed 3. *Incremental change*. Problems are solved incrementally instead of making big changes all at once 4. *Embracing change*, by preserving the most options while solving the present problem 5. *Quality work*. Focus on excellent or even *"insanely"* excellent to support work enjoyment

(creating a structure that organizes the system logic). XP's practices focus on these four basic activities. Next, we introduce each of the 12 practices that are considered in XP's. For a more detailed description of each practice, the reader is referred to Ref. [36].

- *The planning game*. The planning game is used to determinate the scope of the next release. XP calls for implementing the highest priority features first. In this way, any features that slip past the release will be of lower value. Following the Lean principle of value, the customer is responsible to choose the smallest release that makes the most business value, while programmers are responsible for estimating and completing their own work. Particularly, two sources are used to determine the scope of a release: business priorities and technical estimates. Business stakeholders (or customer representatives) decide upon the scope of the release, priorities, composition, and dates. Technical stakeholders make estimations of how long a feature takes to implement, technical consequences of business decisions, work organization (process), and detailed scheduling. *Planning poker* is a common consensus-based technique to make estimations in XP. In Planning poker, estimations are made by using numbered cards that face down to the table. Thus, decision makers (in this case technical stakeholders) do not speak their individual estimations aloud but using the cards. Once everyone has made an estimation using the cards, cards are revealed, and estimates discussed. By hiding estimates in this way, the group can avoid the cognitive bias of anchoring, where the first number spoken aloud sets a precedent for subsequent estimates (Fig. 4).

The planning game creates a plan that is an alive artifact that can be quickly updated when the reality overtakes the plan. Plans are updated at least for each 1–4-week iterations and for 1–3-day tasks, so that the team can solve plan deviations even during an iteration.

- *Small releases.* XP calls for short release cycles, a few months at most. Within a release, XP uses 1–4-week iterations. Short releases cycles include fewer changes during the development of a single release and better ability to adapt to business changes. Key Agile characteristics such as responsiveness and speed are achieved by working in short release cycles.
- *Metaphor.* A metaphor is a simple story of how the system should work that is shared among all programmers. The metaphor helps everyone on

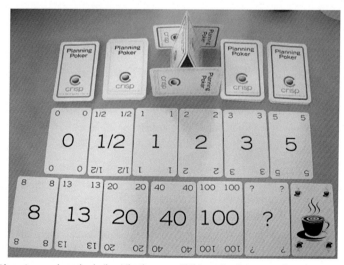

Fig. 4 Planning poker deck (by Hkniberg at Wikimedia Commons).

the project understand the basic elements of the system and their relationships. The metaphor matures as the development proceeds. It is similar to the concept of system architecture but simple enough to be easily communicated between technical and business stakeholders.

- *Simple design.* Design should be as simple as possible to solve today's problem (e.g., runs all tests, does not have duplicated logic, and has the fewest possible classes and methods). Any complexity should be removed as soon as it is discovered. Design is part of the daily business of all programmers in XP, in the midst of their coding.
- *Testing.* Testing and coding are simultaneously conducted in XP. Developers continuously write unit tests that are run for the development to progress. Unit tests written by programmers help convince themselves that their programs work well. In addition, customers write (or at least specify) functional tests to convince themselves that the system as a whole works the way they think it should work. A comprehensive suite of tests is created and maintained, which are run and rerun after each change (several times a day). Automated testing is a key in this context. Test-driven development, the concept of writing first tests for new functionality before the new functionality is added, has acquired a lot of popularity in XP [98].
- *Refactoring.* Refactoring focuses on restructuring a system without changing its behavior in order to improve its internal quality. Refactoring is conducted often in XP to keep the system in shape and remove unnecessary complexity.
- *Pair programming.* As shown in Fig. 5, pair programming refers to two developers programming together at one machine. XP advices for *all* production code to be written in pairs. One member of the pair focuses on writing code while the other thinks more strategically about the solution in terms of tests, complexity, etc. One of the advantages of pair programming is that it allows knowledge sharing among team members.
- *Collective ownership.* In XP, anyone can change anything anywhere in the system at any time. The aim is that anybody who sees an opportunity to add value can do it. Thus, everybody takes responsibility for the whole system.
- *Continuous integration.* Integrate and build the system every time that a new task is completed. As tasks are small, the system gets integrated many times per day. Continuous integration is one of the most important Agile practices nowadays, key for enabling continuous delivery and CD (see Section 5).

Fig. 5 Two coworkers pair programming (by Edward at Wikimedia Commons).

- *40 h per week.* This practice aims to limit working overtime. As a rule, people should not work more than 40 h per week so that they keep fresh, creative, careful, and confident.
- *Onsite customer.* XP asks for a real customer (a person who will really use the system) to be an integral part of the team, available full time to answer questions, resolve disputes, and set small-scale priorities.
- *Coding standards.* Coding standards are followed to maintain the quality of the source code. Coding standards are essential as everyone can change any part of the system at any time, and refactoring each other's code constantly. Moreover, the aim is that code serves also as a communication channel between developers.

4.1.3 Scrum

Nowadays, Scrum is among the most, if not the most, used Agile method [26]. Scrum was first mentioned in 1986, as a new approach to product development in a study in *Harvard Business Review* by Takeuchi and Nonaka [99]. Thus, the ideas behind Scrum did not originate in the software domain either, but were adapted to it. Scrum uses the rugby metaphor of players packing closely together with their heads down and attempting to gain possession of the ball (see Fig. 6) to refer to a holistic product development

Fig. 6 Rugby players forming a Scrum (or Melé) to gain possession of the ball (by Maree Reveley at Wikimedia Commons).

strategy where a development team works as a unit to reach a common goal. In Scrum, a common team goal is defined and shared among all team members, and all team members do their best to reach that common goal (e.g., gaining possession of the ball in rugby). The team is self-organized and can decide the best means to achieve the goal. Thus, the team leader does not *dictate* what to do next, but uses a facilitative style to help the team achieve its goal. Ideally, the team is physically colocated (reminding the concept of *cell* in Lean thinking), or uses close online collaboration, to facilitate communication and create a *team spirit*.

Scrum started to be used for software development in the early 1990s. The origins of Scrum for software development is a topic of frequent debate. Some mistakenly believe that Jeff Sutherland, John Scumniotales, and Jeff McKenna invented Scrum in 1993. However, Sutherland himself has acknowledged that their first team using Scrum at Easel Corporation took Takeuchi and Nonaka's work [99] as their inspiring source [100]. Although the team was already using an iterative and incremental approach to building software, *"the idea of building a self-empowered team where everyone had the global view of the product on a daily basis"* was motivated by Takeuchi and Nonaka's work [100]. Sutherland's team[h] started to develop what we know today as Scrum in software development. In 1995, Sutherland introduced the Scrum team to Ken Schwaber, CEO of Advanced Development Methods, who started to use Scrum at his own company as well. Sutherland and Schwaber

---

[h] Jeff Sutherland was the Chief Engineer (Lean role, see Table 3) for the Object Studio team at Easel Corporation that first started to use Scrum. The team *"defined the Scrum roles, hired the first Product Owner and ScrumMaster, developed the first Product Backlog and Sprint Backlog and built the first portfolio of products created with Scrum"* [101].

worked together to formalize the Scrum development process, which was first presented in a paper describing the *Scrum methodology* at the Business Object Design and Implementation Workshop, held as part of Object-Oriented Programming, Systems, Languages & Applications '95 (OOPSLA '95). Sutherland and Schwaber, as 2 of the 17 initial signatories of the Agile Manifesto, incorporated the ideas behind Scrum into it.

Scrum is a quite simple process framework. Rather than focusing on technical software development practices, Scrum is a method for managing the software development process. It focuses on how a software development team should organize its work to produce software in a flexible way, rather than on the technical practices that the team should apply. Fig. 7 depicts the Scrum process framework. It is composed by three main elements: the scrum team (in the middle of the figure), scrum events (in red), and scrum artifacts (in blue).

• The product owner, the development team, and the scrum master compose the scrum team. The product owner is the responsible person for maximizing the value of the product and the work of the development team. It reminds to the Chief Engineer in Lean thinking. The development team is composed by the professionals who do the work in order to deliver a potentially releasable product increment at the end of each

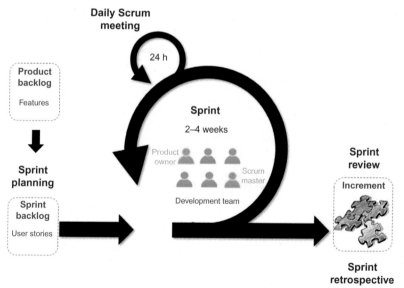

Fig. 7 The Scrum process.

sprint. The scrum master is responsible for making sure that the development work adheres to the Scrum rules (roles, events, and artifacts).

- Scrum considers five well-scheduled events: the sprint, the sprint planning, the daily Scrum meeting, the sprint review, and the sprint retrospective. A sprint is each of the increments in which the software development process is divided. The sprint planning is the meeting held at the beginning of each sprint to agree on the work that will be developed in the following sprint. During the sprint, the Scrum team meets daily during 15 min in the daily Scrum meeting to synchronize activities and create a plan for the next 24 h. At the end of the sprint, the work done during the sprint (potentially shippable increment) is inspected during the sprint review meeting. Finally, the sprint retrospective is a meeting for the Scrum team to inspect its way of working.

- Three main artifacts are considered in Scrum: the product backlog, the sprint backlog, and the increment itself. The sprint backlog is an ordered list that contains all work to be done during the sprint (e.g., user stories, features, functions, requirements, enhancements, and fixes). The product backlog contains all the work to be done for the next product release. The increment is the sum of all product backlog items completed during a sprint (i.e., outcome of the sprint).

Next, we describe how all these pieces work together. The product is defined through *product backlog items*, which are usually known as user stories and features. Product backlog items are stored initially in the product backlog and incrementally transferred to the sprint backlog as the product development progresses through sprints. User stories and features describe the product's functionality from a customer perspective. The underlying strength of these backlog items is that they focus on the value that will be delivered to the user. User stories are often defined following the pattern:

'As a < **User**>, I want to < **Have**> so that < **Benefit**>'

For example, a user story of a library system could be: "*As a librarian I want to have facility of searching a book by different criteria so that I will save time to serve my customers.*" In addition, each user story contains *acceptance criteria* that establish the criteria that the implementation of the user story must meet to be considered as implemented (as done in the Scrum's jargon). User stories contain also an estimation of how much it will take to implement the story, measured as estimated *story points*. The product owner is responsible of defining the user stories and the acceptance criteria for each user story. User stories are usually described at different level of detail.

Particularly in large systems, where high-level product features are initially defined and, as the work progresses, elaborated in more detail through user stories [102]. User stories and features are incrementally created. Thus, it is not expected that a complete set of backlog items is created from the beginning, but the product backlog is an alive artifact that adapts to customer/business's changes. The product owner is the customer's voice to the team and, therefore, responsible for defining features/user stories and prioritizing them to maximize the value of the product. The product owner is also responsible to ensure that the product backlog is visible, transparent, and clear to everyone.

The development is carried out through increments or time boxes, called *sprints* (lasting between 2 and 4 weeks). Thus, products are developed iteratively and incrementally, maximizing opportunities for feedback and, therefore, increasing responsiveness and flexibility. Each sprint has a *sprint goal* that is defined during the *sprint planning meeting*, including the user stories that will be undertaken during that sprint. This is the common goal shared by everyone in the team (as in rugby). The Scrum team is self-organized and has the authority to decide on the necessary actions to complete the sprint goal, which is defined in cooperation with the product owner. Scrum teams are cross-functional, which means that the team has all skills and competences needed to meet the goal. Concerning the size of the team, scrum teams should be small enough to remain nimble but large enough to be able to complete significant work within a sprint. In practice, scrum teams are composed of between 5 and 10 members. Although there are also distributed scrum teams, the ideal case is that all team members are collocated in the same working area, as the *Cell* and *Big-room* concepts in Lean thinking.

The *sprint backlog*, selected from the *product backlog*, contains a prioritized set of user stories (or any other backlog item such as bug fixes) that will be implemented in that sprint. Once the sprint backlog is established during the sprint planning meeting, it remains unchangeable during the sprint. This means that no one is allowed to ask the development team (or any team's member) to work on a different set of requirements or tasks and, similarly, the development team is not allowed to act on what anyone else may request. After every sprint, two meetings are held. On the one hand, the *sprint review meeting*, to analyze the progress, and to demonstrate the current state of the product. The goal of this meeting is to show to all interested stakeholders, and particularly to customers/product owners, the potentially releasable increment that has been developed during the sprint.

The key aspect in Scrum is that each sprint delivers customer value. On the other hand, a *retrospective meeting* is also held at the end of each sprint to reflect about the way of working. In particular, the purpose of the meeting is to reflect on how the last sprint went in order to identify (1) aspects that went well and, therefore, should be kept and (2) potential improvements that need corrections. The outcome of this meeting is an actionable plan for implementing improvements to the way the scrum team does its work. This mechanism allows self-reflection and continuous improvement (kaizen) in Scrum. In addition, every day the scrum team has a *stand-up meeting* of around 10–15 min, usually at the beginning of the day, in which each team member discusses what s/he did since the last stand-up meeting, whether s/he faced any obstacle in that task, and what s/he will do before the next stand-up meeting in the following day. The daily scrum meeting provides transparency on the scrum team daily work and reduces complexity. It allows the team to inspect its progress toward the sprint goal, reveals possible impediments, and promotes quick decision making and adaptation. Fig. 8 shows an example of a daily scrum meeting of a scrum team.

In this process, the scrum master ensures that the rules of Scrum are followed and is in charge to remove impediments in the work of the scrum team.

Fig. 8 Daily Scrum meetings last 15 min every day and are used to provide transparency on the progress toward the sprint goal, reveal possible impediments, and promote quick decision making (by Klean Denmark at Wikimedia Commons).

For example, the scrum master can help the product owner by finding techniques for an effective management of the product backlog, or to the scrum team by facilitating scrum events. Therefore, the scrum master is a servant–leader for the scrum team rather than a manager.

As described earlier, Scrum is a fairly simple method. However, despite its simplicity, Scrum builds upon three important pillars: *transparency, inspection,* and *adaption*. Scrum provides transparency to the development process by making progress visible all the time through events such as daily scrum meetings and sprint reviews. Scrum allows frequent inspection to detect undesirable variances (e.g., changes in customers' needs) and adapt to the new circumstances (e.g., the resulting product will not meet customer's expectations). In words of its founders, *"Scrum implements the scientific method of empiricism [...] to deal with unpredictability and solving complex problems"* [103]. Through small iterations, Scrum improves predictability and control risk by making decisions based on what is known.

Besides the main elements of Scrum described earlier, a number of practices have adhered to Scrum during time. Although, they are not part of the original Scrum framework, they have shown to be very useful as well in certain contexts. For example, Scrum of scrums (or meta-Scrum) are used to synchronize the work of several scrum teams when scaling Scrum from one team to several teams. Backlog grooming is a popular practice to keep the product backlog in shape by reviewing that the backlog contains appropriate items that are properly prioritized. The Agile glossary by the Agile alliance is a good resource to learn more about different terms in the context of Scrum and Agile software development in general (https://www.agilealliance.org/agile101/agile-glossary/).

4.2 Lean Software Development

Nowadays, Lean software development is certainly tied to Agile software development. However, few know that the interest of the software development industry on Lean thinking dates to early 1990s, well before the Agile Manifesto was formulated. Next, we analyze the path followed by Lean thinking within software development before and after the Agile Manifesto, which constituted a breach in the conception of Lean thinking in software development. Then, we focus on Kanban, since it is the most popular Lean technique applied in software development, and list different sources of waste in software development.

4.2.1 Lean Software Development: Pre-Agile Manifesto

The path of Lean thinking within software development started as early as the 1990s, with concepts such as Lean software production and mistake proofing[i] [104,105]. At that time, Lean thinking was understood as a way of making software development processes more efficient and improving their quality. A major focus was on waste reduction. For example, Freeman [104] describes Lean software development as a system of *"achieving ends with minimal means"* by striving to have *"the minimum amount of tools, people, procedures, and so on."* This and similar interpretations of Lean thinking made Lean to be quite unpopular among software engineers, who saw it as a way to get rid of employees. In 1998, Raman [106] analyzed the feasibility of Lean in software development from a wider perspective, considering the five principles of *value, value stream, flow, pull,* and *perfection* as defined by MIT's researchers (see Section 3.1). Raman concluded that Lean could be implemented through contemporary concepts (at that time), such as reusability, rapid prototyping, spiral model, or object-oriented technologies. As we will see next, the conception of Lean software development has considerably evolved after the formulation of the Agile Manifesto.

4.2.2 Lean Software Development: Post-Agile Manifesto

After the formulation of the Agile Manifesto, Lean software development was mainly considered as one of the Agile methods [34]. However, as practitioners started to learn more about Lean principles, Lean software development incrementally acquired an identity in itself [64]. As in manufacturing, Lean software development has been differently interpreted by several authors. Table 5 presents the most popular interpretations of Lean in software development.

Lean software development was initially mainstreamed with an interpretation of Lean thinking led by Poppendieck and Poppendieck [15]. Some argue that this interpretation took the view on software development from the Lean (manufacturing)'s angle. Still, Poppendiecks' principles have been widely used in industry. Poppendieck and Poppendieck have evolved their Lean principles during last years [15,107–109]. Table 5 presents Poppendieck and Poppendieck original principles as they were formulated in 2006 [107] and the current form of these principles. Besides some changes in terminology, the original principles have been complemented

---

[i] See Table 3 for a definition of this concept.

Table 5 Interpretations of Lean Thinking in Software Development

Author	Description
Poppendieck and Poppendieck [15,107–109]	Seven principles guide Lean software development (2006) [107]: 1. *Eliminate waste*, understanding first what value is 2. *Build quality in*, by testing as soon as possible, automation, and refactoring 3. *Create knowledge*, through rapid feedback and continuous improvement 4. *Defer commitment*, by maintaining options open and taking irreversible decisions when most information is available 5. *Deliver fast*, through small batches and limiting WIP (work-in-progress) 6. *Respect people*, the people doing the work 7. *Optimize the whole*, implementing Lean across an entire value stream Seven principles of Lean software development (2017)[a]: 1. *Optimize the whole*, as the synergy between parts is the key to the overall success 2. *Focus on customers* and the problems they encounter 3. *Energize workers* to bring creative and inspired people 4. *Reduce friction* in terms of building the wrong product, building the product wrong, and having a batch and queue mentality 5. *Enhance learning* to respond to future as it unfolds 6. *Increase flow* by creating a steady, even workflow, pulled from a deep understanding of value 7. *Build quality in*, defects should not be tolerated and should be fixed in the moment they occur 8. *Keep getting better*, by changing as fast as the world changes, paying attention to small stuff, and using the scientific method

Continued

Table 5 Interpretations of Lean Thinking in Software Development—cont'd

Author	Description
Middleton and Sutton [110]	Interpretation based on Womack and Jones's five principles of Lean [71]: 1. *Value*, identifying what really matters to the customer 2. *Value stream*, ensuring every activity adds customer value 3. *Flow*, eliminating discontinuities in the value stream 4. *Pull*, production is initiated by demand 5. *Perfection* Middleton and Sutton [110] provide means to implement these principles in practice in a software domain
Larman and Vodde [31]	Larman and Vodde's book [31] focus on scaling Lean and Agile software development with large-scale Scrum. Their interpretation of Lean thinking based on the Lean thinking house, which has two pillars: respect for people and continuous deployment. In addition, Larman and Vodde provided a companion book in 2010 with more concrete practices for scaling Lean and Agile in the software domain [111]
Reinertsen [112]	Set of principles for product development flow, including managing queues, reducing batch size, and applying WIP constraints
Anderson [113]	Kanban as a means to bring Lean thinking into a software development organization. Kanban uses five core properties: (1) visualize workflow, (2) limit WIP, (3) make policies explicit, (4) measure and manage flow, and (5) use models to recognize improvement opportunities. More details on Kanban are provided in Section 4.2.3

[a]http://www.poppendieck.com/ (last accessed 15.08.2017).

with: (1) a deeper *focus on customers* and the problems they encounter; (2) a greater focus on people and creating a Lean mindset [109]; (3) *reducing friction* in terms of reducing building the wrong thing, reducing building the thing wrong, and reducing a batch and queue mentality; (4) a greater focus on *enhancing learning*; and (5) a greater focus on *continuous improvement* (keep getting better).

As time passes by, more diversity was introduced. Thus, besides Poppendieck and Poppendieck's interpretation of Lean software development, other authors have contributed to shape this phenomenon. For example, Middleton and Sutton [110] provide means to implement the original Womack and Jones's (1996) five principles of Lean thinking in the software domain. In a similar vein, Larman and Vodde [111] base their interpretation of Lean software development on the Lean thinking house and provide concrete practices for scaling Lean and Agile based on this interpretation. Other authors, such as Reinertsen [112] and Anderson [113], have focused on more concrete aspects of Lean software development such as flow and Kanban. Moreover, an increasing body of scientific studies reporting experiences when using Lean software development has populated the software engineering literature (e.g., [89,114–121]). Although there is no unified model for the use of Lean and Agile in these studies, some patterns are observable. Most studies converge in the importance of aspects like frequent builds through iterative development, reducing WIP, continuous problem correction and improvement, using cross-functional teams, increasing transparency, and considering human aspects. Table 6 presents some benefits and challenges when applying Lean thinking in software development, as they are mentioned in these scientific studies.

4.2.3 Kanban

Kanban is a popular Lean technique used in software development. It is defined as a signals system to implement the principle of pull[j] in manufacturing. The idea behind Kanban is very simple. It focuses on using a visual signaling device to give authorization and instructions for producing items in a pull system. The signals tell an upstream process the type and quantity of products to make for a downstream process. Thus, products (or product parts) are only manufactured when needed. Kanban cards are the most common example

---

[j] Kanban supports also other principles such as flow and value stream but its main focus was originally on pull.

Table 6 Some Benefits and Challenges When Using Lean Software Development According to the Scientific Literature

Benefits	Challenges
– Increased customer satisfaction – Increased productivity – Decreased lead and development times – Improved progress transparency – Identification of problems' root causes – Identification of bottlenecks	– Creating a Lean and Agile organizational culture to extend Lean thinking beyond software development teams – A deep organizational change is needed to align the entire organization around Lean – Involving management in development tasks – Achieving flow in knowledge work. For example, lack of repetition in knowledge work makes it difficult to specify/standardize tasks – Frustration when starting to use the stop-the-line technique due to the need of stopping-the-line continuously

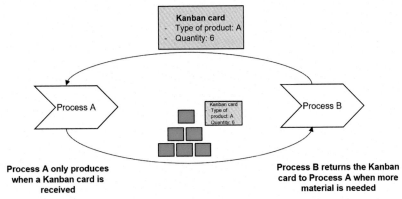

Fig. 9 Kanban as a signaling device in manufacturing.

of these signals. However, Kanban can be also implemented through other devices such as metal plates or colored balls. Fig. 9 shows an example of how Kanban cards support pull. Process A only produces when it receives a Kanban card. Similarly, Process B returns the Kanban card to Process A when it needs more materials.

This same idea has been applied in software development but in the form of a board and sticky notes, as depicted in Fig. 10. The board represents each activity in the development cycle in a column (e.g., definition of backlog items, analysis, development, acceptance testing, and deployment).

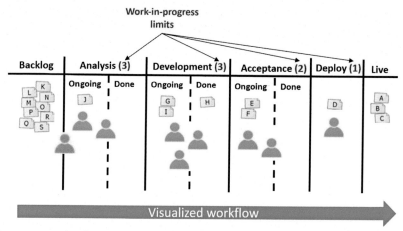

Fig. 10 Example of a Kanban board for software development.

The sticky notes represent user stories or tasks from the product backlog. Thus, the Kanban board visualizes the development flow of these user stories/activities. The number of items under each activity is limited by establishing work-in-progress (WIP) limits. For example, in Fig. 10, three backlog items can be under analysis at the same time, three under development, two in the phase of acceptance testing and one in deployment. Items from the backlog are moved from column to column as the development progresses. An item can be moved to the next development step only if there is room for the new item according to the WIP limits. Thus, teams are not overloaded with work and can focus on the items that provide maximum customer value. Teams can pull more work at any time, triggered by availability of resources. For example, whenever a team member is free, s/he can pull the highest priority ticket from the product backlog. Besides development activities and WIP limits, the Kanban board can include also other information such as the person/s working on each backlog item. It is up to the team to design the Kanban board that better serves its needs.

The board has the same purpose that signals in a manufacturing context, implementing pull in practice by visualizing the workflow (providing visibility to the entire software development process), and enabling the identification of problems, such as bottlenecks in the workflow (a source of waste). Moreover, WIP limits help reduce inventory and focus on the feature that is being developed [113]. Kanban provides five key benefits [113]:

1. *It visualizes workflow.* The Kanban board visualizes the status of each activity in the development process. The team can easily see the big picture of what has been done so far and what needs to be done. Transparency provides a greater control of the workload to deliver in a continuous manner.

2. *It limits work-in-progress.* WIP limits are used to limit the amount of work that can be pulled into each activity. It helps optimize workflow and prevent from team overloading and multitasking. Moreover, WIP limits make bottlenecks visible, helping to a better distribution of resources.

3. *It helps measure and manage flow.* Besides providing information to eliminate bottlenecks in the entire value stream, lead times can also be easily calculated using the board. The shorter the lead-time, the greater is the value for customers and the organization.

4. *It makes process policies explicit.* The process and the policies that the team agrees on how to manage the workflow becomes also visible through the Kanban board. For example, it is clear for everyone what kind of items take priority at a certain time in the development process (e.g., bug fixing vs new user story development).

5. *It supports continuous improvement in a collaborative way.* The Kanban board allows for transparency in the development flow and process monitoring. Inefficiencies and bottlenecks are visible. Kanban teams need to work on those, if they want to maintain a steady flow (Kaizen culture).

Ahmad et al. [122] reviewed the scientific literature on Kanban in software development and found promising benefits when using Kanban, such as better understanding of the whole process by all stakeholders, especially developers, improved team communication, and increased customer satisfaction. However, they also found some challenges, such as integrating Kanban with Agile practices, changing the organizational culture according to Lean and applying Kanban in distributed environments. Regarding integrating Kanban and Agile, Kanban is often used in combination with Scrum in which is known as Scrumban [123]. In Scrumban, Scrum teams employ Kanban boards to visualize their development flow. Depending on the team and the situation at hand, sprints are changed by a continuous delivery flow but other Scrum's elements such as daily scrum meetings or periodic reviews are kept [123].

4.2.4 Waste in Software Development

The main focus of Lean software development is not on reducing costs but on creating value. Still, Table 7 lists some typical sources of waste in software

Table 7 Sources of Waste in Lean Software Development

Author	Sources of Waste	
	Original sources of waste in manufacturing [74]:	Translation to software development [107]:
Poppendieck and Poppendieck [107]	1. *Overproduction*: producing more quantity of product or product's components than needed	1. *Extra features*: producing more features and, therefore, code than what is needed to get the customer's current job done
	2. *Waiting*: waiting for a previous upstream activity to be done	2. *Delays*: waiting for things to happen because of delays in starting a project, decision making, testing, etc.
	3. *Transportation*: unnecessarily moving of materials	3. *Handoffs*: tacit knowledge that is left behind in the mind of the originator
	4. *Overprocessing*: unnecessarily processing by a downstream step in the manufacturing process	4. *Relearning*: forgetting things that were already learned so that knowledge is not preserved
	5. *Inventory*: storing of materials that are not needed for the present	5. *Partially done work*: partially done software that is not integrated into the rest of the environment. For example, uncoded documentation, unsynchronized code, untested code, undocumented code, or undeployed code
	6. *Movement*: unnecessary motion of people	6. *Task switching*: switching time to get into the flow of the new task. Software development requires deep concentrated thinking and switching between tasks takes a lot of effort (e.g., when people are assigned to work on multiple projects)
	7. *Defects*: rework because of quality defects	7. *Defects*: rework because product defects. A primary focus should be on mistake proofing the code and making defects unusual
Sedano et al. [124]	1. *Building the wrong feature or product*: cost of building a feature/product that does not address user or business needs	
	2. *Mismanaging the backlog*: cost of duplicating work, expediting lower value user features, or delaying necessary bug fixes	
	3. *Rework*: cost of altering delivered work that should have been done correctly but was not	
	4. *Unnecessarily complex solutions*: cost of creating a more complicated solution than necessary, a missed opportunity to simplify features, user interface, or code	
	5. *Extraneous cognitive load*: costs of unneeded expenditure of mental energy	
	6. *Psychological distress*: costs of burdening the team with unhelpful stress	
	7. *Waiting/multitasking*: cost of idle time, often hidden by multitasking	
	8. *Knowledge loss*: cost of reacquiring information that the team once knew	
	9. *Ineffective communication*: cost of incomplete, incorrect, misleading, inefficient, or absent communication	

development because we think they can help better understand the concept of waste in software development processes. However, readers should not try to memorize these sources of waste but train their eyes to seeing waste by reflecting upon these typical sources of waste.

Poppendieck and Poppendieck [107] translated the original seven sources of waste in Lean manufacturing [74] into the seven wastes of software development. In their books, they also provide some ways to reduce these wastes. More recently, Sedano et al. [124] provided further insights into waste in software development by empirically identifying sources of waste such as knowledge loss, ineffective communication and psychological distress.

4.2.5 Combining Agile and Lean Software Development in Practice

Principles of Lean software development (see Table 5) are mostly well aligned with the principles and values of the Agile Manifesto. Lean thinking inspired many ideas behind the Agile Manifesto and, therefore, it is not surprising that Agile and Lean share many similarities in the software domain [82,96,125]. Indeed, software companies have traditionally used both in combination. The use of Lean thinking alone is quite uncommon in the software domain [26]. Unlike manufacturing, the transformation toward Agile and Lean in software development has been conducted as a single trip, where the borders between Agile and Lean are not clearly defined. Time and space restrictions, as considered in manufacturing (see Section 3.4), have not been taken into account in the software domain. Wang et al. [64] identified six strategies that the software industry follows to combine Agile and Lean thinking. Those strategies are shown in Table 8 (shorted by popularity).

In our experience, Lean thinking works well as a means for scaling Agile, due to the unique focus of Lean thinking on the organization as a whole [120,121]. For example, in one of our studies with Ericsson R&D Finland [120], we found that Lean thinking was used to support concrete Agile practices. Ericsson R&D transformed its processes from following basic Agile principles to complementing them with Lean principles. The transformation involved around 400 people. In this way, Lean thinking underpins Agile software development. Moreover, Lean is seemed as a means to achieve a learning organization through continuous improvement. The transformation affected the whole development chain, from the earliest product release phases to maintenance, going way beyond

Table 8 Six Categories of Lean Application in Agile Software Development

Strategy	Description
Using Lean principles to improve Agile processes	Selected elements or principles of Lean thinking are used in the context of Agile software development, which is already in place, to improve Agile processes. For example, automating the workflow, using the idea of eliminating waste, or applying Kanban
Using Lean to facilitate Agile adoption	Lean principles serve as bases to decide the concrete Agile practices or methods to be used. Lean principles justify the adoption of certain Agile practices/methods
Using Lean to interact with other business areas	Agile processes are kept in software development while Lean principles are used to interact with neighboring business units such as delivery units, product management, marketing, and customers
From Agile to Lean software development	Agile is in place when the company decides to focus on Lean. A comprehensive application of Lean principles changes Agile processes to a situation that, at the end, Lean processes become dominant and Agile practices play a supporting role
Synchronizing Agile and Lean	Agile teams and Kanban teams work in parallel and in a synchronized manner to address different development aspects of the same product. For example, Scrum teams take care of large-scale changes while Kanban teams focus on small feature requests and bug fixes
Nonpurposeful combination of Agile and Lean	Combination of elements from Agile and Lean with no specific distinction between both paradigms

Based on X. Wang, K. Conboy, O. Cawley, 'Leagile' software development: an experience report analysis of the application of lean approaches in agile software development. J. Syst. Software 85(6), 1287–1299, 2012.

processes and tools and involving a profound change of culture and thinking. Big projects gave room to flexible releases. Agile feature-oriented development, with an end-to-end user story focus, was used to provide flexibility in managing the product and a better ability to serve multiple stakeholders. Development was structured to supports continuous flow. Continuous Integration was a key element in this sense. Testing was done in parallel with development. As a result, the status of the software at all levels became visible and feedback times between traditional test phases were reduced from months to weeks. Seating facilities were also largely modified to support the Lean way of working. Individual offices were gradually given way to team spaces.

We studied the main elements that characterize the combination of Agile and Lean in another example from Elektrobit,[k] a Finnish provider of wireless embedded systems, in Ref. [121]. In general, we found that Lean principles guided also the implementation of Agile methods. Many elements of Agile methods such as Scrum and XP were already in place (e.g., network of products owners, product backlogs, continuous integration, test automation, self-organized and empowered cross-functional teams, and retrospective meeting). In addition, Lean principles, such as creating a culture of customer value in which everyone cares about providing customer value, seeing-the-whole value stream, providing continuous flow through small batches of working software and creating a learning organization to adapt to business and market changes, guided team level activities. Lean practices, such as Kanban, were also used at implementation level.

Overall, Lean helps improve Agile processes and support well-known management issues, such as shorting release cycles and decreasing time-to-market by using flow, defining key performance indicators based on value, improving collaboration with customers by using pull and avoiding short term thinking by using perfection and continuous improvement [40].

5. BEYOND AGILE AND LEAN: TOWARD RAPID SOFTWARE DEVELOPMENT, CONTINUOUS DELIVERY, AND CD

Speed has become increasingly important in current software development. Indeed, it is the key term in the latest advances beyond Agile and Lean software development processes in the form of rapid software development. Rapid software development refers to *develop, release,* and *learn* from software in rapid parallel cycles, often as short as hours or days [47,48]. In this context, software is not seen anymore as an item that evolves through releases that are delivered every certain months; rather, software is seen as a continuous evolution of product functionality that is *quickly* delivered to customers. Value-driven and adaptive real-time business paradigm are key concepts in this context [51]. Value driven because the goal is to create as much value as possible for both, the company and its customers.

---

[k] Elektrobit changes its name to Bittium in 2015.

Adaptive real-time business paradigm because that value is created by activating a continuous—real time—learning capability that allows companies to adapt to business dynamics. In this section, we develop into the importance of speed in current software development processes and describe the concepts of continuous delivery and CD as a means to achieve rapid software development. Particularly, we focus on process elements that are important to achieve this capability.

5.1 Need for Speed

Why does speed matter? The traditional view of software as a static item, which can be bought and owned, is not the current reality of the software industry. Software products and software-intensive systems are increasingly object of a constant transformation. Indeed, customers expect a continuous evolution of product functionality that provides additional value [126]. Speed matters because it provides a competitive advantage to satisfy customers' needs. Speed is concerned with shortening feedback loops between a company and the users of its products. It allows companies to better understand user's behavior and adapt to business/customer changes faster. The risk of predicting what customers want is decreased by increasing feedback loops between users and companies and, thus, *learning* (instead of predicting) what customers want. Need-for-speed (N4S) is the name of the research program in which we have conducted part of our research activities during the last few years (http://www.n4s.fi/en/). Executed by the forefront Finnish software companies and research institutions, N4S focused on making software business in real time. Concretely, during the program, we investigated solutions for adopting a real-time experimental business model to provide capability for instant value delivery, based upon deep customer insight. In the rest of the chapter, we reveal our most relevant findings from the program. Particularly, we discuss key process elements that are important to get speed in software development and the concept of Lean start-up as an experimental model to foster innovation.

Coming back to the reasons why speed is important in software development processes, in his recent book *Speed, data, and ecosystems: excelling in a software-driven world* [127], Bosch reflects on five reasons why speed matters in current software business. According to Bosch,

1. *Speed increases the company's ability to adapt development priorities.* Working in short increments (such as sprints) allows companies set-up priorities

more often. Obviously, the overall product's direction cannot be changed every increment but priorities in small items such as features can be adapted.

2. *Speed allows accelerating the innovation cycle by a faster iteration through the phases of build-measure-learn (BML).* Speed allows companies learn what new ideas would work for their customers faster than the competence. We further develop the concept of BML and the Lean startup movement, in which BML originated, in Section 7.

3. *Speed gives development teams confidence to go into the right direction as it enables believable progress tracking.* Team satisfaction increases as team's members realize that they are able to provide customer value on a continuous basis.

4. *Speed allows continuous feedback on the delivery of value.* It enables fast user feedback to guide the following development steps. Data collection is fundamental in this phase so that decisions are based on data rather than intuition or subjective opinions.

5. *Speed supports product quality.* Faster and more frequent release cycles should not compromise software quality. In fact, although it may seem paradoxical, when properly applied, rapid software development favors quality. In rapid software development coding and testing are done in parallel. Integration and bug fixing are not left to the end of the process, and testing is highly automatized. Overall, speed forces organizations to maintain high quality levels all the time, as new software is continuously delivered to users.

As we described in Section 3.3, speed is one of the four main capabilities of Agility. Speed is embedded in the Agile Manifesto as well. For example, Agile principles, such as *"our highest priority is to satisfy the customer through early and continuous delivery of valuable software"* and *"deliver working software frequently, from a couple of weeks to a couple of months, with a preference to the shorter timescale"* make clear reference to speed. The notion of sprints in Agile methods such as Scrum aims also to increase speed in the software development process. Instead of waiting until the end of development, customers can see progress after every sprint. Based on Agile software development principles, rapid software development increases speed to approaches where the step between development and deployment is minimized and code is immediately delivered to production environment for customers to use (and learn from customer's usage of the product). Iterative Agile and Lean software development are extended toward continuous flow of product functionality.

5.2 Continuous Delivery and Continuous Deployment

Continuous delivery and continuous deployment (CD) are at the heart of rapid software development. Intuitively, both relate to produce new fully evaluated and ready-to-deploy software revisions continuously, even many times per day. Humble and Farley [16], in their book *Continuous Delivery: Reliable Software Releases through Build, Test, and deployment automation* published in 2010, state that continuous delivery provides enterprises with the ability to deliver rapidly, reliably and repeatedly value to customers at low risk with minimal manual overhead. The central concept in Humble and Farley's approach is a deployment pipeline that establishes an automated end-to-end process. Automation is used to ensure that the system works at technical level, executes automated acceptance tests and, lastly, deploys to a production or staging environment. The goal of continuous delivery is to enable the delivery of new software functionality at the touch of a button, using a fully automated process. In the ideal scenario, there are no obstacles to deploying to production; there are no integrations problems, builds can be deployed to users at the push of a button.

There is certain confusion in the use of these terms: continuous delivery and continuous deployment (we refer to the last one as CD). The early scientific literature tended to use both interchangeably, as synonyms [46]. However, there are important differences between them. Humble and Farley [16] describe CD as the automatic deployment of every change to production, while continuous delivery is an organizational capability that ensures that every change can be deployed to production, if it is targeted. However, the organization may choose not to do so, usually due to business reasons. In other words, continuous delivery refers to the organizational capability to keep software always releasable. That is, software is ready to be released to its users at any time but the company decides when it will be released. On the other hand, in CD, software is automatically deployed to production when it is ready. Note that while CD implies continuous delivery, the converse is not true. Fig. 11 graphically illustrates the difference between continuous delivery and CD.

Humble and Farley argue that although CD enables releasing every good build to users, it does not always make sense. In continuous delivery, any build could *potentially* be released to users, but it is the business, not IT, who decides about release schedules.

Leading organizations, such as Facebook, Microsoft and IBM have actively moved toward rapid releases [52]. A plethora of evidence related

Continuous delivery

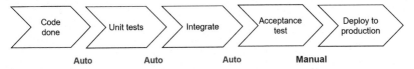

Auto   Auto   Auto   **Manual**

Continuous deployment

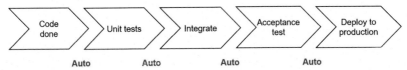

Auto   Auto   Auto   Auto

Fig. 11 Conceptual difference between continuous delivery and continuous deployment.

to continuous delivery and deployment exists in companies' white papers and online blogs, where practitioners have voiced their experiences, expectations and challenges in moving to this direction (e.g., Facebook,[1] Microsoft,[m] Google,[n] Adobe,[o] and Tesla[p]). Next section develops into the main characteristics that allow companies to increase speed in their software development processes and achieve continuous delivery/deployment.

5.3 Key Elements to Get Speed in Your Software Development Process

In the context of the Need-for-Speed project,[q] we reviewed the scientific literature on CD and continuous delivery using the SLR method [46]. Our objective was to determine the underlying factors that characterize this phenomenon as they appear in the scientific literature. Next, we describe the most relevant factors that we found in our study.

1. *Fast and frequent releases*, with preference given to shorter cycles or even continuous flow (weekly or daily deliveries). Release cycles are accelerated when moving toward rapid software development. This implies that the surrounding areas that enable fast releases are also transformed. For example, planning needs to be a continuous activity as well in order

---

[1] https://www.facebook.com/notes/facebook-engineering/ship- early- and- ship- twice- as- often/ 10150985860363920

[m] http://blogs.msdn.com/b/bharry/archive/2012/06/07/announcing-continuous-deployment-to-azure-with-team-foundation-service.aspx

[n] https://air.mozilla.org/continuous-delivery-at-google/

[o] http://steveblank.com/2014/01/06/15756/

[p] http://steveblank.com/2014/01/06/15756/

[q] http://www.n4s.fi/en/

to enable a steady flow of small changes. A tighter integration between planning and execution is needed to ensure alignment between the needs of the business context and software development. The ability to release quickly does not mean that development should be rushed into without a conscious understanding of what is actually being done. The release process needs to be clear, including having a clear release management workflow and delivery workflow.

2. *Flexible product design and architecture.* The software architecture will evolve during the product life-cycle and, therefore, needs to be able to adapt to changing conditions. The architecture must be flexible. Still, it has to be robust to allow organizations to invest resources in offensive initiatives (e.g., new functionality, product enhancements, and innovation) rather than defensive efforts (e.g., bug fixes). The architecture and design have to be highly modular and loosely coupled [128–130]. Moreover, constantly measuring and monitoring source code and architectural quality is important [130,131]. The literature provides some techniques to manage architecture aspects in rapid software development. For example, rapid architecture tradeoff analysis to accommodate rapid feedback and evaluate design options [131], quantifying architectural dependencies by combining Design Structure Matrix (DSM) and Domain Mapping Matrix (DMM) [132] and identifying and assessing risky source code areas based on diverse metrics [133]. In addition, in order to enable continuous and rapid experimentation (see fourth factor), mechanisms in software architectures for run-time variation of functionality and data collection as well as rollback mechanisms to revert changes are required. The literature offers some solutions to achieve this end. For instance, Rally Software suggests A/B testing with Feature Toggle as a technique to manage and support run-time variability of functionality [134].

3. *Continuous testing and quality assurance.* Testing and quality assurance practices have to be performed throughout the whole development process, as new functionality is rolled out, and not just at the end of the development. Transparency on the testing process is important to avoid problems such as duplicated test effort and slow feedback loops [135]. Well-known testing techniques in Agile software development, such as test automation and continuous integration support continuous quality assurance. The concept of technical debt, which is further developed in Section 8, is also important to balance speed and stability. However, besides technical aspects, continuous testing, and quality assurance require also a culture of quality. Developers bear the responsibility for writing

good code, perform thorough tests as well as support the operational use of their software [119,136,137]. As systems become larger and more complex, such culture complements test automation and allows quality to be maintained at scale. The principle is that more frequent release cycles should not compromise software quality.

The latest advances toward managing quality in rapid software development focus on supporting quality-awareness during rapid cycles through continuous monitoring and visualization of product and process quality. For example, our work on the Q-Rapids project[r] focuses on a data driven, quality-aware rapid software development framework. In this framework, quality and functional requirements (e.g., user stories or features) are managed together [138]. Thus, quality requirements are incrementally elicited and refined based on data gathered at both development time and runtime. Using data-mining techniques, project, development, runtime, and system usage data is aggregated into quality-related strategic indicators to support decision makers in steering future development cycles. A strategic dashboard aggregates the collected data into strategic indicators related to factors such as time to market, team productivity, customer satisfaction, and overall quality to help decision makers making data-driven, requirements-related decisions in rapid cycles [139].

4. *Rapid and continuous experimentation.* In rapid software development, systematic design and execution of small field experiments guides product development and accelerates innovation. Continuous experimentation allows companies to learn customer needs and base business and design decisions on data rather than on stakeholder opinions [140]. For example, Facebook uses *A/B (split) testing* as an experimental approach to identify user needs [136]. In A/B testing, randomized experiments are conducted over two options (e.g., two similar features) to compare, through statistical hypothesis testing, how end-users perceive each feature [136,141]. Experimental results are continuously linked with product roadmap in order to provide guidance for planning activities [142]. Rapid experiments are at the heart of the BML innovation cycle (see Section 7).

5. *User involvement.* The role of users is essential in rapid software development. Everything goes around providing value to users. Thus, users are continuously monitored to learn what will provide value to them.

[r] Project funded from the European Union's Horizon 2020 research and innovation program under grant agreement No. 732253 (2016–19), http://q-rapids.eu/

Mechanisms to involve users in the development process and collect user feedback from deliveries as early as possible (even near real-time) are essential in rapid software development. They guide design decisions and innovation. However, techniques need to be nonintrusive so that users are not stressed with continuous feedback requests. Several options can be used in this sense. For instance, similar to Facebook, Rally Software [134] uses also A/B testing with Feature Toggle as a technique to manage and support run-time variability of functionality. Facebook uses a fast upgrade of database systems to deploy experimental software builds on a large scale of machines without degrading the system uptime and availability [143]. Besides involving customers in experiments, customers are also involved in testing activities. Again, strategies applied with such purpose have to be nonintrusive. For example, *dark deployment* is used to deliver new features or services, which are invisible to customers and have no impact on the running system, to test system quality attributes and examine them under simulated workload in a real production environment [134,136]. Another relevant practice, *canary deployment*, delivers a new version of the system to a limited user population to test it under real production traffic and use. The new version is then delivered to the whole user population once it reaches a high enough quality level [134,136].

6. *Automation.* Computers perform repetitive tasks, people solve problems. In the context of continuous delivery and deployment, the focus is on automating the entire delivery pipeline as much as possible. The goal is that deployments are predictable, routine tasks that can be performed on demand in a matter of seconds or minutes. Continuous integration automates tasks such as compiling code, running unit and acceptance tests, monitoring and validating code coverage, checking compliance with coding standards, static code analysis and automatic code review. The continuous integration practice of Agile software development is extended to release and deploy automation. For example, Facebook [136] has a tool for deployment called Gatekeeper that allows developers to turn features on and off in the code, and to select which user groups see which features. Similarly, Agarwal [144] presents an automation system to integrate configuration management, build, release and testing processes. The system enables migrating changes from one environment to another (e.g., development to production), performing a release with selected content, and upgrading software in a target environment. Humble et al. [145] recommend automating build, testing,

and deployment in the early stage of the project and evolving the automation along with the application.

7. *DevOps*. DevOps is another key element in continuous delivery and deployment because it allows achieving rapid end-to-end software development. DevOps integrates development and operations. It enables transparency and understanding of the whole development pipeline to overcome corporate constraints that often cause delays in product deliveries (e.g., handover delays and communication gaps). The concept of DevOps is further developed in next section.

6. DevOps

When moving beyond Agile and Lean software development toward CD, the software development process becomes characterized by increasing collaboration of cross-organizational functions and automation, particularly, in software testing, deployment and operations. In this section, we introduce the concept of DevOps, whose core principles center around cross-discipline collaboration, automation and the use of Agile principles between software development and operations. The main goal of DevOps is to improve the delivery speed of high-quality software changes while the system is in operational state.

6.1 The Origins of DevOps

Compared to early years of its inception, the concept of DevOps today is less ambiguous owing to the increased number of scientific studies describing its underlying characteristics and its adoption in software-intensive organizations [146–150]. DevOps emerged in 2009 as an approach for implementing CD [151]. For several years, it had remained unclear what the concept entails, despite its popularity in the software industry.

The term "DevOps" is a combination of two words, "*Development*" and "*Operations*." As a concept, DevOps stresses on building an Agile relationship and collaboration between software development and operations, whereby operational requirements are considered early into, and throughout the software development process. The main aim of DevOps is to speed up the delivery of quality software changes more reliably and repeatedly in production environment. DevOps proposes some practices and the use of tools to automate the management of software infrastructures, which, over the years, have become complex, heterogeneous and of large-scale.

Historically, the term DevOps was coined by Patrick Debois when organizing a practitioners' event called DevOpsDays.[s] Prior to DevOpsDays, in 2008, Debois gave a presentation at the 2008 Agile conference titled: *"Agile infrastructure and operations"* [152] that focuses on employing Agile principles to infrastructure-related activities and projects, such as upgrading application server that has multiple infrastructural component dependencies. After his presentation, Debois observed a growing interest on the topic among other software practitioners. For instance, few months prior to the DevOpsDays event, two other pioneers of DevOps, John Allspaw and Paul Hammond, shared their experiences of multiple deployments in a presentation titled *"10 deploys per day: Dev & ops cooperation at Flickr"* at the 2009 Velocity conference. Together with other software practitioners, the pioneers introduced a DevOps movement that today embodies a diverse set of practices and tools crucial for speeding up the delivery process of software changes to end-users.

Much of what is addressed by DevOps movement deals with problems resulting from organizational silos between software development and operations. For many medium and large software-intensive organizations, software development and operations activities are traditionally performed by separate organizations. Typically in such context, software is designed, implemented and tested by software developers and testers, and, upon completion, the developed software is handed over to operations personnel, such as system administrators. Operations personnel then deploy and operate software in production environment. Such a setup is beneficial to facilitate required specialization and prevent mix up of software development and operations tasks in daily work of engineers. However, the separation, exacerbated by poor communication between software development and operations, has been reported to bring several serious problems, including *unsatisfactory testing environments and systems being put into production before they are complete* [153]. The problems intensified when software developers started to use Agile methods, but without their adoption among operations personnel [154,155]. As a result, conflicting goals of *"stability vs. change"* began to appear. Operations personnel, being skeptical of the impacts of frequent changes to production environment, tend to seek stability by avoiding frequent system changes to production, which software developers tend to seek with Agile methods. To facilitate effective collaboration between software development and operations, several approaches, ranging from ad-hoc

---

[s] DevOps days is a worldwide series of technical conference covering topics of software development and IT infrastructure operations (http://devopsdays.org/about/).

and informal to formal approaches, were suggested [156]. However, more recently DevOps emerged as a solution that focuses on automating the deployment pipeline and extending Agile methods to operations, as visualized in Fig. 12 [146,147]. The deployment pipeline, visualized in Fig. 13, is the path taken by all software changes to production, including changes of software features and configuration parameters. In the deployment pipeline, though it may manifest differently across software-intensive organizations, similar patterns can be observed. A common pattern is that, the software developer performs development tasks on his/her local development environment and may performs some tests before deciding to commit the changes to team's shared development environment, where the changes are made visible to others. Upon commit, software changes are automatically compiled, tested

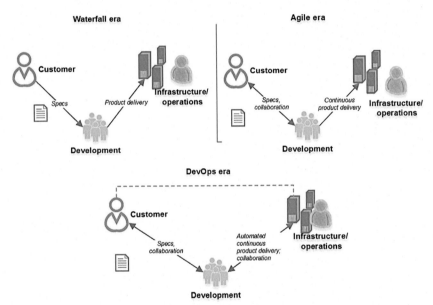

Fig. 12 DevOps extending Agile collaboration to operations engineers. *Adapted from T. Rautonen, DevOps—evolution of developers' role, https://gofore.com/devops-sovelluskehittajan-roolin-evoluutio/ [Online; accessed 3.10.2017].*

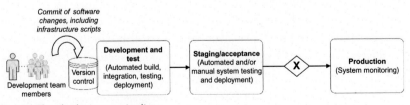

Fig. 13 The deployment pipeline.

for integration and with successful tests a build is created. The newly created software build is then passed for additional tests such as acceptance tests before being deployed to production environment, where software changes are made visible to end-users. Specific details in the level and efforts needed for test and deployment automation vary across software-intensive organizations, depending on the domain and type of application being developed.

6.2 DevOps Practices

The adoption of DevOps in software-intensive organizations is not homogenous due to factors such as target objective and starting baseline. DevOps adoption entails a series of organizational and sociotechnical practices to be employed. The organizational practices focus on organizational structures and ways in which an organization deals with the relationships and responsibilities between software development and operations. Meanwhile, the sociotechnical practices focus on engineering practices related to the development of software changes, deployment pipeline and interaction of software developers and operations personnel.

From the organizational perspective, DevOps has four common practices [157,158]:

a. *separating development and operation with high collaboration;*
b. *separating development and operation with a facilitator between the two groups;*
c. *development team gaining more responsibilities of operations and operations existing in a small team and;*
d. *no visible operations team.*

Among the four practices, the practice of separating software development and operations with high collaboration is most common in large software-intensive organizations [157,158]. The practice of separating software development and operations with a facilitator between them, and assigning more responsibilities to developers are more common in small and medium sized organizations [157,158]. However, regardless of the organization size, the practice of having developers who gain more responsibilities of operations is observed to increasingly becoming common [157,158]. This has huge implications to the working culture and division of responsibilities [136,148,159,160] A complimentary practice consists of establishing of a separate team that acts as a facilitator between the two groups. [148,159,161,162] argue for such a team because it can significantly reduce required efforts and minimize disruptions for development team. Particularly, in large organizations, the established team consists of a small number of members with multidisciplinary background including that of software development and operations. The team is responsible for standardizing deployment mechanisms such as building a deployment pipeline platform, which is to be used by software developers [148,161,162]. An important aspect to notice is that when knowledgeable software developers are given the responsibility to perform operations tasks, they tend to be given access to critical environments [159,160]. Such access allows software developers to provision by themselves different software development environments.

DevOps practices from a sociotechnical perspective focus on automating the deployment pipeline, including automatic management of software infrastructure, which consists of development and production environments. The sociotechnical DevOps practices include technical infrastructure where software development and operations activities take place and its interaction with human agents. Therefore, the deployment pipeline is the technical infrastructure where sociotechnical DevOps practices are observed. In specific, a survey of software practitioners by Cito et al. [157] reports on incorporation of practices in the following areas for DevOps: *(1) software deployment and provisioning; (2) software design and implementation; (3) software troubleshooting; and (4) software monitoring and production data usage.* Automation practices in the deployment process, which focus on automatic provisioning of environments, are more prominent in cloud-native software [157]. One reason is the progress in virtualization technology, which is made prevalent through cloud computing. Cloud computing provides access to computing resources that, through DevOps tools such as Puppet and Chef, software developers can use to provision and manage environment repeatedly and fast. The tools advocated in DevOps, which are used to automatically provision and manage environments, apply the concept of *infrastructure-as-code*, which is central to DevOps. The *Infrastructure as Code* concept entails a practice of using software engineering practices for managing source code such as programming language and version control, to implement and manage scripts used for environment provisioning, configuration management and system deployment [163]. In practice, the *infrastructure-as-code* concept is facilitated by DevOps tools like Puppet and Chef, which feature a domain-specific language used to write scripts for environment provisioning and software deployment. The benefit of using *infrastructure-as-code* is that it helps minimize inconsistencies and complexity of software infrastructure, which in turn helps shorten the release cycle time. We further elaborate on two sociotechnical DevOps practices as these were the focus areas in our research.

Automation of deployment process. The automation of deployment process in DevOps involves two main activities. First, an automatic provisioning of development environments, e.g., test, that are similar to the production environment. Second, the application of deployment strategies such as blue/green deployment, rolling upgrade and canary deployment, to deploy software changes to desired environments. The latter is meant to replace a manual process for deploying software and configuring environment as well as to minimize environment inconsistencies with production, as these constitute to common antipatterns in software delivery process and have impacts to the release cycle time of changes [16]. Additionally, a deployment strategy is selected and facilitated by the use of deployment automation tools. Blue/green deployment and rolling upgrade are two commonly used deployment strategies for cloud application [147]. In blue/green deployment, a number of virtual machine (VM) instances, which contain old version of the software, are maintained while in parallel the same number of VM instances, which contain an updated version of the software, are provisioned and only when they are ready, traffic is routed to the new instances [147]. In rolling upgrade deployment strategy, a small number of VM with new software version are upgraded directly to environment containing VM instances with old version and in parallel a number of old VM instances are switched off [147]. However, each strategy may also pose some reliability issues, including the amount of time needed to prepare and place a new VM to the desired environment [147]. Recently, microservice architecture has been found beneficial when considering deployment automation. Microservice architecture loosely decouples the software into collaborating services whereby each service implements a set of related functions that can be easily developed and frequently deployed, using technology such as Docker, independent of other services [164]. Despite its many benefits, microservice architecture still presents several challenges that need to be addressed in research, including complexities introduced in managing and monitoring system comprised of many different service types [165].

Monitoring. In DevOps, monitoring becomes important not just for operations personnel but also software developers. For instance at Facebook, developers in a team are responsible for monitoring and teams schedule. They rotate among themselves an "oncall" person who receives alerts of urgent issues through automated phone calls but also shields others to focus on development work [166]. Software monitoring tools, such as New Relic, allow developers to install monitoring agents to the software in order to gather real-time software and server monitoring data and metrics specified as relevant by the development team [147]. Insights from monitoring data can be used to inform and guide developers identify software failures and performance improvements as well as gain feedback from users based on software usage among other things [147]. Recent research contributions on this aspect have focused on finding effective ways to correlate runtime data to software changes during development time for such purpose, as to automate troubleshooting activity among others [167]. This is contrast to the traditional approach whereby developers and operators employ manual steps to diagnosis faults by navigating and inspecting relevant information from log statements, software metrics, and using mental models [167]. For this purpose, Cito et al. [167] have proposed an analytical approach to established explicit links that carry semantic relations between related fragments that can be individual logs, metrics, or excerpts of source code.

Finally, it is relevant to mention that DevOps practices are championed and well established in software-intensive organizations that are involved in cloud and web development [136,148,157,162]. DevOps practices are increasingly being evaluated for their feasibility in other contexts, such as embedded systems, particularly because several factors make the adoption of DevOps in such contexts challenging, e.g., unavailability of technology to enable flexible and automatic system redeployment repeatedly [159,168,169].

6.3 The Benefits and Challenges of DevOps

Benefits and challenges of DevOps have been reported in empirical studies. For example, among the most popular benefits are *a reduction in release cycle time; improved quality and eased communication between software development and operations* [148,161,162]. However, successful adoption of DevOps requires a mindset change, management support and shared responsibility among all stakeholders involved in the delivery of software. Support from senior management strengthens the value of and the adoption of DevOps across an organization [170]. However, the organization may encounter several challenges during the adoption of DevOps, including the need to rearchitect the software to better support DevOps in deployability, testability and loggibility [171]. Microservice architecture is one architectural design having such benefits as scalability, flexibility, and portability, and is in support of DevOps [164]. Furthermore, since in DevOps, the development team is given the responsibility to develop and operate the system that they develop, this requires the team to be equipped with necessary skills, thus, increasing the learning curve among development team members [160]. Alternatively, for some processes, their inner working may be abstracted through automation achieved in the deployment pipeline, thus, demanding a low learning curve from development team members.

7. THE LEAN STARTUP MOVEMENT

This section focuses on a movement that has acquired particular relevance during last years, the Lean startup movement. Indeed, applying the principles of Lean thinking has been broadened from the areas described in the earlier sections. One new area is the emergence and early stages of new enterprises, startups. Lean thinking is widely applicable where something new is built—manufacturing, software development, or startups. In this section, applying Lean thinking on software startups is described.

7.1 Origin and Key Concepts

In 2011, Ries [172] applied the principles of Lean thinking into startup companies in his book *"The Lean startup: How today's entrepreneurs use continuous innovation to create radically successful businesses."* In the book, he focused on the question how to tune a successful business case from a product idea. In his thinking, leanness was crystallized by defining waste as a product without market potential: the biggest waste is to create a product that nobody is willing to buy. Ries presented how to work in a Lean way and avoid waste by introducing a process of early learning. The key characteristics of that process are (1) earliness, (2) close customer cooperation, (3) rapid progress, (4) measurement of the value, (5) learning, (6) corrective actions, and (7) iteration.

- *Earliness* refers to the need to follow Lean thinking from the very beginning of a startup. In the opposite case, the course of the company may already be fixed toward a wrong direction, and the efforts done turn out to be waste.
- *Close customer cooperation* refers to the principle that the customer(s) need to be linked to the startup's doings as early and closely as possible. The measurement of the value of the product is fully based on the customers' opinions. A real customer is the source of the feedback needed when measuring the business potential of the innovation. Ries' book classifies customers by defining different groups of customers being relevant for a startup in different phases of its evolution path. The customer groups

are defined mostly in the context of American business environment and products aimed to mass markets. In a practical situation, however, close customer relationship is easier to create in case of products aimed to specific customers instead of mass markets [173]. In any case, early and close customer cooperation is very beneficial for a startup figuring out its first product. Reasonable amount of effort should be put in finding potential customers willing to cooperate with a startup—willingness to work with a startup having only a vague idea of a product and a MVP (minimum viable product), may be low among the potential customers.

- *Rapid progress* refers to a startup's need to validate the value of the product idea as rapidly as possible. Rapid progress along with earliness is a key means to avoid waste in Lean startup thinking, but the need to speed up the product development is a common trend in the software industry, as discussed earlier in Section 5. The need for a rapid progress and close customer cooperation are the drivers behind one of the key concepts of Ries' Lean startup, building the MVP.

- *Measurement of the value* of the business case is done by the customer on the basis of that MVP—and only in that way. An MVP should be at the same time easy and fast to create, but functional enough to be a valid basis for an evaluation by the targeted customer(s). What the MVP is at the end, is left fairly open. Ries' book addresses MVPs by introducing examples. That is understandable when aiming at covering a big variety of startup with the same principles, but in a practical situation the MVP must be well figured out to fit to the innovation, to the measurements purposes, and to the overall context.

- *Learning* refers to the experiences and opinions collected from the customer(s) concerning the value of the product idea. How the learning is gained, is covered in Ries' original book by defining such terms as validated learning and actionable metrics. Validated learning is a result of validated experiments with the latest version of the MVP and the real customer(s). As the experiments are the source of validated learning, a key issue, besides the definition of the MVP, is careful planning of the experiments in which the value of the MVP is measured.

- *Corrective actions* refer to the conclusions drawn from learning. The corrective actions are crystallized in Lean startup thinking into two clear alternatives, changing the idea or continuing its development, called in Ries' terminology pivoting or persevering. The criteria for pivoting or persevering are summarized by Ries as so-called actionable metrics as an opposite of vanity metrics. The actionable metrics should be accessible and audible. The point in putting such attributes as actionable and vanity on the metrics used for decision making is to highlight courage of a startup's key persons to explore the innovation in question in a critical and objective way. The positive feelings or bride of the innovator(s) should not lead to development of products that turn out to be waste. Measuring and learning in Lean startup thinking can be seen as a means to identify difficulties and find reasons not to persevere but pivot, as studied in Ref. [174].

- *Iteration* refers to the principle that the whole process of building MVP(s), measuring, and learning, has to be repeated until the aimed product gains the functionality and characteristics that guarantee customer acceptance. This principle represents the old idea of continuous and incremental development, applied in Lean startup thinking on the customer-centric validation of an innovation. The same continuous and incremental way of doing is proposed also for the growth stage of a startup, when the first product(s) has successfully been launched to the markets and the startup is seeking growth. A broader view on the practices of continuous and incremental development was presented in earlier sections.

Ries encapsulates the earlier characteristics in a simple, cyclic process containing three steps: (1) build, (2) measure, and (3) learn, BML cycle (see Fig. 14). After the three steps, the startup evaluates the results of the steps and makes a decision whether to keep the idea, persevere, or to modify it, pivot. In the former case the development of the idea to a product continues,

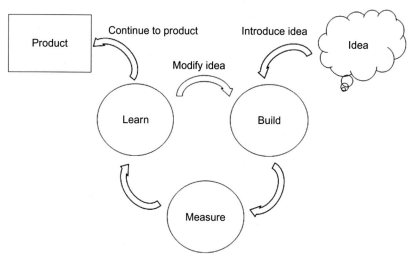

Fig. 14 The BML cycle.

while in the latter case the idea is modified and the build–measure–learn steps are repeated. Following this process should prevent a startup wasting its resources in developing products without market demand.

7.2 Derivatives and Further Developments

Ries' Lean startup thinking was rapidly accepted not only by practitioners but also by the academia too. A set of variants based on similar thinking were presented. Stainert and Leifer presented fairly similar ideas in their hunter–gatherer model [175], which was further tuned by Nguyen-Duc et al. [176]. While the hunter–gatherer model was at a fairly high and philosophical level, Bosch et al. [177] and Björk et al. [178] presented in 2013 a development model called Early Stage Software Startup Development Model (ESSSDM), claimed to be an operationalization of Lean thinking on software startups (see Fig. 15).

The ESSSDM model was developed to bring the ideas of Ries closer to the practical work in software startups by addressing such areas as managing multiple product ideas in a product portfolio, validation criteria of the product ideas, and techniques to use for validating them. The ESSSDM model, with its focus on product portfolios, is not principally different from the original Lean startup thinking. A product portfolio including alternative product ideas, means validating several ideas in parallel or in series. However, a smaller startup may not have resources for parallel evaluation of several product ideas, as strongly recommended in the ESSSDM model. An idea portfolio may also be difficult from the focus and motivation perspectives: founding a company and taking the responsibility and risks tied to it requires courage [179]. A strong belief on a single idea may be a better catalyzer for the needed courage than a set of ideas, which possibly are not as credible.

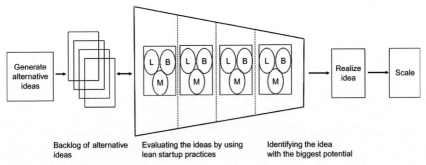

Fig. 15 The ESSSDM model.

7.3 Lean Startup and Other Startup Research

Starting fairly early, Lean startup thinking has been combined with other research paths on innovation and entrepreneurship. Müller and Thoring studied in 2012 differences and similarities of Lean startup model and design thinking [180]. The comparison highlighted, in an interesting way, certain characteristics of Lean startup thinking: (1) it assumes that a product vision exists, e.g., a Lean startup is driven by a product vision of the startup founders, (2) it is customer centered instead of user centered, and (3) its focus is on the business model for a product idea—not on user desirability or technical feasibility of the product.

The Lean startup thinking focuses on creating the business case and on validating product ideas from the business perspective. The customer acceptance of the idea is defined as the criterion of the waste. Practitioners having experience in product development know that the business feasibility is only one dimension of the whole. The technical feasibility and user desirability must be evaluated, too. Putting too much effort on an idea that is too expensive or too difficult to realize with existing technology, or difficult to use, is waste even when the idea has a real market potential [173]. Technical feasibility and user desirability are broad areas where the weight of different criteria varies according to the context and purpose of the product: a data communications product must provide the user with speed and reliability, an interactive product with desirable user experience, understandability and easiness of use, and a banking system with data security.

An example of a broader focus validation of the idea is the work of Hokkanen et al., combining the lean startup principles with the user experience of the targeted product [181–183]. Hokkanen's work complements the Lean startup from the user-centered view by highlighting the importance of a good user experience in an early stage—as a part of the MVP. According to her, an MVP must be mature enough and its user experience desirable enough to help selling the idea and collecting meaningful feedback from the customer(s). Following the original Lean startup thinking, she recommends early building of communication channels to real users, implementing usability tests before measuring, and applying planned methods to collect the feedback necessary for correct learning.

7.4 Summary of the Lean Startup Movement

As mentioned earlier, Lean startup thinking is vision driven—there is an innovation, the value of which has to be measured. The startup is founded to realize the innovation to a product. Research [173] has shown that many times it really is the case. A typical story of a startup is that a single person gets an idea that he/she believes in, gathers possibly a team, preferring old friends and other trusted people, and founds a company [179]. Sometimes, founding is preceded by a long period of deliberation before the final decision for a startup is made. Of course, there are also different cases, where the need for a certain product or service is clear and the business case is easier to figure out. That does not, however, decrease the value of the basic idea of the Lean startup thinking, better to evaluate than to develop waste.

The works by Bosch et al. [177], Björk et al. [178], Müller and Thoring [180], and Hokkanen et al. [181–183] are examples of how the Lean startup thinking, defined at a high abstraction level, and trying to address a big variety of different startups, needs additional stuff around the principles. However, the principles, validating before progressing, iterating, and improving continuously, seeking learning from real customers, being critical not vain, and avoiding waste, are good guidelines for any product development. The principles are broadly applicable also outside the original business-related focus, and their value is intuitively understandable. The basic message is: *keep on learning—do not believe that you know the value of your idea without measuring it.*

8. MISCELLANY

Before closing this chapter, we want to discuss two areas that have been the focus of our recent research in Agile and Lean software development processes, and we believe are useful for the readers of this chapter. On the one hand, we focus on software metrics that are usually employed in the context of Agile and Lean software development. These metrics are important not only for understanding the situation of the product under development but also for guiding the transformation toward Agile and Lean. On the other hand, we discuss on the topic of technical debt, which has become popular in the context of Agile software development.

8.1 Metrics in Agile and Lean Software Development

Our SLR about the use of software metrics in Agile software development [68] revealed both similarities and differences between Agile and traditional software development. The top goals of both development approaches are similar, that is, to improve productivity and quality. The majority of the metrics are in one way or the other tied to those goals. However, we found value-based differences in means how those goals are to be reached. Traditional software development utilizes software metrics from the viewpoint of scientific management by Taylor, i.e., controlling software development from outside. The traditional measurement programs use ideas stemming from industrial product lines and manufacturing industry such as Six Sigma. Traditional metrics programs were often large-scale corporate initiatives, which emphasized following a pregiven plan. Agile principles, such as empowering the team, focusing on rapid customer value, simplicity, and willingness to embrace changes, form a contrast to the traditional measurement programs.

Fig. 16 shows the top six metrics used in Agile software development. We selected the top metrics based on the number of occurrences and the perceived importance of each metric in the sources (for more details, see Ref. [68]).

Velocity was found as the most important software metric in Agile development. Velocity indicates how fast is the team working or how much a team can achieve in a certain time-boxed period, e.g., in a sprint of 4 weeks. Velocity is used in sprint planning. Velocity is also used to determine different scopes for the sprint such as having a minimum scope of features that can be delivered for sure and having a stretch goal for the iteration. Having two different scopes can help in customer communication as the minimum can be promised and the maximum indicates all the changes that may happen during a sprint. Velocity enables finding bottlenecks in a software process where progress is slower than desired. Making process improvement or automation to these parts results in an excellent cost benefit ratio for the

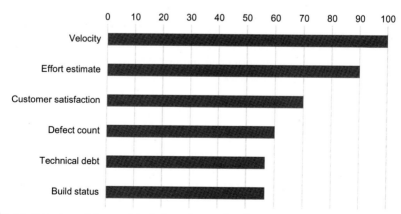

Fig. 16 Top six metrics used in Agile software development.

improvement. However, our study also found cases where attempts to maintain velocity led to cutting corners and lowered the product quality.

Effort estimate, which indicates the anticipated work cost of a particular feature or story, was found as the second most important metric. Effort estimates are important in feature prioritizations as all features with equal value can be sorted by the effort estimates to provide a sprint scope with the highest possible value. However, *No Estimates* movement started by Ron Jefferies, approximately says that making effort estimates can be counterproductive as the accuracy of the estimates is too poor to make the effort spent in estimation worthwhile. Nevertheless, effort estimates were often used and found important in the sources of our literature review.

Customer satisfaction was found as the third ranked metric, although it could be claimed that it should be the most important due to the customer centric nature of Agile software development. Yet, it was less frequently mentioned in our sources than Velocity or Effort estimate. Customer satisfaction is a metric with many possible operationalizations. We found several measures of customer satisfaction such as the number of change requests by the customer, net promoter score (the probability that a customer would recommend the product or producer to another potential customers), Kano analysis (quality model that distinguishes between must-be, one-dimensional, attractive, indifference, and negative quality), and postrelease defects found by the customers. We found that most successful projects measured customer satisfaction more often than other projects.

We find that the core books giving advice on Agile software development are missing some core software engineering metrics such as the *defect count*. In our review of industrial empirical studies, it was ranked in the fourth place indicating its importance in Agile software development. Defect count can have many uses. It may be a measure of software testing as the number of defects found per time unit in testing measures in-house software quality. Furthermore, the defects experienced in customer usage can be a measure of the software quality in the wild.

Technical debt, which is covered in more detail next, was among the top metrics used by Agile teams. Technical debt hinderers day-to-day software development and reduces velocity. In one company, a technical debt board was used to make technical debt visible in order to ensure that enough resources and motivation were given to removing the debt. The board was also used to make plans how to remove technical debt. The board made sure that developers picked the highest priority technical debt items instead of cherry picking lower priority but more interesting tasks.

Build status is an important measure in Agile software development. It is an indication of the agile principle working code. If builds are not passing, this indicates that no progress according to Agile principles has been made. Build status often encompasses quality gates in addition to the simple assertion that the software has been successfully built. It can include the results of automated unit testing. Frequently, successful build requires that certain end-to-end automated acceptance test is passed. When working on software with very high reliability requirements, even violations from static code analysis tools, such as Findbugs or Coverity, may be treated as build breakers.

We also studied the purposes of using metrics in Agile software development, which are depicted in Fig. 17.

We found that over 70% sources used metrics for planning and progress tracking of sprints or projects. Understanding and improving quality were found as a reason for using metrics in roughly one-third of the studies. Fixing process problems inspired roughly half of the studies in using metrics. Finally, motivating people was found as the reason for using metrics in 25% of the studies, for example displaying a technical debt board can motivate people to reduce technical debt. The percentage does not add up to 100% because many studies had multiple motivations in using metrics.

8.2 Technical Debt in Agile and Lean Software Development

The Agile Manifesto highlights *"continuous attention to technical excellence and good design enhances agility."* Similarly, XP has *"coding standards"* as one of its key practices. Still, the scientific literature highlights that Agile software

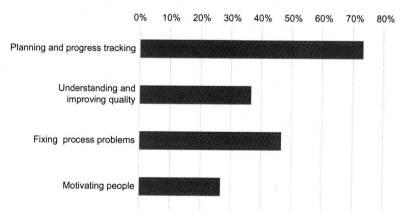

Fig. 17 Reasons for using metrics in Agile teams.

development appears to be more prone to technical debt compared to traditional software development [184]. This section develops into the concept of technical debt and describes different strategies that can be applied to manage it in Agile software development.

8.2.1 What Is Technical Debt?

Technical debt is a metaphor used in software development to communicate the consequences of poor software development practices to nontechnical stakeholders. Ward Cunningham first introduced it in 1992 [185]. The concept works like a financial debt that needs to be paid back with some extra interest but applied to the software development process. Technical debt happens, for example, when developers focus only on delivering new functionality at the end of each sprint and lack time in refactoring, documentation, or other aspects of the project infrastructure. If these aspects are not considered in the sprint backlog and ignored, the software will end up growing less understandable, more complex, and harder to modify. We can view this system degradation as a debt that developers owe to the system [186]. Cunningham [185] explained that a little debt can speed up software development in the short run. However, this benefit is achieved at the cost of extrawork in the future, as if paying interest on the debt. The debt increases when it is not promptly repaid. Time spent on *not-quite-right* code counts as interest on that debt.

Technical debt has been usually associated to poorly written code. For example, code that is unnecessarily complex, poorly organized or includes unnecessary duplications will require future work (e.g., refactoring). However, as time passes by, technical debt has evolved to embrace different aspects of software development that range from architectural deterioration and design problems to testing, documentation, people, and deployment issues [187]. Another example of technical debt in the area of testing is when tests scripts are not properly maintained, leading to additional manual testing. In general, technical debt can be manifested in any aspect of the software development process that relates to internal system qualities, primarily maintainability and evolvability [188].

8.2.2 Technical Debt: Debt or Investment?

Technical debt seems to be, therefore, something negative, which often results in undesirable consequences such as system degradation, increased maintenance costs, or decreased productivity. However, technical debt is not only about technical shortcuts because pressures of the moment or accumulation of unfinished work. It can also be used constructively as a framework to reflect about different alternatives and make informed decisions about what to pass up [186]. As everything in life, there are always trade-offs in software development, which, unfortunately, is not a world of infinite resources. Software companies need to decide where to invest their finite resources to get maximum value for their business and their customers. Not doing something good today will allow doing something else that is also good and, at the end, will pay back better in the future [186]. For example, we may decide to sacrifice certain quality level in the next release because we prefer to include a feature that will provide a competitive advantage. Getting to the market to catch an important business opportunity may influence the decision to take on technical debt. Overall, technical debt is seen as a balance between software quality and business reality [189].

According to Fowler [190], technical debt can be classified in two groups: reckless and prudent debt. Reckless debt is the "*bad*" or unhealthy debt; the debt that is considered negative. It is incurred as the result of a shortage, sloppy programming, or poor development. It can be inadvertent, for example, because of poor technical knowledge, or deliberated, if the debt is a result of sloppy or careless programming. However, there is also a healthy or "*good*" debt, the prudent debt. Prudent technical debt is proactively incurred in order to get some benefit. For example, achieving a quicker release. In this situation, the team deliberately decides to take on technical debt and has plans to deal with consequences. Prudent debt is, therefore, related to intentional decisions to trade-off competing investments during development. Fowler argues that prudent debt could be also inadvertent, in the sense that software development is usually a learning activity. Software engineers may learn different alternatives during the development process that would have been better options from a technical point of view.

Overall, technical debt effects are not always negative. The problem with technical debt comes when it is not properly managed or gets uncontrolled. The key is to make informed decisions and be prepared for the consequences. When it is properly managed and controlled, technical debt can be utilized as an Agile strategy to gain business opportunities and be paid back later. Next section describes different strategies to manage technical debt in an Agile context.

8.2.3 Strategies to Manage Technical Debt in Agile Software Development

Besides the emphasis on quick delivery, which is frequently identified as one of the most important causes for accumulating technical debt in Agile [69], other aspects can also generate technical debt in Agile. For example (1) architecture and design issues including inflexibility in architecture, poor design, and suboptimal up-front architecture and design solutions, (2) inadequate test coverage and issues with test automation, (3) lack of understanding of the system being built, (4) oversight in estimations of sprints and, in general, unrealistic estimations, and (5) inadequate refactoring [69]. The consequences are quite straightforward, system quality degradation, which increases maintenance costs and reduces productivity, market loss/hurt business relationships, and even complete redesign or rework of a system [69]. Therefore, it is important to learn strategies that help manage technical debt in Agile and use it in a healthy way. Our SLR on technical debt in Agile software development found the following solutions to manage technical debt and balance rapid delivery and quality development [69]:

- *Refactoring* is the most common technique applied to reduce technical debt in Agile software development. Refactoring consists in restructuring the code base or system architecture without altering the external behavior of the system. The key idea is to redistribute classes, variables, and methods across the class hierarchy to simplify the underlying logic of the software and eliminate unnecessary complexity. Refactoring should be a continuous activity in Agile and rapid software development. Nowadays, many development environments include automated support for refactoring, at least for mechanical aspects of refactoring. A comprehensible overview of software refactoring, including an illustrative example on how it is conducted in practice, can be found in Ref. [191].
- *Measuring/visualizing technical debt and communicating its consequences.* Measuring technical debt is fundamental to make it visible. Nonvisible technical debt goes usually unnoticed, being a clear risk for the development (unhealthy debt). As presented earlier, technical debt is among the top metrics used by Agile teams. Unmeasured technical debt makes it hard for development teams to make a strong case for business stakeholders to invest in technical debt fixes. Overall, technical debt should be transparent. Design decisions related to technical debt and architecture dependences should be clear for everyone. The literature suggests different strategies to monitor technical debt and make it visible, such as keeping track of a list of architectural and design decisions in a backlog [130,192], using technical debt visualization boards [193–196], visualizing technical debt using the "code Christmas tree" [197], pie and bar charts to visualize and manage technical debt [198], and technical visualization tools like Ndpend to detect code violations [199].

- *Planning for technical debt and prioritizing technical debt when needed.* Using technical debt proactively means that we know that we are taking technical debt and we plan how to deal with the consequences generated by our decision. Advance planning for technical debt involves assigning preemptive efforts to address it when needed by allocating resources in advance for tasks such as code cleanup, removal of design shortcuts, and refactoring. Dedicated teams, which are particularly responsible for technical debt reduction, can also be allocated when needed. Measuring is essential here in order to objectively determinate "when needed." Optimistic estimations by Agile teams lead to technical debt. Technical debt items should be explicitly included in the product backlog so resources are explicitly allocated for these tasks. Moreover, increasing teams' estimation ability leads to reduce reckless technical debt.
- The *Definition of Done* (DoD) supports also controlling technical debt in Agile. Employing a common DoD in different levels, such as story, sprint, and release, helps achieve a common understanding on technical debt and manage it strategically. For example, definition of the *right-code* can be include in the DoD. The DoD can be also used during sprint planning meetings and sprint reviews to reveal technical debt issues and plan how to deal with consequences. Overall, the DoD helps with establishing an acceptable level of technical debt.
- *Automation*, and particularly, test automation helps also reduce technical debt and increase test coverage.

A more detailed analysis of the scientific literature on technical debt in Agile software development is reported in our SLR on the topic [69].

9. CONCLUSIONS AND FUTURE DIRECTIONS

Agile software development is well established in software industry. In this chapter, we have gone through the main elements of Agile and Lean software development processes, from the very fundamental principles and values that guide these approaches to the concrete methods and practices existing in the literature to implement the fundamentals in practice. Particularly, we have discussed the main characteristics of "agility" and "leanness" as they emerged in the manufacturing domain (Section 3). The five original principles of Lean thinking [71] are: (1) *value*, understood from a customer's perspective; (2) *value stream*, as the stream of production activities in which every activity delivers customer value; (3) *flow*, which provides continuity to the value stream; (4) *pull*, to make sure that production is guided only by demand; and (5) *perfection*, the pillar for continuous improvement. Agility, on the other hand, is characterized by four capabilities [85]: (1) *responsiveness*, to identify and respond fast to changes; (2) *competence*, to have all capabilities to achieve productivity, efficiency, and effectiveness; (3) *flexibility*, to make changes in products and achieve different objectives using similar facilities; and (4) *speed*, to reduce time-to-market.

The closeness of ideas between both paradigms has favored its combination (known as leagility). Although this combination was taken with caution in manufacturing—as it is described in Section 3—the journey of Agile and Lean in the software domain has followed a quite different path; a path characterized by a symbiosis between Agile and Lean in which limits are not clearly established. The foundations of Agile software development processes were established in the Agile Manifesto through 4 values and 12 principles. These foundations created the basis for Agile software development methods that implement those values and principles in practice (e.g., Scrum and XP). Agile software development emerged as the best possible solution to face the challenges posed by a turbulent and dynamic software market. However, although Agile processes provided important benefits at development team level, they were not enough to operate a whole organization. Thus, Lean software development emerged as a way of scaling Agile and, in general, complementing Agile software development processes. More recently—as discussed in Section 5—Agile and Lean software development processes have been extended to continuous delivery and deployment. The idea is that software development becomes a continuous process that is able to dynamically adapt to business changes and discover new business opportunities.

If we had to highlight two essential characteristics of current software processes, those would be *speed* and *learning*. Speed is essential because it does not only allow obtaining a competitive advantage, but also it favors short feedback loops. Short feedback loops are the basis for the second essential characteristic of current software processes, learning, and more concretely, *validated learning*. Learning is important because it is the key element that allows software companies to adapt to business changes. The goal is to learn (instead of predicting) where value is through practices such as continuous experimentation, which guide product development and accelerate innovation. DevOps (discussed in Section 6) and Lean start-up (discussed in Section 7) are key elements to achieve speed and validated learning.

How do we see the future of software development processes and, particularly, Agile software development? As the software development industry evolves, versatile, and flexible software development processes are becoming essential to cope with current market and customers' demands. Old-fashioned software development processes, such as the waterfall model, worked well under the software industry conditions where they were developed. However, they are too complex to surface innovation in current software markets, as they are not flexible and fast enough for many software companies. Thus, Agile is turning into a mainstream methodology for software development. Although, it was initially taken with care by many software sectors, in which Agile was seen just as the latest fad that, as any fad, would have a short life, it is quite clear now that Agile won the recognition battle and it is here to stay. Nowadays, Agile is not only used by pioneers in innovative projects, but also it has spread across a broad range of industries. This is particularly true in web, mobile applications, or services domains, where applications are frequently developed in weeks or months, rather than years.

Years ago, people were not as exposed to software products as they are now. Nowadays, we are using software every single day of our lives. In our houses, our jobs, hospitals, schools, … Smartphones, watches, fridges, and cars are obvious devices containing software. However, smart microprocessors are embedded in almost every device that you can imagine: thermostats, garage doors, front doorbells, … Did you know that smart microprocessors are starting to be installed in walls and foundations of houses to monitor stress fractures, watering, and weakening in the structure of buildings? It is clear that the software development industry is growing fast. Many innovations are emerging and will continue to emerge. For example, we think that trends like the IoT will shape the software development processes of more traditional industries in the near future. The IoT is much more than hardware and connectivity. Software is needed to run and connect all these smart devices collecting and exchanging data, and connecting every aspect of our lives. Moreover, with IoT there are increasing demands in reliability and in particular to security. When software controls physical world products, security holes must be quickly and continuously patched. Agile, Lean, and rapid software development will play a key role in such development.

Both the software development industry and the research community are very active in shaping and extending Agile and Lean to processes such as rapid software development. For example, continuous processes are under constant exploration to identify ways to safely get speed in software development. Ways to activate a continuous—real time—learning capability to allow companies to adapt to business dynamics, such as continuous experimentation, are also increasingly explored. Software processes must be value driven. Better understanding on the concept of value will provide the means to improve value-based decision making as well.

REFERENCES

[1] T. Wasserman, in: Low ceremony processes for short lifecycle projects, Keynote at the 2013 International Conference on Software and System Process, ACM, 2013.

[2] J. Münch, O. Armbrust, M. Kowalczyk, M. Soto, Software Process Definition and Management, Springer, 2012.

[3] W.W. Royce, in: Managing the development of large software systems, Proceedings of IEEE WESCON, Los Angeles, CA, USA, 1970, pp. 1–9.

[4] M. Oivo, A. Birk, S. Komi-Sirviö, P. Kuvaja, R.V. Solingen, in: Establishing product process dependencies in SPI, Proceedings of European Software Engineering Process Group Conference, 1999.

[5] B. Boehm, A spiral model of software development and enhancement, Computer 21 (5) (1988) 61–72.

[6] ISO/IEC 122007:2008(E), Systems and Software Engineering—Systems Life Cycle Processes, second ed., ISO/IEC, Geneva, Switzerland, 2008.

[7] ISO/IEC 90003:2004(E), Software Engineering—Guidelines for the Application of ISO 9001:2000 to Computer Software, 2004. ISO/IEC 90093:2004(E).

[8] CMMI, Capability Maturity Model Integration 1.3, Carnegie Mellon Software Engineering Institute, 2010. http://www.sei.cmu.edu/cmmi/.

[9] ISO/IEC 15504-1:1998, Information Technology—Process assessment—Part 1: Concepts and Introductory Guide, 1998. WG10N222.

[10] P. Kuvaja, A. Bicego, BOOTSTRAP—a European assessment methodology, Softw. Qual. J. 3 (3) (1994) 117–127.

[11] K. Beck, M. Beedle, A. van Bennekum, A. Cockburn, W. Cunningha, M. Fowler, J. Grenning, J. Highsmith, A. Hunt, R. Jeffries, J. Kern, B. Marick, R.C. Martin, S. - Mellor, K. Schwaber, J. Sutherland, D. Thomas, Manifesto for Agile Software Development, http://www.agilemanifesto.org/, 2001.

[12] V.R. Basili, A.J. Turner, Iterative enhancement: a practical technique for software development, IEEE Trans. Softw. Eng. 4 (4) (1975) 390–396.

[13] W.W. Agresti, New Paradigms for Software Development: Tutorial, IEEE Computer Society Press, 1986.

[14] I. Jacobson, G. Booch, J.E. Rumbaugh, The Unified Software Development Process—The Complete Guide to the Unified Process from the Original Designers, Addison-Wesley, 1999.

[15] M. Poppendieck, T. Poppendieck, Lean Software Development: An Agile Toolkit, Addison-Wesley Professional, 2003.

[16] J. Humble, D. Farley, Continuous Delivery: Reliable Software Releases through Build, Test, and Deployment Automation, Pearson Education, 2010.

[17] I. Sommerville, Software Engineering, ninth ed., Addison-Wesley, 2010.

[18] L. Osterweil, in: Software Processes are Software Too, Proceedings of the 9th International Conference on Software Engineering, IEEE Computer Society Press, 1987, pp. 2–13.

[19] M. Castells, The Rise of the Network Society: The Information Age: Economy, Society, and Culture, Wiley-Blackwell, 2011.

[20] K.M. Eisenhardt, J.A. Martin, Dynamic capabilities: what are they? Strateg. Manag. J. 21 (10 − 11) (2000) 1105–1121.

[21] P. Abrahamsson, O. Salo, J. Ronkainen, J. Warsta, Agile Software Development Methods: Review and Analysis, VTT Publications, 2002, p. 478.

[22] K. Conboy, S. Coyle, X. Wang, M. Pikkarainen, People over process: key challenges in agile development, IEEE Softw. 28 (4) (2011) 48–57.

[23] A. Cockburn, J. Highsmith, Agile software development, the people factor, Computer 34 (11) (2001) 131–133.

[24] P. Kruchten, in: A plea for lean software process models, Proceedings of the 2011 International Conference on Software and Systems Process, ACM, 2011, pp. 235–236.

[25] VersionOne, Inc, 10th Annual "State of Agile Development" Survey, https://versionone.com/pdf/VersionOne-10th-Annual-State-of-Agile-Report.pdf, 2016.

[26] P. Rodríguez, J. Markkula, M. Oivo, K. Turula, in: Survey on agile and lean usage in Finnish software industry, 2012 ACM-IEEE International Symposium on Empirical Software Engineering and Measurement (ESEM), IEEE, 2012, pp. 139–148.

[27] S. Rakitin, Manifesto elicits cynicism, IEEE Comput. 34 (12) (2001) 4.

[28] B. Boehm, Get ready for agile methods, with care, Computer 35 (1) (2002) 64–69.

[29] K. Beck, B. Boehm, Agility through discipline: a debate, Computer 36 (6) (2003) 44–46.

[30] D. Rosenberg, M. Stephens, Extreme Programming Refactored: The Case Against XP, Apress, 2003.

[31] C. Larman, B. Vodde, Scaling Lean & Agile Development: Thinking and Organisational Tools for Large-Scale Scrum, Addison-Wesley Professional, 2009.

[32] K. Vilkki, When agile is not enough, in: Lean Enterprise Software and Systems, Springer, 2010, pp. 44–47.

[33] M. Laanti, Agile methods in large-scale software development organisations, in: Applicability and Model for Adoption, University of Oulu, 2012, Doctoral thesis.

[34] T. Dybå, T. Dingsøyr, Empirical studies of agile software development: a systematic review, Inf. Softw. Technol. 50 (9) (2008) 833–859.

[35] K. Schwaber, M. Beedle, Agile Software Development with Scrum, Prentice Hall, Upper Saddle River, 2002.

[36] K. Beck, C. Andres, Extreme Programming Explained: Embrace Change, Addison-Wesley Professional, 2004.

[37] D. Turk, R. France, B. Rumpe, in: Limitations of agile software processes, Third International Conference on eXtreme Programming and Agile Processes in Software Engineering (XP 2002), 2002, pp. 43–46.

[38] P. Abrahamsson, K. Conboy, X. Wang, 'Lots done, more to do': the current state of agile systems development research, Eur. J. Inf. Syst. 18 (4) (2009) 281–284.

[39] C. Maples, in: Enterprise agile transformation: the two-year wall, Agile Conference, 2009. AGILE'09, IEEE, 2009, pp. 90–95.

[40] A. Maglyas, U. Nikula, K. Smolander, Lean solutions to software product management problems, IEEE Softw. 29 (5) (2012) 40–46.

[41] J.P. Womack, D.T. Jones, D. Roos, The Machine that Changed the World: The Story of Lean Production: How Japan's Secret Weapon in the Global Auto Wars Will Revolutionize Western Industry, Rawson Associates, New York, NY, 1990.

[42] B. Bremner, C. Dawson, K. Kerwin, C. Palmeri, P. Magnusson, Can anything stop Toyota? Bus. Week 117 (2003).

[43] M.A. Cusumano, Reflections on the Toyota debacle, Commun. ACM 54 (1) (2011) 33–35.

[44] N. Tokatli, Global sourcing: insights from the global clothing industry—the case of Zara, a fast fashion retailer, J. Econ. Geogr. 8 (1) (2008) 21–38.

[45] L.B. de Souza, Trends and approaches in lean healthcare, Leadersh. Health Serv. 22 (2) (2009) 121–139.

[46] P. Rodríguez, A. Haghighatkhah, L.E. Lwakatare, S. Teppola, T. Suomalainen, J. Eskeli, T. Karvonen, P. Kuvaja, J.M. Verner, M. Oivo, Continuous deployment of software intensive products and services: a systematic mapping study, J. Syst. Softw. 123 (2017) 263–291.

[47] M.V. Mäntylä, B. Adams, F. Khomh, E. Engström, K. Petersen, On rapid releases and software testing: a case study and a semi-systematic literature review, Empir. Softw. Eng. 25 (2) (2015) 1384–1425.

[48] B. Fitzgerald, K.J. Stol, Continuous software engineering: a roadmap and agenda, J. Syst. Softw. 123 (2017) 176–189.

[49] H.H. Olsson, H. Alahyari, J. Bosch, in: Climbing the "Stairway to Heaven"—a multiple-case study exploring barriers in the transition from Agile development towards continuous deployment of software, 38th Euromicro Conference on Software Engineering and Advanced Applications, IEEE, 2012, pp. 392–399.

[50] B. Fitzgerald, K.J. Stol, in: Continuous software engineering and beyond: trends and challenges, Proceedings of the 1st International Workshop on Rapid Continuous Software Engineering, ACM, 2014, pp. 1–9.

[51] J. Järvinen, T. Huomo, T. Mikkonen, P. Tyrväinen, From agile software development to mercury business, in: Software Business. Towards Continuous Value Delivery, Springer, 2014, pp. 58–71.

[52] G.G. Claps, R. Berntsson Svensson, A. Aurum, On the journey to continuous deployment: technical and social challenges along the way, Inf. Softw. Technol. 57 (2015) 21–31.

[53] J. Highsmith, A. Cockburn, Agile software development: the business of innovation, Computer 34 (9) (2001) 120–127.

[54] D.J. Reifer, How good are agile methods? IEEE Softw. 19 (4) (2002) 16–18.

[55] E. Arisholm, H. Gallis, T. Dyba, D.I. Sjoberg, Evaluating pair programming with respect to system complexity and programmer expertise, IEEE Trans. Softw. Eng. 33 (2) (2007).

[56] I.M. Chen, Rapid response manufacturing through a rapidly reconfigurable robotic workcell, Robot. Comput. Integr. Manuf. 17 (3) (2001) 199–213.

[57] M. Perepletchikov, C. Ryan, Z. Tari, The impact of service cohesion on the analyzability of service-oriented software, IEEE Trans. Serv. Comput. 3 (2) (2010) 89–103.

[58] E. Hossain, M.A. Babar, H.Y. Paik, in: Using scrum in global software development: a systematic literature review, Fourth IEEE International Conference on Global Software Engineering (ICGSE), IEEE, 2009, pp. 175–184.

[59] D. Port, T. Bui, Simulating mixed agile and plan-based requirements prioritization strategies: proof-of-concept and practical implications, Eur. J. Inf. Syst. 18 (4) (2009) 317–331.

[60] J. Vähäniitty, K.T. Rautiainen, Towards a conceptual framework and tool support for linking long-term product and business planning with agile software development, in: Proceedings of the 1st international Workshop on Software Development Governance, ACM, 2008, pp. 25–28.

[61] R. Hoda, J. Noble, S. Marshall, The impact of inadequate customer collaboration on self-organizing agile teams, Inf. Softw. Technol. 53 (5) (2011) 521–534.

[62] C. Delen, H. Demirkan, Data, information and analytics as services, Decis. Support. Syst. 55 (1) (2013) 359–363.

[63] J.M. Rivero, J. Grigera, G. Rossi, E.R. Luna, F. Montero, M. Gaedke, Mockup-driven development: providing agile support for model-driven web engineering, Inf. Softw. Technol. 56 (6) (2014) 670–687.

[64] X. Wang, K. Conboy, O. Cawley, 'Leagile' software development: an experience report analysis of the application of lean approaches in agile software development, J. Syst. Softw. 85 (6) (2012) 1287–1299.

[65] E. Scott, G. Rodríguez, A. Soria, M. Campo, Are learning styles useful indicators to discover how students use scrum for the first time? Comput. Hum. Behav. 36 (2014) 56–64.

[66] J. Babb, R. Hoda, J. Norbjerg, Embedding reflection and learning into agile software development, IEEE Softw. 31 (4) (2014) 51–57.

[67] M. Leppänen, S.V. Mäkinen, M.E. Pagels, V.P. Eloranta, J. Itkonen, M.V. Mäntylä, T. Männistö, The highways and country roads to continuous deployment, IEEE Softw. 32 (2) (2015) 64–71.

[68] E. Kupiainen, M.V. Mäntylä, J. Itkonen, Using metrics in agile and lean software development—a systematic literature review of industrial studies, Inf. Softw. Technol. 62 (2015) 143–163.

[69] W.N. Behutiye, P. Rodríguez, M. Oivo, A. Tosun, Analyzing the concept of technical debt in the context of agile software development: a systematic literature review, Inf. Softw. Technol. 82 (2017) 139–158.

[70] V. Garousi, M.V. Mäntylä, Citations, research topics and active countries in software engineering: a bibliometrics study, Comput. Sci. Rev. 19 (2016) 56–77.

[71] J.P. Womack, D.T. Jones, Lean Thinking: Banish Waste and Create Wealth in Your Organisation, Rawson Associates, New York, 1996.

[72] T. Dybå, H. Sharp, What's the evidence for lean? IEEE Softw. 29 (5) (2012) 19–21.

[73] T. Fujimoto, Evolution of Manufacturing Systems at Toyota, Oxford University Press, Inc., New York, 1999.

[74] T. Ohno, Toyota Production System: Beyond Large-Scale Production, Productivity press, 1988.

[75] J.K. Liker, The Toyota Way, Esensi, 2004.

[76] J.M. Morgan, J.K. Liker, The Toyota Product Development System, Productivity Press, New York, 2006.

[77] M. Poppendieck, M.A. Cusumano, Lean software development: a tutorial, IEEE Softw. 29 (5) (2012) 26–32.

[78] R. Shah, P.T. Ward, Defining and developing measures of lean production, J. Oper. Manag. 25 (4) (2007) 785–805.

[79] C. Marchwinski, J. Shook, Lean Lexicon: A Graphical Glossary for Lean Thinkers, fourth ed., Lean Enterprise Institute, 2008.

[80] R.N. Nagel, R. Dove, 21st Century Manufacturing Enterprise Strategy: An Industry-Led View, DIANE Publishing, 1991.

[81] M. Christopher, D.R. Towill, Supply chain migration from lean and functional to agile and customized, Supply Chain Manag. Int. J. 5 (4) (2000) 206–213.

[82] K. Conboy, Agility from first principles: reconstructing the concept of agility in information systems development, Inf. Syst. Res. 20 (3) (2009) 329–354.

[83] L. Donaldson, The Contingency Theory of Organizations, Sage, 2001.

[84] B. Sherehiy, W. Karwowski, J.K. Layer, A review of enterprise agility: concepts, frameworks, and attributes, Int. J. Ind. Ergon. 37 (5) (2007) 445–460.

[85] H. Sharifi, Z. Zhang, A methodology for achieving agility in manufacturing organizations: an introduction, Int. J. Prod. Econ. 62 (1) (1999) 7–22.

[86] J.B. Naylor, M.M. Naim, D. Berry, Leagility: integrating the lean and agile manufacturing paradigms in the total supply chain, Int. J. Prod. Econ. 62 (1) (1999) 107–118.

[87] R. Mason-Jones, J.B. Naylor, D. Towill, Engineering the leagile supply chain, Int. J. Agil. Manag. Syst. 2 (1) (2000) 54–61.

[88] R.I. Van Hoek, The thesis of leagility revisited, Int. J. Agil. Manag. Syst. 2 (3) (2000) 196–201.

[89] B.R. Staats, D.J. Brunner, D.M. Upton, Lean principles, learning, and knowledge work: evidence from a software services provider, J. Oper. Manag. 29 (5) (2011) 376–390.

[90] V. Mandic, M. Oivo, P. Rodriguez, P. Kuvaja, H. Kaikkonen, B. Turhan, in: What is flowing in lean software development?, Proceedings of the 1st International Conference on Lean Enterprise Software and Systems (LESS), 2010, pp. 72–84.

[91] E. Mendes, P. Rodriguez, V. Freitas, S. Baker, M.A. Atoui, Towards improving decision making and estimating the value of decisions in value-based software engineering: the VALUE framework, Softw. Qual. J. (2017) 1–50. https://link.springer.com/article/10.1007%2Fs11219-017-9360-z#citeas.

[92] S. Biffl, A. Aurum, B. Boehm, H. Erdogmus, P. Grünbacher (Eds.), Value-Based Software Engineering, Springer Science & Business Media, 2005.

[93] J. Stapleton, DSDM: Business Focused Development, Addison-Wesley Professional, 2003.

[94] A. Cockburn, Crystal Clear: A Human-Powered Methodology for Small Teams, Pearson Education, 2004.

[95] S.R. Palmer, M. Felsing, A Practical Guide to Feature-Driven Development, Pearson Education, 2001.

[96] J.A. Highsmith, Agile Software Development Ecosystems, Addison-Wesley Professional, 2002.

[97] C.G. Cobb, Making Sense of Agile Project Management: Balancing Control and Agility, Wiley, 2011.

[98] K. Beck, Test-Driven Development: By Example, Addison-Wesley Professional, 2003.

[99] H. Takeuchi, I. Nonaka, The new new product development game, Harv. Bus. Rev. 64 (1) (1986) 137–146.

[100] J. Sutherland, Agile development: lessons learned from the first scrum, Cutteer Agile Project Management Advisory Service: Executive Update 5 (20) (2004) 1–4.

[101] J. Sutherland, K. Schwaber, The Scrum Papers: Nut, Bolts, and Origins of an Agile Framework, SCRUM Training Institute, 2011, p. 152.

[102] D. Leffingwell, Agile Software Requirements: Lean Requirements Practices for Teams, Programs, and the Enterprise, Addison-Wesley Professional, 2010.

[103] K. Schwaber, J. Sutherland, The Scrum Guide. The Definitive Guide to Scrum: The Rules of the Game, http://www.scrumguides.org/, 2016. last accessed 09.08.2017.

[104] P. Freeman, in: Lean concepts in software engineering, IPSS-Europe International Conference on Lean Software Development, Stuttgart, Germany, 1992, pp. 1–8.

[105] J. Tierney, in: Eradicating mistakes from your software process through Poka Yoke, Proceedings of 6th International Conference on Software Quality Week, 1993, pp. 300–307.

[106] S. Raman, in: Lean software development: is it feasible?, Digital Avionics Systems Conference, 1998. Proceedings of 17th DASC, The AIAA/IEEE/SAE, IEEE, 1998. 1:C13/1–8.

[107] M. Poppendieck, T. Poppendieck, Implementing Lean Software Development: From Concept to Cash, Addison-Wesley Professional, 2006.

[108] M. Poppendieck, T. Poppendieck, Leading Lean Software Development: Results Are Not the Point, Pearson Education, 2009.

[109] M. Poppendieck, T. Poppendieck, The Lean Mindset: Ask the Right Questions, Pearson Education, 2013.

[110] P. Middleton, J. Sutton, Lean Software Strategies: Proven Techniques for Managers and Developers, Productivity Press, 2005.

[111] C. Larman, B. Vodde, Practices for Scaling Lean and Agile Development: Large, Multisite and Offshore Product Development With Large-Scale Scrum, Addison-Wesley Professional, 2010.

[112] D.G. Reinertsen, The Principles of Product Development Flow: Second Generation Lean Product Development, Celeritas Redondo Beach, Canada, 2009.

[113] D.J. Anderson, Kanban, Blue Hole Press, 2010.

[114] P. Middleton, Lean software development: two case studies, Softw. Qual. J. 9 (4) (2001) 241–252.

[115] P. Middleton, A. Flaxel, A. Cookson, Lean software management case study: Timberline Inc, in: Anonymous Extreme Programming and Agile Processes in Software Engineering, Springer, 2005, pp. 1–9.

[116] K. Petersen, Implementing Lean and Agile Software Development in Industry, Blekinge Institute of Technology, 2010. Doctoral Dissertation Series No. 2010:04.

[117] P. Middleton, D. Joyce, Lean software management: BBC worldwide case study, IEEE Trans. Eng. Manag. 59 (1) (2012) 20–32.

[118] M. Mehta, D. Anderson, D. Raffo, Providing value to customers in software development through lean principles, Softw. Process Improve. Pract. 13 (1) (2008) 101–109.

[119] J. Trimble, C. Webster, in: From traditional, to lean, to agile development: finding the optimal software engineering cycle, System Ssciences (HICSS), 2013 46th Hawaii International Conference on, IEEE, 2013, pp. 4826–4833.

[120] P. Rodríguez, K. Mikkonen, P. Kuvaja, M. Oivo, J. Garbajosa, Building lean thinking in a telecom software development organization: strengths and challenges, in: Proceedings of the 2013 international Conference on Software and System Process, ACM, 2013, pp. 98–107.

[121] P. Rodríguez, J. Partanen, P. Kuvaja, M. Oivo, in: Combining lean thinking and agile methods for software development: a case study of a Finnish provider of wireless embedded systems detailed, 2014 47th Hawaii International Conference on System Sciences (HICSS), IEEE, 2014, pp. 4770–4779.

[122] M.O. Ahmad, J. Markkula, M. Oivo, in: Kanban in software development: a systematic literature review, Proceedings of the 39th Euromicro Conference Series on Software Engineering and Advanced Applications (SEAA) Santander, Spain, 2013.

[123] C. Ladas, Scrumban—Essays on Kanban Systems for Lean Software Development, Modus Cooperandi Press, 2009.

[124] T. Sedano, P. Ralph, C. Péraire, in: Software development waste, Proceedings of the 39th International Conference on Software Engineering, 2017, pp. 130–140.

[125] K. Vilkki, H. Erdogmus, Point/counterpoint, IEEE Softw. 29 (5) (2012) 60–63.

[126] J. Bosch, Building products as innovation experiment systems, in: Software Business, Springer, 2012, pp. 27–39.

[127] J. Bosch, Speed, Data, and Ecosystems: Excelling in a Software-Driven World, CRC Press, 2017.

[128] A. MacCormack, How internet companies build software, MIT Sloan Manag. Rev. 42 (2) (2001) 75–78.

[129] H.H. Olsson, J. Bosch, H. Alahyari, in: Towards R&D as innovation experiment systems: a framework for moving beyond agile software development, IASTED Multiconferences–Proceedings of the IASTED International Conference on Software Engineering, SE 2013, 2013, pp. 798–805.

[130] S. Bellomo, R.L. Nord, I. Ozkaya, A study of enabling factors for rapid fielding combined practices to balance speed and stability, in: 35th international Conference on Software Engineering (ICSE), IEEE, 2013, pp. 982–991.

[131] S. Bellomo, R.L. Nord, I. Ozkaya, in: Elaboration on an integrated architecture and requirement practice: prototyping with quality attribute focus, 2013 2nd International Workshop on the Twin Peaks of Requirements and Architecture (TwinPeaks), IEEE, 2013, pp. 8–13.

[132] A.W. Brown, S. Ambler, W. Royce, in: Agility at scale: economic governance, measured improvement, and disciplined delivery, Proceedings of the 2013 International Conference on Software Engineering, IEEE Press, 2013, pp. 873–881.

[133] V. Antinyan, M. Staron, W. Meding, P. Osterstrom, E. Wikstrom, J. Wranker, A. Henriksson, J. Hansson, in: Identifying risky areas of software code in agile/lean software development: an industrial experience report, 2014 SoftWare Evolution Week-IEEE Conference on Software Maintenance, Reengineering and Reverse Engineering (CSMR-WCRE), IEEE, 2014, pp. 154–163.

[134] S. Neely, S. Stolt, in: Continuous delivery? Easy! Just change everything (well, maybe it is not that easy), Agile Conference (AGILE), 2013, IEEE, 2013, pp. 121–128.

[135] A. Nilsson, J. Bosch, C. Berger, Visualizing testing activities to support continuous integration: a multiple case study, in: Agile Processes in Software Engineering and Extreme Programming, Springer, 2014, pp. 171–186.

[136] D.G. Feitelson, E. Frachtenberg, K.L. Beck, Development and deployment at Facebook, IEEE Internet Comput. 17 (4) (2013) 8–17.

[137] M. Marschall, in: Transforming a six month release cycle to continuous flow, Agile Conference (AGILE), 2007, IEEE, 2007, pp. 395–400.

[138] L. Guzmán, M. Oriol, P. Rodríguez, X. Franch, A. Jedlitschka, M. Oivo, in: How can quality awareness support rapid software development? A research preview, International Working Conference on Requirements Engineering: Foundation for Software Quality, Springer, 2017, pp. 167–173.

[139] X. Franch, C. Ayala, L. López, S. Martínez-Fernández, P. Rodríguez, C. Gómez, A. Jedlitschka, M. Oivo, J. Partanen, T. Räty, V. Rytivaara, in: Data-driven requirements engineering in agile projects: the Q-rapids approach, Proceedings of the 2nd International Workshop on Just-in-Time Requirements Engineering: Dealing With Non-functional Requirements in Agile Software Development, Lisbon, Portugal, 2017, 2017, pp. 1–4.

[140] U. Eklund, J. Bosch, in: Architecture for large-scale innovation experiment systems, 2012 Joint Working IEEE/IFIP Conference on Software Architecture (WICSA) and European Conference on Software Architecture (ECSA), IEEE, 2012, pp. 244–248.

[141] R. Benefield, in: Agile deployment: lean service management and deployment strategies for the SaaS enterprise, 42nd Hawaii International Conference on System Sciences, 2009, HICSS'09, IEEE, 2009, pp. 1–5.

[142] F. Fagerholm, A.S. Guinea, H. Mäenpää, J. Münch, in: Building blocks for continuous experimentation, Proceedings of the 1st International Workshop on Rapid Continuous Software Engineering (RCoSE 2014), Hyderabad, India, 2014.

[143] A. Goel, B. Chopra, C. Gerea, D. Mátáni, J. Metzler, F. Ul Haq, J. Wiener, in: Fast database restarts at Facebook, Proceedings of the 2014 ACM SIGMOD International Conference on Management of Data, ACM, 2014, pp. 541–549.

[144] P. Agarwal, in: Continuous Scrum: agile management of SaaS products, Proceedings of the 4th India Software Engineering Conference, ACM, 2011, pp. 51–60.

[145] J. Humble, C. Read, D. North, in: The deployment production line, Agile Conference, 2006, IEEE, 2006, p. 6.

[146] L.E. Lwakatare, P. Kuvaja, M. Oivo, C. Lassenius, T. Dingsøyr, M. Paasivaara (Eds.), Dimensions of DevOps, 16th international Conference on Agile Software Development (XP), vol. 212, Springer International Publishing, Cham, 2015, pp. 212–217.

[147] L. Bass, I. Weber, L. Zhu, DevOps: A Software Architect's Perspective, Addison-Wesley Professional, 2015.

[148] M. Callanan, A. Spillane, DevOps: making it easy to do the right thing, IEEE Softw. 33 (3) (2016) 53–59.

[149] R. Penners, A. Dyck, Release Engineering vs. DevOps—An Approach to Define Both Terms, Full-Scale Software Engineering, 2015. Retrieved from, https://www2.swc.rwth-aachen.de/docs/teaching/seminar2015/FsSE2015papers.pdf#page=53.

[150] J. Smeds, K. Nybom, I. Porres, in: DevOps: a definition and perceived adoption impediments, 16th International Conference on Agile Software Development (XP), Springer International Publishing, Helsinki, 2015, pp. 166–177.

[151] J. Humble, J. Molesky, Why enterprises must adopt DevOps to enable continuous delivery, Cutter IT J. 24 (8) (2011) 6–12.

[152] P. Debois, in: Agile infrastructure and operations: how infra-gile are you?, Agile 2008 Conference, IEEE, 2008, pp. 202–207. https://doi.org/10.1109/Agile.2008.42.

[153] J. Iden, B. Tessem, T. äivärinta, Problems in the interplay of development and IT operations in system development projects: a Delphi study of Norwegian IT experts, Inf. Softw. Technol. 53 (4) (2011) 394–406.

[154] A. Elbanna, S. Sarker, The risks of agile software development: learning from adopters, IEEE Softw. 33 (5) (2016) 72–79.

[155] O. Gotel, D. Leip, Agile software development meets corporate deployment procedures: stretching the agile envelope, in: Agile Processes in Software Engineering and Extreme Programming, Springer, Berlin, Heidelberg, 2007, pp. 24–27.

[156] B. Tessem, J. Iden, in: Cooperation between developers and operations in software engineering projects, Proceedings of the 2008 International Workshop on Cooperative and Human Aspects of Software Engineering, 2008.

[157] J. Cito, P. Leitner, T. Fritz, H.C. Gall, in: The making of cloud applications: an empirical study on software development for the cloud, Proceedings of the 2015 10th Joint Meeting on Foundations of Software Engineering—ESEC/FSE 2015, 2015, pp. 393–403.

[158] M. Shahin, M. Zahedi, M.A. Babar, L. Zhu, in: Adopting continuous delivery and deployment: impacts on team structures, collaboration and responsibilities, 21st Evaluation and Assessment in Software Engineering Conference (EASE), 2017.

[159] T. Schneider, in: Achieving cloud scalability with microservices and DevOps in the connected car domain, CEUR Workshop Proceedings on Continuous Software Engineering, 2016, pp. 138–141. CEUR-WS.org. Retrieved from, http://ceur-ws.org/Vol-1559/.

[160] K. Nybom, J. Smeds, I. Porres, On the Impact of Mixing Responsibilities Between Devs and Ops, Springer; Cham, 2016, pp. 131–143.

[161] L. Chen, Continuous delivery: overcoming adoption challenges, J. Syst. Softw. 128 (2017) 72–86. https://doi.org/10.1016/j.jss.2017.02.013.

[162] F. Elberzhager, T. Arif, M. Naab, I. Süß, S. Koban, From Agile Development to DevOps: Going Towards Faster Releases at High Quality—Experiences from an Industrial Context, Springer, 2017, pp. 33–44.

[163] W. Hummer, F. Rosenberg, F. Oliveira, T. Eilam, Testing idempotence for infrastructure as code, in: ACM/iFIP/USENIX International Conference on Distributed Systems Platforms and Open Distributed Processing, Springer, Berlin, Heidelberg, 2013, pp. 368–388.

[164] A. Balalaie, A. Heydarnoori, P. Jamshidi, Microservices architecture enables DevOps: migration to a cloud-native architecture, IEEE Softw. 33 (3) (2016) 42–52.

[165] R. Heinrich, A. van Hoorn, H. Knoche, F. Li, L.E. Lwakatare, C. Pahl, J. Wettinger, in: Performance engineering for microservices: research challenges and directions, Proceedings of the 8th ACM/SPEC on International Conference on Performance Engineering Companion, ACM, 2017, pp. 223–226.

[166] C. Tang, T. Kooburat, P. Venkatachalam, A. Chander, Z. Wen, A. Narayanan, R. Karl, in: Holistic configuration management at Facebook, Proceedings of the 25th Symposium on Operating Systems Principles—SOSP '15, 2015, pp. 328–343.

[167] J. Cito, F. Oliveira, P. Leitner, P. Nagpurkar, H.C. Gall, in: Context-based analytics: establishing explicit links between runtime traces and source code, Proceedings of the 39th International Conference on Software Engineering: Software Engineering in Practice Track, IEEE Press, 2017, pp. 193–202.

[168] T. Laukkarinen, K. Kuusinen, T. Mikkonen, in: DevOps in regulated software development: case medical devices, Proceedings of the 39th International Conference on Software Engineering: New Ideas and Emerging Results Track, 2017, pp. 15–18.

[169] L.E. Lwakatare, T. Karvonen, T. Sauvola, P. Kuvaja, J. Bosch, H.H. Olsson, M. Oivo, in: Towards DevOps in the embedded systems domain: why is it so hard?, 49th Hawaii International Conference on Systems Science, IEEE, 2016, pp. 5437–5446.

[170] S. Jones, J. Noppen, F. Lettice, in: Management challenges for DevOps adoption within UK SMEs, Proceedings of the 2nd International Workshop on Quality-Aware DevOps—QUDOS 2016, 2016, pp. 7–11.

[171] M. Shahin, in: Architecting for devops and continuous deployment, Proceedings of the ASWEC 2015 24th Australasian Software Engineering Conference, ACM, 2015, pp. 147–148.

[172] E. Ries, The Lean Startup: How today's Entrepreneurs Use Continuous Innovation to Create Radically Successful Businesses, Random House LLC, 2011.

[173] P. Seppänen, N. Tripathi, M. Oivo, K. Liukkunen, in: How are product ideas validated? The process from innovation to requirement engineering in software startups, 8th International Conference on Software Business (ICSOB), 2017.

[174] S.S. Bajwa, X. Wang, A.N. Duc, P. Abrahamsson, "Failures" to be celebrated: an analysis of major pivots of software startups, Empir. Softw. Eng. 22 (5) (2017) 2373–2408.

[175] M. Steinert, L.J. Leifer, Finding One's Way': re-discovering a hunter–gatherer model based on wayfaring, Int. J. Eng. Educ. 28 (2) (2012) 251.

[176] A. Nguyen-Duc, P. Seppänen, P. Abrahamsson, in: Hunter–gatherer cycle: a conceptual model of the evolution of software startups, Proceedings of the 2015 International Conference on Software and System Process, ACM, 2015, pp. 199–203.

[177] J. Bosch, H.H. Olsson, J. Björk, J. Ljungblad, The early stage software startup development model: a framework for operationalizing lean principles in software startups, in: Lean Enterprise Software and Systems, Springer, Berlin, Heidelberg, 2013, pp. 1–15.

[178] J. Björk, J. Ljungblad, J. Bosch, in: Lean product development in early stage startups, IW-LCSP@ ICSOB, 2013, pp. 19–32.

[179] P. Seppänen, M. Oivo, K. Liukkunen, in: The initial team of a software startup, narrow-shouldered innovation and broad-shouldered implementation, 22nd ICE/IEEE International Technology Management Conference, 2016.

[180] R.M. Müller, K. Thoring, Design thinking vs. lean startup: a comparison of two user-driven innovation strategies, in: international Design Management Research Conference, Boston, MA, 2012, pp. 151–161.

[181] L. Hokkanen, K. Kuusinen, K. Väänänen, Minimum viable user experience: a framework for supporting product design in startups, in: international Conference on Agile Software Development, Springer, Cham, 2016, pp. 66–78.

[182] L. Hokkanen, K. Kuusinen, K. Väänänen, in: Early product design in startups: towards a UX strategy, International Conference on Product-Focused Software Process Improvement, Springer International Publishing, 2015, pp. 217–224.

[183] L. Hokkanen, M. Leppänen, in: Three patterns for user involvement in startups, Proceedings of the 20th European Conference on Pattern Languages of Programs, ACM, 2015, p. 51.

[184] Y. Guo, R.O. Spínola, C. Seaman, Exploring the costs of technical debt management—a case study, Empir. Softw. Eng. 21 (1) (2016) 159–182.

[185] W. Cunningham, The WyCash portfolio management system. SIGPLAN OOPS Mess. 4 (2) (1992) 29–30, https://doi.org/10.1145/157710.157715.

[186] F. Shull, Perfectionists in a world of finite resources, IEEE Softw. 28 (2) (2011) 4–6.

[187] E. Tom, A. Aurum, R. Vidgen, An exploration of technical debt, J. Syst. Softw. 86 (6) (2013) 1498–1516.

[188] P. Kruchten, Refining the definition of technical debt, https://philippe.kruchten.com/2016/04/22/refining-the-definition-of-technical-debt/, 2016. Accessed 29 June 2017.

[189] E. Lim, N. Taksande, C. Seaman, A balancing act: what software practitioners have to say about technical debt, IEEE Softw. 29 (6) (2012) 22–27.
[190] M. Fowler, Technical debt quadrant, http://martinfowler.com/bliki/TechnicalDebt Quadrant.html, 2009. Accessed 29 June 2017.
[191] T. Mens, T. Tourwé, A survey of software refactoring, IEEE Trans. Softw. Eng. 30 (2) (2004) 126–139.
[192] P. Abrahamsson, M.A. Babar, P. Kruchten, Agility and architecture: can they coexist? IEEE Softw. 27 (2) (2010) 16–22.
[193] R.L. Nord, I. Ozkaya, R.S. Sangwan, Making architecture visible to improve flow management in lean software development, IEEE Softw. 29 (5) (2012) 33–39.
[194] R.L. Nord, I. Ozkaya, P. Kruchten, M. Gonzalez-Rojas, in: Joint Working IEEE/IFIP Conference on Software Architecture (WICSA) and European Conference on Software Architecture (ECSA), IEEE, 2012, pp. 91–100.
[195] J.L. Letouzey, in: The SQALE method for evaluating technical debt, 2012 Third International Workshop on Managing Technical Debt (MTD), IEEE, 2012, pp. 31–36.
[196] P.S.M. dos Santos, A. Varella, C.R. Dantas, D.B. Borges, Visualizing and managing technical debt in agile development: an experience report, in: international Conference on Agile Software Development, Springer, Berlin, Heidelberg, 2013, pp. 121–134.
[197] M. Kaiser, G. Royse, in: Selling the investment to pay down technical debt: the code Christmas tree, Agile Conference (AGILE), 2011, IEEE, 2011, pp. 175–180.
[198] K. Power, in: Understanding the impact of technical debt on the capacity and velocity of teams and organizations: viewing team and organization capacity as a portfolio of real options, 2013 4th International Workshop on Managing Technical Debt (MTD), IEEE, 2013, pp. 28–31.
[199] G. Hanssen, A.F. Yamashita, R. Conradi, L. Moonen, in: Software entropy in agile product evolution, 2010 43rd Hawaii International Conference on System Sciences (HICSS), IEEE, 2010, pp. 1–10.

ABOUT THE AUTHORS

Pilar Rodríguez (PhD, Software Engineering) is a postdoctoral researcher of the M3S research unit at the University of Oulu (Finland). Prior to moving to Finland, she was a research assistant at the System and Software Technology Group, Technical University of Madrid (Spain). Her research interests include empirical software engineering with a focus on software processes, agile and lean software development, value-based software engineering, and software quality. She is a member of the Review Board of Empirical Software Engineering (2015, 2016, 2017), and has served as reviewer in leading academic forums in Software Engineering (e.g., TSE, EMSE, IST, IEEE Software, and ESEM).

Mika Mäntylä is a professor of Software Engineering at the University of Oulu, Finland. He received a D.Sc. degree in 2009 in software engineering from the Helsinki University of Technology, Finland. His research interests include empirical software engineering, software testing, software maintenance, mining software repositories, and behavioral software engineering. He has previously worked as a postdoc at the Lund University, Sweden and as an assistant professor at the Aalto University, Finland. His studies have appeared in journals such as IEEE Transaction on Software Engineering, IEEE Software, Empirical Software Engineering, and Information and Software Technology. For more information http:// mikamantyla.eu/.

Markku Oivo (PhD, eMBA) is professor and head of the M3S research unit at the University of Oulu, Finland. During 2000–2002 he was Vice President and director of R&D at Solid Co. He held several positions at VTT in 1986–2000. He had visiting positions at the University of Maryland (1990–91), Schlumberger Ltd. (Paris 1994–95), Fraunhofer IESE (1999–2000), University of Bolzano (2014–15), and Universidad Politécnica de Madrid (2015). He has worked at Kone Co. (1982–86) and at the University of Oulu (1981–82). He has initiated and managed 100+ projects with tens of millions of euros funding. He has 150+ international publications.

Lucy Ellen Lwakatare, PhD, is a researcher of the M3S research unit at the University of Oulu. She received her doctoral degree in the field of science in 2017 at the same university with a major in Information Processing Science. Her research interests include (but not limited to): Portfolio management of software product and services, software development processes and approaches specifically DevOps and Continuous Deployment. She is currently conducting research on DevOps software engineering approach.

Pertti Seppänen (MSc., 1993 Electrical Engineering) is a PhD student at the Empirical Software Engineering in Software, Systems and Services (M3S) research unit of the University of Oulu (Finland). Prior to joining the M3S unit he did over 30 years' career in ICT industry in different managerial positions. His research interests include early-stage software startups, entrepreneurship, human resources in enterprises, innovation, software startup processes and practices.

Pasi Kuvaja is a professor of Software Engineering at the University of Oulu, Finland. He received PhD degree in 2012 in software engineering in the University of Oulu, Finland. He is also Vice Head of M3S Research Unit in Faculty of ITEE at the University of Oulu. His research interests include empirical software engineering, software processes, software development approaches like DevOps and continuous deployment, and software product line engineering. He has long history of cooperation with European industry in numerous European research projects (EU framework, Eureka, etc.). He has worked as director of BOOSTRAP institute and work package leader in SPICE project. His studies have been published in journals such as IEEE Software, Software Process, Software Quality, Software and Knowledge Engineering, Systems Architecture, and Information and Software Technology.

CHAPTER FIVE

Advances in Symbolic Execution

Guowei Yang*, Antonio Filieri†, Mateus Borges†, Donato Clun†, Junye Wen*

*Texas State University, San Marcos, TX, United States
†Imperial College London, London, United Kingdom

Contents

1.	Introduction	226
2.	Background	228
3.	Constraint Solving	230
	3.1 Simplification, Reuse, and Caching	231
	3.2 Strings	232
	3.3 Nonlinear Constraints and Bitvectors	233
	3.4 Other Theories	235
4.	Path Explosion	236
	4.1 Heuristics-Guided Path Exploration	236
	4.2 Pruning Paths	239
	4.3 Merging States or Paths	240
	4.4 Handling Loops	242
	4.5 Parallel Exploration	243
5.	Compositional Analysis	244
	5.1 Improving Compositional Symbolic Execution	244
	5.2 Applying Compositional Symbolic Execution	245
6.	Memory Modeling	246
	6.1 General Advancements	247
	6.2 Managing Heap Input	248
	6.3 Handling Symbolic Memory Access	249
	6.4 Advancements in Static Analysis	249
7.	Concurrency	250
	7.1 Detecting Concurrency Bugs	251
	7.2 Improving Scalability	252
8.	Test Generation	253
	8.1 Finding Bugs and Improving Coverage	254
	8.2 Regression Testing, Data Flow Testing, and Load Testing	255
	8.3 Testing Nontraditional Programs	257
	8.4 Test Optimization	259
9.	Security	259
	9.1 Hybrid Techniques	259
	9.2 Embedded Systems	260

Advances in Computers, Volume 113
ISSN 0065-2458
https://doi.org/10.1016/bs.adcom.2018.10.002

9.3   Underconstrained Symbolic Execution   261
9.4   Ad Hoc Solutions and Enhancements   262
10. Probabilistic Symbolic Execution   264
10.1 Statistical Symbolic Execution   266
10.2 Beyond Linear Integer Arithmetics   267
11. Tools Support   269
12. Conclusion   271
References   271
About the Authors   286

Abstract

Symbolic execution is a systematic technique for checking programs, which forms a basis for various software testing and verification techniques. It provides a powerful analysis in principle but remains challenging to scale and generalize symbolic execution in practice. This chapter reviews the cutting-edge research accomplishments in addressing these challenges in the last 5 years, including advances in addressing the scalability challenges such as constraint solving and path explosion, as well as advances in applying symbolic execution in testing, security, and probabilistic program analysis.

1. INTRODUCTION

Symbolic execution is a program analysis technique enabling the exploration and summarization of a large number of execution paths by replacing program inputs with symbolic parameters and studying the conditions on these parameters that determine the execution of each element of the program [1].

A symbolic execution engine implements two complementary features. First, it maintains a symbolic representation of the program state, which is updated after the execution of each instruction according to the semantics of the programming language used to implement the program; depending on the language, the program state may be include complex information, like a symbolic representation of the heap or the state of multiple threads and their scheduler. Second, it uses a constraint solver to determine if the evaluation of a conditional branch in the current symbolic state can be satisfied; this is critical to decide if there exists a concrete assignment to the symbolic variables of a program that can actually exercise the current execution path. Both these features place critical challenges to the applicability of symbolic execution in practice.

In this chapter, we review the main results in symbolic execution achieved in the last 5 years. By focusing on the current edge of research, we aim at providing an overview of the latest challenges elicited by the research community to improve the scalability and generality of symbolic execution.

After introducing the necessary definitions and background in Section 2, we report on recent results in constraint solving that directly enabled significative advance in symbolic execution. These include both reasoning techniques for domains and theories specifically relevant for symbolic execution, e.g., strings or nonlinear numerical constraints, and techniques aiming at improving the practical scalability of constraint solver, from caching and reuse to heuristics enabling faster solutions for specific constraint patterns arising in symbolic execution.

We discuss the problem of *path explosion* and the most recent mitigation strategies in Section 4. These strategies include, among the others, the selective concretization of certain execution paths, the use of loop summarization to prevent unnecessary loop unwinding of loop, and the use of directed and incremental search strategies to maximize the exploration effectiveness of partial symbolic execution given the available computational resources and time.

Section 5 reviews recent advance in compositional symbolic execution techniques, which allow a principled trade-off between the accuracy of local analyses and the reuse of partial results, which may also enable the parallelization of several analyses.

Symbolic data structures and heap representations are reviewed in Section 6, where we overview both the formalization of heap objects' structure and properties for automatic reasoning and the challenges of capturing the semantics of memory accesses via symbolic object references.

Section 7 concludes the overview of the main across-the-board challenges of symbolic execution by discussing the recent advances on modeling concurrency, both for detecting concurrency bugs and for improving the scalability of symbolic execution engines in presence of the combinatorial explosion of execution paths due to the interleaving of multiple threads.

Sections 8–10 present the recent advances in three application areas of symbolic execution that have seen significant advances in the last 5 years. In particular, Section 8 discusses how automated test case generation and test suite optimization benefited from advances in constraint solving, hybrid concrete/symbolic execution, directed symbolic execution, and the symbolic modeling of nonstandard programming artifacts like databases.

Section 9 presents a set of symbolic/hybrid execution techniques for the detection, automatic exploitation, and automated repair of security vulnerabilities. Finally, Section 10 summarizes a new research thread on probabilistic symbolic execution, which extends classic symbolic execution to quantify the probability of executing each program element instead of simply assessing the possibility of such element being executed. This paves the way to new applications of symbolic execution for the analysis of quantitative program properties, including performance, information leakage, or reliability.

Finally, in Section 11 we include a list of currently maintained symbolic execution tools, sketching for tool the languages it supports and the problems it is optimized for. We conclude with some final remarks in Section 12.

2. BACKGROUND

Symbolic execution [1, 2] is a powerful program analysis technique for systematic exploration of a large number of program execution paths. It provides a basis for various software testing and verification techniques. The key idea is to use symbolic values in place of concrete values as inputs to execute the program, and to compute resulting output as a function of the symbolic inputs.

A symbolic program state includes the (symbolic) values of program variables and a path condition (PC). The path condition is a (quantifier free) Boolean formula over the symbolic inputs, collecting constraints on the inputs in order for an execution to follow the associated path. Path conditions are checked for satisfiability using off-the-shelf decision procedures [3] during symbolic execution; each time the path condition is updated, it is checked to determine the feasibility of the path; if a path condition becomes unsatisfiable, which means the corresponding path is infeasible, symbolic execution stops exploration of that path and backtracks. A *symbolic execution tree* characterizes all the paths explored during the symbolic execution. Each node represents a symbolic program state and each arc represents a transitions between two states.

We illustrate symbolic execution with the example program shown in Fig. 1, where the method compute has three integer inputs: curr (current), thresh (threshold), and step; it calculates the relationship between the current and the threshold, in increments given by the step value. Its corresponding symbolic execution tree is shown in Fig. 2. The path

```
1 int compute(int curr, int thresh, int step){
2   int delta = 0;
3   if (curr < thresh){
4     delta = thresh - curr;
5     if ((curr + step) < thresh)
6       return -delta;
7     else
8       return 0;
9   } else {
10    int counter = 0;
11    while (curr >= thresh) {
12      curr = curr - step;
13      counter++;
14    }
15    return counter;
16  }
17 }
```

Fig. 1 Example program.

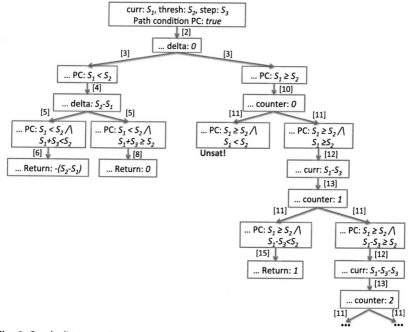

Fig. 2 Symbolic execution tree.

condition PC is initialized as *true*, and the three input variables curr, thresh, and step have symbolic values S_1, S_2, and S_3, respectively. Program variables are then represented by expressions in terms of these symbolic inputs; e.-g., after executing statement 4, delta becomes $S_2 - S_1$. At each branch point, there is a *choice* in the execution and *PC* is updated with assumptions about the inputs, to choose between alternative paths. For example, after statement 3 is executed, both `then` and `else` alternatives of the `if` statement are possible, and *PC* is updated accordingly and checked for satisfiability. If *PC* becomes false, which means there are no inputs to satisfy it and thus the state is unreachable, symbolic execution does not continue for that path. For instance, when the `while` statement at line 11 is executed the first time, the *PC* corresponding to exiting the loop is unsatisfiable, and symbolic execution does not continue on that path. The *PC*s for the explored program paths during symbolic execution can be solved by a constraint solver and the solutions can be formed as test inputs—the execution of the program on these concrete inputs will follow the same path as the symbolic execution.

For programs with loops or recursion, symbolic execution may result in an infinite symbolic execution tree. For example, in Fig. 2, the expansion of the right-most leaf in the tree may continue forever. To address this problem, a limit (e.g., a depth bound or a time bound) is typically put on the search for symbolic execution.

Classic symbolic execution statically analyzes programs without any concrete executions, and in practice it can be challenging to apply it, for example, for programs with invocations to untraceable libraries/native code or with constraints that are hard to solve. Dynamic symbolic execution [4–7], which is also called concolic execution, combines concrete executions and symbolic analysis to mitigate this problem. It starts with a concrete execution on some given or random inputs and collects the *PC* along the executed path at the same time. The constraints in the *PC* can be negated so that the resulting *PC*s are solved by a constraint solver to find inputs to explore alternative paths. This process is repeated until the desired testing coverage has been achieved.

3. CONSTRAINT SOLVING

Symbolic execution relies on *constraint solvers* to decide the satisfiability of path conditions and thus avoid exploring unfeasible paths. Although satisfiability is a hard (NP-complete) [8] problem, algorithmic and engineering advances in constraint solving over the last few decades made it more manageable. One of the main developments is the creation of efficient *Satisfiability Modulo Theories* (SMT) solvers. SMT solvers combine together specialized decision procedures for specific *theories* (i.e., a well-defined interpretation of a set of symbols in a first-order logic formula) with a core SAT solver. This combination allows to express and decide constraints that depend on multiple theories. Developers of symbolic execution engines must choose carefully the theory, and consequently the solver, used to model and decide the constraints encountered during the exploration of the program. For example, the code in Fig. 1 only performs addition and subtraction over the integers, which can be handled by the theory of linear integer arithmetic. However, this representation may miss edge-case behaviors like arithmetic overflows since ints are represented in hardware as 32-bit values. Modeling the constraints in the theory of *bitvectors* will ensure that the formula semantics will match the actual execution in the hardware.

3.1 Simplification, Reuse, and Caching

Symbolic execution of large and/or complex programs will produce similarly large/complex constraints, which may take a long time to be solved. The solving cost may be ameliorated by splitting the constraint into independent subproblems that can be analyzed faster and independently. Visser et al. [9] go a step further with Green, a framework designed to reduce the number of constraint solver/model counter calls by reusing previous results across multiple symbolic execution runs. Green maintains an in-memory database with the constraint solving results. Incoming PCs will be sliced into independent subproblems, which are transformed (canonicalized) into a normal form to increase reuse opportunities. The canonicalized subproblems that have never been seen before are solved using off-the-shelf solvers, and their solution is stored on the database; otherwise Green will return the respective stored solution. Jia et al. [10] improve Green in the context of linear integer arithmetic by taking in account *logical implications* between constraints. For example, a solution for the constraint $x > 2$ is also a solution for $x! = 0$, which means that the second formula is redundant and can be discarded if both formulas appear in the same PC. Aquino et al. [11] leverage a canonical matrix representation of linear integer arithmetic constraints to build an efficient search index of solutions for equivalent constraints.

Romano and Engler [12] propose an expression optimizer for bitvector constraints that can learn reduction rules automatically. The optimizer learns new rules in two stages: first, it looks for possibly equivalent constraints using the hash of a sequence of evaluations with specific values. To further improve the performance, the PCs are also split into distinct sets depending on the number/width of the variables. Later, the equivalence map is refined into a set of reduction rules: expressions are simplified and canonicalized, rules that do not reduce the size of the constraint are discarded, and finally equivalence is confirmed using a SMT solver. Lloyd and Sherman [13] simplify path conditions by removing redundant inequalities through polyhedral algorithms, but the approach is limited to linear integer constraints.

Zhang et al. [14, 15] propose a simple approach to reduce the number of calls to constraint solvers, *speculative symbolic execution*. As the name says, speculative symbolic execution delays the feasibility check for the current branch until a (user-defined) number of new clauses is added to the path condition. If the current branch is not satisfiable, the execution backtracks to the last feasible branch, otherwise the execution progresses normally. The authors also propose an optimization to improve performance in scenarios with lots of unfeasible branches: given a reachable branch, if the constraint solver returns UNSAT for one of its children, the other must be satisfiable. Kausler and Sherman [16] propose user-defined backtracking policies for symbolic execution. The rationale is that default backtrack policies, such as a fixed timeout on a solver call, might waste resources due to its strictness. The authors suggest that more flexible policies, like timing out solver calls that exceed the average of previous calls by a certain factor, can lead to improved exploration coverage.

3.2 Strings

Symbolic execution of user-facing applications, such as web apps, frequently requires the ability to reason over string constraints. Unfortunately, the general theory of strings is undecidable [17]; current research efforts focus on discovering theory fragments which may have efficient decision procedures. Early approaches used either automatons to model and decide standard regular expression constructs (e.g., concatenation/union/Kleene star) [18], or encoded string constraints to bitvector formulas [19]. The latter approach can handle mixed integer-string formulas (e.g., using their operator `length` or `indexOf`), but it must decide the size of the final string before attempting to solve it.

The lack of a general theory for strings also poses modeling challenges for symbolic execution: Kausler and Sherman [20] showed that existing string solvers cannot model precisely many of the string operations available in traditional programming languages (e.g., Java or Python). Furthermore, the authors evaluated the modeling cost, performance, and accuracy of a selection of string solvers. The experiment used a set of PCs extracted (using dynamic symbolic execution) from the execution of the test suite of string-manipulating applications. Results show that the performance/accuracy/modeling cost varies between solvers, and as such users must consider the characteristics of the expected path conditions before choosing a solver.

Zheng et al. [21] proposed Z3-str, an extension to the Z3 SMT solver capable of solving mixed string constraints. As opposed to previous bitvector-based string solvers, Z3-str treats strings as a primitive datatype, and thus it does not need to decide the length of the string before solving. This is made possible by the Z3 plugin system: the core modules (DPLL (T) and congruency closure engines) handle the exploration of the formula and the coordination between the specific theory solvers. Trinh et al. [22] extends the S3 solver [23] with a *progressive* search algorithm for the fragment theory of strings with recursive replacement, length and Kleene Star. The impact of nontermination due to the recursive nature of the operations is mitigated by pruning the recursion tree when the current branch implies that a shorter subtree is satisfiable. The authors also implemented a variant of conflict-driven clause learning (CDCL) for string constraints.

Both Liang et al. [24] and Abdulla et al. [25, 26] propose new solvers for the fragment theory of unbounded strings with regular language membership and length. The decidability of this fragment is an open problem and thus termination cannot be ensured. However, both authors claim that a practical solver for the fragment can be built. Liang's solver is a proof procedure for a nondeterministic proof calculus, created by combining a off-the-shelf linear integer solver and a EUF (theory of equality plus uninterpreted functions) solver that has been extended to support string and regular language constraints. This allows the solver to be integrated with the DPLL(T) framework of SMT solvers, and not just be a plugin that discharges to other theories. Nevertheless, the procedure was implemented in CVC4 and benchmarked favorably against well-known string solvers; it returns correct solutions very quickly for most of the problems, while it timeouts for a larger share in comparison with Z3-str. Abdulla's approach is very similar: the work introduces a new logic for strings together with a proof calculus based on a set of inference rules. The proof procedure is guaranteed to terminate if the formula is *acyclic*, that is, each variable appears only once in an equality.

3.3 Nonlinear Constraints and Bitvectors

Nonlinear constraints are ubiquitous in software that makes use of complex mathematical operations, such as hybrid control systems and the aerospace domain. In the worst case, nonlinear constraint over finite-width values (e.g., IEEE754 floats) can be modeled as bitvector operations: the constraints can be converted (i.e., *bitblasted*) to a set of propositional formulas that match the binary manipulations done in a CPU.

Borges et al. [27] propose a solver for complex mathematical constraints over floating-point variables that combines metaheuristic search algorithms and interval constraint propagation. Infeasible regions of the domain are removed with interval constraint propagation, which improves the quality of the initial (*seed*) candidate solutions. Experimental results show that the combination can increase the chances of solving a constraint for a small time overhead for the ICP calls. Dinges and Agha utilize a similar approach in [28]: the algorithm, *Concolic Walk*, removes infeasible regions of the domain that are outside of the polyhedra formed by the subset of linear clauses of the constraint. Afterward, Concolic Walk uses a combination of tabu search algorithm and root-finding numerical methods to find to find solutions inside the polyhedra.

Bagnara et al. introduced FPSE, a Constraint Programming (CP) solver for constraints over floating-point numbers and the four basic arithmetic operations [29]. The performance of CP tools is directly related to how well their filtering algorithms can eliminate infeasible points from the domain. FPSE improves upon previous CP approaches by reformulating existing filtering algorithms to support multiplication and division, as well as implementing several heuristics to speed up convergence.

Fu and Su [30] propose XSat, a solver that models the solving process as a mathematical optimization problem. XSat builds a function F from the original constraint C such that finding the minimum point of F is equivalent to solving C. Informally, F measures how far a model is from a potential solution. The "representing function" F is translated into a C program, and the optimization process is performed using Markov-Chain Monte Carlo. The tool stops if the optimum is not found after a certain (user-specified) number of iterations. XSat is capable of handling both basic arithmetic and arbitrary floating-point functions/rounding modes. Experiments show that the approach is faster than traditional SMT solvers (Z3/MathSAT) and existing metaheuristic approaches [27]. A similar approach using machine learning optimization algorithms is proposed by Li et al. [31].

Tiwari and Lincoln [32] propose a pseudo decision procedure for nonlinear arithmetic constraints over the reals. The procedure always returns correct results (i.e., it is a decision procedure) if the constraint is a conjunction of *multilinear polynomial equations*, e.g., $xy + 4y + 1$. It works by decomposing subsets of clauses of the constraint into a matrix form where one of the variables must be one of the eigenvalues of a constant matrix. This rewriting rule is applied recursively for each variable; if the constraint is fully decomposed, it can be decided if a model exists. Otherwise, the procedure returns unknown. To support more complex polynomials, the authors implement a few (equisatisfiable) transformations.

Hadarean et al. [33] investigated the impact of delaying bitblasting when solving bitvector formulas. According to Hadarean, most bitvector solvers attempt to simplify the formula first, and then bitblast it. This *eager* approach discards information that might be used during solving because the word-level structure is lost after preprocessing the formula, thus denying any opportunities for cooperation with solvers for other theories. The authors propose a *lazy* approach: before bitblasting, other algebraic solvers specialized in different fragments of the bitvector theory, are called in sequence (from the least to the most expensive) to analyze the formula. Experiments show that the lazy approach is complementary to the eager one: problems that are easily solved by the lazy solver take a long time with the eager solver, and vice versa.

3.4 Other Theories

Separation logic [34] is a novel, nonclassical logic, that gained a lot of popularity in recent years due to its power to reason about heap-manipulating programs. However, separation logic requires specialized decision procedures that are not easily integrable into existing tools. Piskac et al. [35] attack this problem by proposing a novel fragment of first-order logic, called the "logic of graph reachability and stratified sets (GRASS)," to which separation logic formulas can be reduced to in linear time. GRASS can be decided by existing SMT solvers, which allows all the reasoning tasks to be performed at the solver level and thus leverage existing optimizations and theories.

Daca et al. [36] propose "Arrays-Fold Logic" (AFL), an extension of quantifier-free array integer theory that is able to express *counting* constraints, such as "the number of elements smaller than x is the same in arrays $A1$ and $A2$." Applications of AFL include summarizing loops with branches, checking whether the histogram of an array content matches a discrete distribution, and speeding concolic execution by encoding the statements of a parser as AFL formulas. The decision procedure for AFL encodes the formulas as symbolic counter machines. Decidability for those machines reduce to satisfiability of existential Presburger arithmetic.

Bansal et al. [37] leverage SMT solvers to decide constraints that use domain-specific user-defined axioms (i.e., *local theory extensions*). Developers can specify those axioms to "encode" domain-specific properties about theories that are not supported by the host SMT solver. However, this is done through the use of universal quantifiers; since SMT solvers use incomplete heuristics to guide instantiation, they may never know when to stop enumerating new instances. Furthermore, those heuristics are not easily customizable; developers must encode their (potentially complete) strategy by preprocessing the input file. Bansal shows that if the axiom/theory extension is local, i.e., you only need to check instances with ground terms that appear in the formula, a decision procedure for this extension can be generated in existing SMT solvers by leveraging E-matching algorithms.

Cristi and Rossi [38] propose a new decision procedure for sets and binary relations. According to the authors, existing decision procedures for set theory are incomplete because they rely on a "lossy" encoding to other theories, such as arrays or predicate calculus. The proposed solver, SAT_{br}, implements a complete decision procedure by attempting to reduce the input formula to an equisatisfiable *solved form*. The reduction rules are designed to eliminate problematic expressions, such as constraints over the range of relations, that would result in nontermination. The solver applies the reduction rules nondeterministically until a fixpoint is reached: the result will be either *false* if the original formula is not satisfiable, or its solved form which is guaranteed to be satisfiable.

4. PATH EXPLOSION

Symbolically analyzing all feasible paths cannot scale to large and complex programs, since the number of feasible paths grows exponentially with the increase of branching instructions and can even be infinite in the case of programs with loops or recursion. Therefore, path explosion is a key problem to address in symbolic execution.

4.1 Heuristics-Guided Path Exploration

Many approaches have been proposed to mitigate path explosion by exploring paths of interest guided by heuristics. One approach was introduced by Li et al. [39] to steer symbolic execution to less traveled paths. The main idea is to exploit "length-n subpath program spectra" to systematically approximate full path information for guiding path exploration. In particular, it uses frequency distributions of explored length-n subpaths to prioritize "less traveled" parts of the program to improve test coverage and error detection. This technique recursively update the frequency of traveling through subpaths of a certain length, and the paths traveled less frequently will be explored first as it is tend to be more possible for a potential problem to exist.

Seo and Kim used a context-guided search (CGS) strategy to relieve path explosion for concolic execution [40]. CGS looks at preceding branches in execution paths and selects a branch in a new context for the next input. As symbolic execution reached a branch, a *k-context* of this branch will be generate and checked. For instance, in case a symbolic execution reaches a branch *b6* following path *b1, b4, b6*, a 2-context of b6 will be generate as *b4, b6*. CGS will then check whether this 2-context is new or not by checking all 2-context stored in the cache. If it is new, this branch will be select to generate an input following this path, and this 2-context will be stored in cache as well. If it is not new, this branch will be skipped since we assume the corresponding path has been checked before. Meanwhile, in order to handle the situation when k is too small, this framework also calculates the *dominators* to make sure all branches in the k-context is meaningful. In other words, for a certain branch *b*, CGS will calculate the importance of all branches that leads to it. Thus, the k-context of branch *b* will ignore branches that all execution path *must* go through to reach it, only keeping those relevant branches that can generate significantly different paths in its k-context.

A more recent approach developed by Christakis et al. [41] guides dynamic symbolic execution toward unverified program executions. This approach annotates programs to reflect which executions have been verified under which assumptions. Such annotations are used in this approach to guide dynamic symbolic execution, so that tests that lead symbolic execution to verified executions are pruned to avoid exploring parts of the state space, and tests that lead to not yet fully verified properties are prioritized. To be more specific, a program with a property P, which is in the shape of assertion, under an assumption A will be annotated in the form of *assumed A* whereas property A is considered true without checking at this point in the program and *assert P verified A* which means when assumption A is considered true, assertion P should hold. However, since assumption A is potentially unsound, dynamic symbolic execution should generate test cases in which A is violated to check if assertion P still stands true under such condition. With these annotations, this framework is able to guide the symbolic execution in order to prune the redundant tests where A still stands.

Coughlin et al. developed an approach [42] to find "validation scopes," which is the minimal parts of code that needs to be checked to verify a property. This technique uses "enforcement windows" which are nonfaulty sequence of establish–check–use operations found with symbolic trace interpretation, in order to improve the efficiency of proving properties of interest with an acceptable number of false alarms using symbolic trace interpretation. Enforcement window distances are dynamically measured based on two dimensions, namely control and data reasoning, to be used for describing and quantifying the complexity of the validation scopes to provide information needed by designers to observe the spots for improvement.

Zhang et al. [43] introduced regular property guided dynamic symbolic execution to find the program paths satisfying a regular property as soon as possible. They argue that only the paths with specific sequences of events can satisfy the regular property, and the number of such paths is often very small. It is desirable not to explore the irrelevant paths and the relevant paths not satisfying the property. What this application proposes is to explore the off-path-branches, or unchecked branches in current path, along which the paths are most likely to satisfy the property. It combines symbolic execution with model checking on finite-state machine (FSM). For each off-path-branches b, two FSM models will be extracted: one indicates the path leading to b, and one includes the states that could be reached from b. The author argued that if the interception of these FSM is empty, the likelihood of a path satisfying the regular property can be considered very low. Thus, such branches should be less prioritized in the following dynamic symbolic execution process. In other words, the symbolic execution would choose other branches along paths that can satisfy the regular property, rather than doing it in a traditional depth first search or breadth first search style.

It is expensive for symbolic execution to systematically analyze functional correctness properties of programs, e.g., using assertions or executable contracts. iProperty [44] facilitates incremental checking of programs based on a property differencing technique. The key insight is that if a property Φ about program p holds at a control point l in p and Φ implies another property Φ', Φ' also holds at l in p. Based on this observation, the algorithm for computing property differences checks for implications between corresponding old and new properties to minimize the set of properties that must be checked after a property change is made and hence reduces the overall cost of checking. Furthermore, for programs with both changes to properties and changes to code that implements the functionality, iProperty uses the property differences in conjunction with code-based change impact analysis techniques, e.g., DiSE [45] to guide checking onto relevant parts of code and properties, thus reducing much redundancy otherwise present in reapplying symbolic execution.

4.2 Pruning Paths

Since the number of paths and states is potentially large, if we can find a way to prune uninterested states and paths, we can reduce the number of states we actually need to explore and thus relieve the impact of path explosion. One of the useful approaches is using incremental checking. The insight of using incremental checking for path explosion problem is twofold: First, for the same version of a program, we can leverage the state space in fragments, e.g., by the depth boundary, and incrementally check different parts of the state space step by step in multiple symbolic execution runs. Second, we can use incremental checking to avoid checking unchanged states and paths as we conduct regression test on different versions of a program.

One of the incremental checking techniques is Directed Incremental Symbolic Execution (DiSE) [45, 46]. DiSE applies symbolic execution and static analysis in synergy to enable more efficient symbolic execution of programs as they evolve.

The static analysis is based on intraprocedural data and control flow dependences. It identifies instructions in the source code that define program variables relevant to changes in the program. Conditional branch instructions that use those variables, or are themselves affected by the changes, are also identified as affected. The information generated by the static analysis is used to direct symbolic execution to explore only the parts of the programs affected by the changes, potentially avoiding a large number of unaffected execution paths. DiSE generates, as output, path conditions only on conditional branch instructions that use variables affected by the change or are otherwise affected by the changes. Based on the development of DiSE, some applications of symbolic execution are developed to increase the efficiency of regression testing [47].

Another approach about how to scale symbolic execution for efficiently analyzing program increments is developed by Makhdoom et al. [48]. This approach focuses more on test cases rather than source code analysis. It patches automated test suites based on code changes. Using the test suite of a previous version, this approach eliminates constraint solving for unchanged code. This technique identifies ranges of paths, each bounded by two concrete tests from the previous test suite, and by exploring them, all paths affected by code can be covered.

Yi et al. investigated postconditioned symbolic execution [49] by pruning paths based on subsumption checking. At each branching location, this approach checks whether the branch is subsumed by previous explorations by checking the summarized previously explored paths by weakest precondition computations. Postconditioned symbolic execution can identify path suffixes shared by multiple runs and eliminate them during test generation when they are redundant, results in a reduction in the number of explored paths.

Li et al. [50] proved that the subsequent symbolic analysis of two z-equivalent states traverses the same set of paths and gives the same answers to validity and satisfiability queries. Thus, by analyzing one state in a set of z-equivalent states, redundant path exploration can be avoided. They introduced an algorithm that can detect z-equivalent states in a linear scalability, identifying the abstract syntax tree that represents symbolic states in question as unconstrained expressions.

Yang et al. developed memoized symbolic execution (Memoise) [51] to leverage the similarities across successive symbolic execution runs to reduce the total cost by maintaining and updating the state of a symbolic execution run. A trie—an efficient tree-based data structure—is used for a compact representation of the symbolic paths generated during a symbolic execution run. It is maintained during successive runs in order to reuse of previously computed results of symbolic execution without the need for recomputing them again. Constraint solving is turned off for the paths that were previously explored and the search is guided by the choices recorded in the trie. Moreover, the search is pruned for the paths that are deemed to be no longer of interest for the analysis. Memoise reuses state and path information in previous symbolic execution runs to avoid reexploring state space that is unchanged or of no interest. Thus, the overall cost of exploring the whole state space is reduced.

4.3 Merging States or Paths

Some approaches aim to reduce the number of states or paths to explore in general, instead of addressing path explosion problem in certain contexts. One way to reduce the number of states or paths is to merge them. Such a method is introduced by Kuznetsov et al. [52]. This approach first presents a method for statically estimating the impact that each symbolic variable has on solver queries that follow a potential merge point. It merges states only when such merging promises to be advantageous. It then presents another technique for merging states that interacts favorably with search strategies in automated test case generation and bug finding tools. One more recent approach developed by Scheurer et al. [53] reduces number of states by merging symbolic execution branches. This work devises a general lattice-based framework for joining operations and proves soundness of these operations. It has been extensively evaluated with the highly complex TimSort case study and it was demonstrated that significant improvements can be gained.

Jaffar et al. developed an approach [54] to boost concolic testing via interpolation. First, they assume that bug conditions of the program is in the form if C then bug. Which is to say, if a condition C is satisfied, the path is buggy. Then, the solver will generate an interpolant at each program point along the unsatisfiable path. The interpolant at a given program point can be seen as a formula that succinctly captures the reason of infeasibility of paths at the program point, or in other words, the reason why paths are not buggy. As a result, if the program point is encountered again through a different path such that the interpolant is implied, the new path can be subsumed, because it can be guaranteed to not be buggy.

MultiSE [55] is an approach developed by Sen et al. using symbolic execution on JavaScript programs. This technique merges states incrementally during symbolic execution, without using auxiliary variables. The key idea of MultiSE is to a set of guarded symbolic expressions called a value summary based on an alternative representation of the state. MultiSE does not introduce auxiliary symbolic variables, and updates value summaries incrementally at every assignment instruction.

MergePoint introduced by Avgerinos et al. [56] is a symbolic execution technique that operates on Linux binaries. This approach uses a technique named veritesting, which alternates between static and dynamic symbolic execution to take advantage of each of them. The authors argue that as dynamic symbolic execution (DSE) is a path-based technique for testing purpose, it is suffering from path explosion problem by its natural. On the other hand, however, static symbolic execution (SSE) is treating the program as a whole formula since it is a verification technique to detect potential vulnerabilities. In other words, it only checks the satisfiability of certain formulas in which program problems are encoded as part of it, and since there is no dynamic path exploration, it avoided path explosion problem. Veritesting combines these two symbolic execution techniques basing on the control flow graph of the target program. Starting with DSE process, this technique will not always fork into several substates when a symbolic branch is reached. Instead, an analyze will first check if the following program fragment could be statically treated, in which case it will switch to SSE mode and summarize a formula to represent the corresponding program fragment, and switches back to DSE mode for the parts that are difficult to be handled. By treating fragments of the whole program as a "subprograms" statically, the overall path explosion for DSE is relieved.

4.4 Handling Loops

Strejček and Trtík developed an approach on abstracting path conditions [57], which leads the exploration to a certain location while avoiding the path explosion caused by loops. The algorithm is based on computation of loop summaries for loops along acyclic paths, which are called "backbones," that lead to the target location. These backbones are detected by recursively removing the leftmost repeating nodes using loop summaries in the complete path until all cycle states are removed. Given a program and a location within it, the approach produces a nontrivial necessary condition to reach it.

Another approach aims to improve the code coverage in a loop body [58]. This algorithm increases the amount of user-control over symbolic execution of loops by implementing a k-bounded loop unwinding, as the authors argue that loop bounding and search space bounding which are widely used by default could miss important paths and thus test cases generated cannot reach ideal branch coverage in loop bodies. This approach semiautomatically concretizes variables used in a loop by symbolically execute the loop out of context and extract a model for the generated path conditions, following by adding concrete values to these constraints to concretize the variables used in loop bodies or as loop guards. In other words, a complex path constraints inside a loop body will be simplified based on the conditions to cover the branches inside the loop rather than recursively update the path condition by the potential infinite iterations.

S-looper is an approach developed by Xie et al. [59], aiming to deal with loops with multipaths inside the loop body. For multipath loops, the key challenge for summarizing is that there is a large number of possibilities of a loop traversal for the different execution orders of the paths in it.

Typically for a loop condition related to string, it is highly possible that tracking every character in it for loop summarization is necessary but expensive. S-looper automatically summarizes a type of loops related to a string traversal. This approach is to identify patterns of the string based on the branch conditions along each path in the loop. It then generates loop summary based on this patterns. The summary describes the path conditions of a loop traversal as well as the symbolic values of each variable at the exit of a loop.

Le developed another analysis method named segmented symbolic analysis [60]. This hybrid technique addresses not only challenging problem of loops, but also library calls. It performs symbolic and concrete executions based on demand, similar to a concrete execution. Dynamic executions are performed on the unit tests constructed from the code segments to infer program semantics needed by static analysis.

4.5 Parallel Exploration

Applying parallel algorithm is another great tool to deal with path explosion. Instead of exploring the whole state space by only one process, we can divide the state space trie into small sub-trees and launch multiple workers to explore them simultaneously. Ranged symbolic execution [61] is such an approach developed by Siddiqui and Khurshid. This approach embodies this insight and uses two test inputs to define a range for a symbolic execution run. It is built base on the idea that the state of a symbolic execution run can be encoded succinctly by a test input. By defining a fixed branch exploration ordering, the symbolic execution is divided into several ranged symbolic execution subproblems and solved separately.

Synergise [62, 63] extends ranged symbolic execution and introduces two special kinds of ranges—*feasible ranges* and *unexplored ranges*, which lay the foundation of a twofold synergistic approach: improved distributed symbolic execution and seamless integration with complementary search-based testing tools. A feasible range compactly encodes constraint satisfaction results. Given a feasible range, ranged symbolic execution can be used to quickly populate a constraint database by just building the path conditions for all paths that are within the range—all such paths are feasible by definition—without requiring any additional constraint solving. The constraints results stored in the database can then be efficiently reused when performing symbolic execution in other, unexplored, parts of the program. Unexplored ranges enable symbolic execution to efficiently reuse existing tests, providing a natural integration between any type of test generation tool and symbolic execution. Previously generated tests using other tools, such as random testing tools, can be ordered with respect to the specific search order used by the symbolic execution tool employed to define a set of unexplored ranges, which only contain program paths that none of the existing test covers, and these unexplored ranges can then be efficiently explored by symbolic execution in parallel.

SCORE framework [64] distributes concolic execution [5] so that whole execution paths are generated one by one on distributed nodes in a systematic manner while preventing redundant test case. In concolic testing, a symbolic path formula is extracted on the path traversed during each concrete execution, and further symbolic path formulas are then generated by negating path condition. The concrete executions based on the values generated by solving these symbolic path formulas can traverse different new paths in the program. Rather than exploring these paths sequentially in regular concolic testing, the *SCORE* framework employs distributed nodes to explore these paths in parallel.

5. COMPOSITIONAL ANALYSIS

5.1 Improving Compositional Symbolic Execution

Since computation is highly demanded by symbolic execution, scalability is one of the major challenges in practice. One of the methods to address this challenge is compositional analysis, a general-purpose methodology that capable of scaling up multiple static analysis and software verification techniques, including symbolic execution. The main idea of compositional analysis is to encode the input–output behavior of each elementary unit (i.e., a method or a procedure) by analyzing them separately. The results of the analysis are stored in a summary for each elementary unit, and by incrementally composing and utilizing these summaries, the whole-program analysis results are obtained. In other words, the preconditions and postconditions of a path are stored in the summary, and for all the later calls of the path that satisfy the preconditions, we can prune applying symbolic execution on the path again, and return the stored post conditions directly instead [65].

Compositional analysis is first proposed by Godefroid in 2007 [66] to improve symbolic execution performing along concrete paths [67]. This method is further extended [68] to reach a better scalability, and the latest approach, SMASH [69], employs both "may" and "must" summaries, expressed with logical formulae.

One common problem in compositional symbolic execution techniques that use logical formulas as summaries is that in presence of heap updates, compositional symbolic execution could be less efficient due to the lack of a natural way to compose in presence of heap operations. One approach addressing this problem is developed by Rojas and Păsăreanu [70]. This approach is based on partial evaluation (PE) [71]. PE, also known as program specialization, is a well-known technique for automatically specializing a program with respect to some of its inputs. Through this technique, method summaries that consist of path condition and heap constraints for a particular symbolic execution path are obtained, and can be reexecuted with reconstructing the heap naturally without keeping explicit representations in them.

Based on this approach, Memoized Reply is introduced by Qiu, et al. [72]. This approach summarizes each analyzed method as a memoization tree. The memoization tree captures crucial elements of symbolic execution such as path choices and path conditions for all complete paths. Instead of explicitly encoding the method's outputs, a composition operation is defined to replay the symbolic execution of the methods in different calling methods efficiently, using the memoization trees in a bottom-up fashion. To reduce the number of solver calls, constraint solving is only used to determine which paths in the method summary are feasible at a method call site, and turned off when exploring these paths.

Lin et al. improved compositional symbolic execution by using fine-grained summaries [73]. Different from summarization on function level used in the conventional compositional symbolic execution, fine-grained summaries are the summarizing blocks of code within functions. This change makes summaries more possible to be reused to reduce path constraint solver calls. By reducing the number of solver calls, fine-grained summaries can improve the efficiency of symbolic execution in terms of time cost. The evaluation shows the improvement can be affected by the usage of summaries, which means the more opportunities for summaries to be used, the better improvement we can get.

5.2 Applying Compositional Symbolic Execution

Besides improving the scalability of symbolic execution, some researchers inspired by the analyze result from compositional symbolic execution itself, and try to take advantage of this information to apply different approaches efficiently. One such approach is Modular And Compositional analysis with

KLEE Engine (MACKE) [74] developed by Ognawala, et al. This tool enables testers to detect low-level vulnerabilities in a program using symbolic execution in a reasonable amount of time. This solution first performs symbolic execution on individual components in isolation, in order to find out low-level vulnerabilities in these components. The result is reasoned about from a compositional perspective, and the vulnerabilities detected are then scored and report to the user.

DrE is a framework recently introduced by Pustogarov et al. [75] that puts directed compositional symbolic execution into more specific usage. This framework targets software for the popular MSP430 family of microcontrollers. While conventional symbolic execution does not perform good enough on lower-end embedded architecture due to path explosion, directed compositional symbolic execution can be conducted to mitigate this problem and served well as an approach to automatically discover sequences of digital sensor readings that drive the firmware to an adversarially chosen state, especially states of potential vulnerabilities.

6. MEMORY MODELING

Symbolic execution is playing an increasingly important role in proving memory safety properties or checking correctness of programs manipulating complex data structures, however, reasoning on the content of memory brings its own specific challenges and issues. Separation logic [34] has emerged as a major tool to reason about heap manipulation, however, there are still restrictions affecting its practical usability that are being addressed in the ongoing research, like its lack of support from current SMT solvers, or the heavy interdependence between how heap properties are expressed and what can be practically verified, while these issues must be tackled within a further exacerbation of the state space explosion due to the large number of possible memory configurations.

This section will focus on the most recent advances in the tools, methodologies, and approaches pertaining symbolic execution and memory modeling. Regarding separation logic, since it proved to be useful to tackle also other symbolic execution related problems (e.g., analyzing concurrent programs), this section will include only results that are strictly focused on memory modeling, while we refer the reader to Section 3.4 for additional material which is still relevant but more broadly applicable.

6.1 General Advancements

The verification of heap-manipulating programs is a challenging task. Separation logic allows to describe concisely program states that holds in separate regions of the memory; however, this resource oriented logic is not supported by current SMT solvers, since the satisfiability of separation logic formulae with inductive predicates and Presburger arithmetic is undecidable.

In [76] the authors identified a fragment of separation logic more expressive than what is used in other current state-of-the-art approaches (e.g., it is not limited to shape-only inductive predicates), and developed an approach to solve satisfiability problems. This fragment is recursive, which may result in the decision procedure taking an infinite amount of time. To handle that, the authors propose an over/under-approximation procedure that unfolds the formula until it finds a model or proves that all leaves are unsatisfiable.

Verification tasks involving data structures that are traversed in various ways (e.g., graphs) is problematic because the recursive predicates generally used to denote their properties effectively restrict the access patterns that can be effectively verified. To overcome this issue in [77] the authors added support to *iterated separating conjunctions* (ISC) in a symbolic execution engine. Iterated separating conjunctions are an alternative way to express properties of a set of heap location, that do not constraint the traversal order, therefore handling an unbounded number of locations at the same time. The proposed approach uses quantifiers over heap locations, however, the quantifier instantiation is controlled in a way that maintains a low performances impact. An implementation has been evaluated and showed encouraging results on different verification scenarios.

The verification of code involving the manipulation of data structures is particularly challenging because of the large state space of the data that must be explored. In [78] Geldenhuys et al. focused on bounded exhaustive bug finding, and proposed an approach that increases the scalability of a current state-of-the-art tool (Symbolic PathFinder [79], an extension of Java Path-Finder) combining two different techniques aimed at the reduction of the size of the search space. The first is the precomputation of tight field bounds using the structural invariants of the classes under analysis, which prunes invalid states from the search space. The second is the reduction of the number of partially initialized structures that are considered, by using symmetry-breaking to avoid the generation of isomorphic ones.

Rosner et al. [80] also focused on developing techniques to obtain an early pruning of invalid or redundant data structures and proposed two mechanisms: bound refinement, which improves over current bounded lazy initialization approach [78], and introducing auxiliary feasibility checks relying on a SAT solver.

Belt et al. [81] focused on algorithms for the symbolic execution of programs operating with statically allocated value-based data structures that are found in the development and verification of high assurance applications. Their work focused on Spark [82], a subset of Ada that lacks constructs that are difficult to reason about and includes a notation allowing the specification of pre- and postconditions, assertions, and information flow relationships. They provide a formal operational semantic for the symbolic representation of Spark programs, with two popular approaches: a logical representation and a graph-based symbolic representation. Interestingly, the graph-based representation proved to be strictly faster than its logical counterpart.

6.2 Managing Heap Input

The symbolic analysis of a program to generate relevant test cases, or to verify certain properties, is particularly challenging when the input is constituted by a dynamic heap structure, as in general the space of possible heap configurations is extremely large and only a small portion of it represents valid input values.

The introduction of the *lazy initialization* approach, presented in [83] by Khurshid et al. in 2003, improved significantly the applicability of symbolic execution in this context, however, within the current state of the art there are still methodologies that do not take into account structural properties of the input, performing an extensive exploration that includes isomorphic structures and invalid program executions (thus increasing the analysis cost and raising false alarms) and methodologies that *do* take into account these constraints, but either enumerate in advance all the possible valid structures or perform a consistency check of the whole heap after every new assumption.

Braione et al. [84] propose a methodology based on domain invariants specifically designed to define constraints over the lazy initialization process, rather than executable predicates over the shape of the data structure, simplifying the verification of the data structure integrity and avoiding other issues, like the fact that the structure validation procedure itself may trigger the lazy initialization of some fields. This idea is further developed in [85] where the authors explicitly define a language to specify structural constraints of partially initialized data structures, along with a decision procedure that is used to evaluate the constraints in an incremental fashion as the symbolic execution progresses, resulting in a significant scalability improvement over the current state of the art. An evaluation of the resulting tools against multiple experiments over the TSAFE air traffic control application, the Google Closure compiler project, and various programs that manipulates classic recursive data structures, is described in [86], confirming the benefits of the incremental verification.

6.3 Handling Symbolic Memory Access

The symbolic execution of code containing pointer handling and dereferencing is a valuable tool to detect software bugs involving invalid memory access operations (e.g., access to uninitialized or previously freed memory locations); however, the symbolic evaluation of operations including pointers is a challenging task, since a symbolic pointer may represent a large number of concrete addresses for which it is necessary to evaluate the validity.

In [87] Romano and Engler developed a dispatcher to resolve symbolic memory access that can be employed on unmodified machine code including also demand-allocated buffers. Their approach defines a separation between the mechanism used to issue a symbolic memory access, and the policy used to resolve it. This allowed to easily support a variety of policies, which is an interesting strength compared with other state-of-the-art solutions, as different policies produce different performance and completeness results, and no single policy can be considered the best. Examples are the *fork on address* policy, which has a high cost but explores all feasible accesses, and the much faster *prioritized concretization* policy, which searches for single a feasible address containing a symbolic value. The library has been evaluated in test case generation and memory fault detection applications, showing an improvement over the current state of the art tools.

In [88] Fromherz et al. describe an enhancement of the Symbolic Path-Finder (SPF) for Java bytecode that is capable of performing symbolic execution of programs handling arrays of symbolic length. By using the array theory supported by a variety of SMT solvers, the approach is capable of handling both arrays of primitive types and, thanks to a novel combination with lazy initialization, also of reference types.

6.4 Advancements in Static Analysis

Analyzing heap reachability (i.e., checking if an object can be reached through a sequence of pointer dereferences) is a notable static analysis task that is performed to statically check assertions concerning, for example, the lifetime of an object or the correctness of field encapsulation and is also essential for various other relevant static analysis problems such as escape analysis or taint analysis. In [89] the authors describe a technique for reasoning about heap reachability that is flow-, context-, and path-sensitive, and that provides location materialization. The approach consists in an initial flow-insensitive points-to-analysis that is refined on-demand: the flow-insensitive points-to facts are integrated into a mixed symbolic-explicit representation of the program state, that, together with an algorithm to infer loop invariants for heap reachability queries, is used to perform the analysis, which demonstrated to scale well while maintaining a reasonable precision.

Zhu et al. described in [90] a novel methodology and a tool to prove the linearizability of a concurrent data structure implementations, interpreting it as a property checking problem using separation logic. The tool tries to identify *witness states* of the linearization of the operation with respect to the abstract specification: symbolic execution is used to identify candidate witness states that are then validated, and if every possible path presents a witness state the linearizability is guaranteed.

In [91] the authors focused on a particular class of dangling reference issues that are often found web applications. They observed that these applications often rely on code written in different languages and executed on different physical machines, where parts of the code are generated dynamically; a common configuration (which is the also the reference situation in the paper) involves a server executing PHP code to generate dynamically HTML and JavaScript code that will be then executed on the client side. The lack of cohesiveness between the various components arising in this circumstance may leads to dangling references: certain versions of the generated code may contain variables that have not been appropriately initialized. The authors proposed a static analysis method that uses symbolic execution to build a model that approximates all the possible versions of the generated output, and then uses it to verify that all the necessary initializations are performed in every scenario.

7. CONCURRENCY

Analyzing concurrent programs is notoriously hard due to their inherent nondeterminism. Much research projects have been conducted on applying symbolic execution to analyze concurrent programs. While some are focused on detecting concurrency bugs such as data races and deadlocks, others aim to scale up the analysis of concurrent programs.

7.1 Detecting Concurrency Bugs

A key question for analysis of concurrent programs is how to deal with data races and deadlocks, which can cause many concurrency bugs. A data race occurs when different threads concurrently access the same memory location, and at least one of these accesses is a write. A deadlock occurs as a thread is holding a lock (L1) and waiting for another one (L2), while another thread is holding L2 and waiting for L1. Thus, none of these threads can continue and the whole process gets stuck. Ideally, data race could be avoided by using proper synchronization between threads, e.g., using concurrent data structures or data-locking mechanisms. However, in practice it is difficult to guarantee data race free for a concurrent program.

Symbolic execution has been applied for predictive analysis [92], which aims at predicting executions that expose races by mutating the schedule order of an execution trace. In particular, Wang et al. [93] and Said et al. [94] proposed symbolic predictive analysis approaches that use symbolic predictive model to precisely capture feasible interleavings and encode the concurrency errors detection problem as an SMT formula. Liu et al. [95] proposed IPA, which allows the schedule mutation to change the location referenced by a shared access, and thus enables more schedule mutations and improves the data race detection. IPA also applies a hybrid symbolic encoding scheme to achieve practical applicability. For operation not supported by the solver, IPA requires all the operands to take the concrete values observed in the seed execution.

Razavi et al. proposed a test generation technique for concurrent programs [96]. It extended sequential concolic execution [4, 97] with predictive analysis [92] for the concurrent setting. The approach alternates between sequential concolic execution and concolic multitrace analysis (CMTA) to generate tests that will increase structural coverage. Concolic execution is first applied to increase branch coverage on individual threads until coverage is saturated; CMTA is then used to generate new test inputs and thread schedules to cover previously uncovered branches in one of the threads; concolic execution is used again based on these new test inputs and thread schedules to cover more previously uncovered branches. Farzan et al. proposed con2colic testing [98] for performing concolic execution [4, 97] for concurrent programs. Different from traditional concolic execution that is applied for sequential programs, con2colic introduces interference scenario as a representation of a set of interferences among the threads, and extends the traditional sequential concolic execution to model interferences as interference constraints and navigates the space of all interference scenarios in a systematic way. Con2colic testing executes a concurrent program based on a given schedule and stores the observed execution in a forest data structure that keeps track of the already explored interference scenarios, and then decides what new scenario to try next, aiming to cover previously uncovered parts of the program, based on the set of interferences that have already been explored. Deligiannis et al. proposed WHOOP [99] focusing on detecting data races typically in device drivers. Compared to traditional data race detection techniques that are based on happens-before and typically attempt to explore as many thread interleavings as possible (and thus has code coverage and scalability issues), WHOOP uses over-approximation and symbolic pairwise lockset analysis, which scales well.

In addition, several techniques were developed recently to apply symbolic execution on concurrent programs in specific fields. For example, Parallelized Compiled Symbolic Simulation [100] is for verification of cooperative multithreading programs available in the Extended Intermediate Verification Language (XIVL) format. The XIVL extends the SystemC IVL, which has been designed to capture the simulation semantics of SystemC programs, with a small core of OOP features to facilitate the translation of C++ code. Li et al. proposed GKLEE [101], a framework that uses symbolic execution to analyze C++ GPU programs to detect races, deadlocks, as well as performance bugs such as noncoalesced memory accesses, memory bank conflicts, and divergent warps. They introduced a symbolic virtual machine (VM) to model the execution of GPU programs on open inputs.

7.2 Improving Scalability

Analysis of multithreaded programs must reason about possible thread interleavings for each input, in addition to reasoning about program behaviors over all possible inputs. In practice, the number of interleavings grows exponentially with the length of a programs execution, and the interleaving space is too massive to be explored exhaustively [102, 103], thereby exaggerating the well-known path explosion problem of symbolic execution. Despite the fact that much research has been done to relieve the impact of path explosion in general, several studies have been conducted to address such problem in particular for concurrent programs.

Although the number of possible thread interleavings for each input could be large for real-world programs, not all of these interleavings are necessary to check for testing. For example, if there are two threads for a certain program without any data race, it does not matter which thread run first. Kahkonen et al. [104] developed an algorithm that combines dynamic symbolic execution and unfoldings for testing concurrent programs, with a goal of exploring only the interleavings that can lead to a change in software behavior. Unfoldings [105] fights path explosion using a "compression approach" by constructing a symbolic representation of the possible interleavings that is more compact than the full execution tree. Unfoldings naturally capture the causality and conflicts of the events in concurrent programs in a way that allows test cases to be efficiently generated.

Guo et al. [106] developed an assertion guided pruning framework that identifies executions guaranteed not to lead to an error and removes them during symbolic execution. This approach uses a generalized interleaving graph (GIG) to capture the set of all possible executions of a concurrent program. By analyzing GIG and summarizing the reasons why previously explored executions cannot reach an error, redundant executions are pruned in future runs. Guo et al. [107] developed ConciSE which reduces the state space exploration by leveraging change impact information and exploring only the executions affected by code changes between two program versions.

Another idea to limit path explosion for concurrent programs is to symbolically execute relatively small fragments of a program in isolation—this reduces path length, which in turn reduces the potential for path explosion. This idea is first introduced by Bergan et al. as input-covering schedules [108]. This approach implements many optimizations to avoid the inherent combinatorial explosion, like bounding epochs: loops containing synchronization code are analyzed separately and in lockstep for all threads. A further-developed approach [109] is given later on performing symbolic execution of concurrent programs from arbitrary program contexts, rather than starting it at one of a few natural starting points, such as program entry (for whole-program testing) or a function call (for single-threaded unit testing). This approach proposes a way to integrate data flow analysis with symbolic execution. This data flow analysis computes a cheap summary that is used as the starting point for symbolic execution by combining reaching definitions, which summarize the state of memory, with locksets and barrier matching.

8. TEST GENERATION

The generation of test input is a major application of symbolic execution, which is currently tackling relevant problems like the identification of the input values necessary to cover parts of the code that are particularly difficult to reach. In recent years, multiple techniques have been developed for more efficient and effective test generation using symbolic execution.

8.1 Finding Bugs and Improving Coverage

Batg [110] combines features of static analysis and bounded symbolic execution for test case generation. An initial static analysis identifies a list of potential errors on the control flow graph (CFG): potentially buggy nodes are identified, and the set of paths leading to these nodes is passed to the "path processor," where symbolic execution is performed. All the paths with satisfiable path constraints are considered as real bugs, and one test case is generated for each of them; otherwise, these paths are unwound from its "CFG form" where the loops are considered as only one iteration, and the satisfiability of the new generated constraint is checked again. This process is done iteratively until a satisfiable context is found, which means a test case for the bug can be generated, or the unwinding bound is reached.

Following the guiding principle of RANDOOP [111], which relies on feedback-directed random input generation, Garg et al. [112] propose an automatic unit test generation technique to improve the coverage obtained by feedback–directed random test generation method. It employs dynamic symbolic execution on the generated test drivers and uses nonlinear solvers in a lazy manner to generate new test inputs for programs with numeric computations. It conducts analysis on unsatisfied cores returned by SMT solvers, checking whether a branch can be reached with another input or can be used to tell whether certain test suits cannot reach a target branch.

Bardin et al. [113] improve the coverage of generated test cases from a different perspective. They argue that although behaviors related to control are well handled, interesting behaviors related to data can still be missed in path-oriented criteria. Thus, they propose *label coverage*, a new coverage criterion based on *labels*, a well-defined and expressive specification mechanism for coverage criteria associated to program instructions. By appropriately changing the labeling function, multiple coverage criteria can be emulated such as instruction coverage, decision coverage, and condition coverage. A dynamic symbolic execution approach is developed for covering labels with no exponential blowup of the search space.

You et al. [114] propose an incremental test generation strategy using dynamic symbolic execution for generating test cases to satisfy the

Observable Modified Condition/Decision Coverage (OMC/DC) [115]. OMC/DC is an improved coverage criterion based on Modified Condition/Decision Coverage (MC/DC), which is a condition-based criterion that is widely used in safety-critical systems. They use dynamic symbolic execution to generate tests that, first, satisfy a condition-based coverage criterion to exercise a particular part of the code (MC/DC) and, second, satisfy a dataflow criterion propagating the possibly corrupted state to a point where it is used by an observable output. A tag is assigned to each condition and the propagation of tags to outputs approximates observability. Incremental test generation starts with concrete test inputs that satisfy an obligation and then invokes a model checker at each test step repeatedly to solve path conditions in an attempt to propagate tags through nonmasking paths toward outputs.

8.2 Regression Testing, Data Flow Testing, and Load Testing

Software evolves and many techniques have been developed recently using symbolic execution for regression testing.

Jamrozik et al. [116] propose an augmented dynamic symbolic execution (ADSE) for better regression testing. ADSE augments path conditions with additional conditions that enforce target criteria such as boundary or mutation adequacy, or logical coverage criteria.

Qi et al. [117] propose *semantic signature* to partition program paths based on program output. Semantic signature is more concise than a complete enumeration of all possible execution paths and conveys enough information to reason effectively about changes between different program versions. It analyzes the output of the symbolic execution on-the-fly, and groups together paths with the same symbolic output. The signature can be further used to detect changes between different versions with higher efficiency, which is frequently done in regression testing.

Marinescu and Cadar [118] propose ZESTI (Zero-Effort Symbolic Test Improvement), which leverages symbolic execution to check all possible inputs on the same path toward a certain operation in the program, while using regression test suits to guide symbolic execution. The slightly divergent but potentially buggy paths are explored by this guided symbolic execution, detecting more potential bugs in the program without any extra effort from programmer or testing team.

Cadar and Palikareva [119] propose Shadow Symbolic Execution: in one symbolic execution instance, two different program versions will be run together at the high level, while the information of the old version is gathered and used to "shadow" the new one. By doing so, the changes between versions are effectively detected by the precise dynamic value information gathered, and symbolic execution is driven based on these differences. In other words, symbolic execution on the new version will generate test inputs addressing actual changes between two versions, instead of only generating test cases that make both versions behave identically.

Le and Pattison [120] propose hydrogen for patch verification. It performs symbolic analysis on multiversion interprocedural control flow graph (MVICFG), which represents the differences among various versions of a program and is the central element used to verify relevant aspects of a patch, namely whether it correctly fixes an existing bug without introducing new bugs, and whether it can be applied on all software releases impacted by the bug.

Yoshida et al. [121] propose FSX, a technique that uses a diagnostics engine to guide and refine test drivers and symbolic execution, and dynamically analyzes the dependency with Reduced Ordered Binary Decision Diagrams (ROBDDs) [122]. It conducts dynamic symbolic execution incrementally: it starts with concrete input values as a minimal test suite and obtains diagnostic information during the symbolic execution; based on the diagnostic information, relevant variables made symbolic for further symbolic analysis.

Su et al. [123] introduce a hybrid data flow testing framework based on dynamic symbolic execution with a guided path search strategy. It first statically analyzes a program to find def–use pairs, and then applies dynamic symbolic execution to generate test inputs that cover as many def–use pairs as possible within a time boundary. All uncovered pairs are then checked for feasibility using CEGAR based approach. Afterward, dynamic symbolic execution with a higher time boundary is applied to cover feasible pairs. This cycle repeats with the time boundary been increased for each cycle to cover feasible def–use pairs.

Zhang et al. [124] propose compositional load test generation to address the limited effectiveness of load test generation on large software pipelines. It analyzes system components separately with symbolic execution to collect path constraints that induce performance problems on each system component, selects constraints and find a compatible path across the whole system, generates channeling constraints, and then solves conjunction of constraints to derive test inputs.

8.3 Testing Nontraditional Programs

8.3.1 Web Applications

Symbolic execution face specific challenges in the domain of JavaScript-based Web applications which are quite different from traditional programs. Web applications are usually event driven, user interactive, and string intensive. Also, unlike traditional programming languages, JavaScript is a dynamic, untyped and functional language. To address these challenges, some techniques have been developed to apply symbolic execution on web applications using JavaScript.

Li et al. [125] propose SymJS, a comprehensive technique for automatic testing of client-side JavaScript Web applications. It consists of two major parts: a symbolic execution engine designed for JavaScript language and an automatic event explorer tool for Web pages. Web events are discovered through the symbolic execution of the associated JavaScript. After this process, dynamic feedback from all these execution is refined, and SymJS generates test cases with high coverage based on it. The whole process can be performed automatically, without the need of users supervision and intervention.

Dhok et al. [126] propose type-aware concolic testing of JavaScript programs to reduce the number of generated inputs for JS programs. It is developed based on the observation that the number of inputs generated for JavaScript programs can increase dramatically due to a naive type-agnostic extension of concolic testing to JS programs, where many executions operate on undefined values and repeatedly explore same paths resulting in redundant tests.

Loring et al. [127] propose ExpoSE, a technique that is designed to generate test cases with a high path coverage for JavaScript programs. The ECMA standard for JavaScript specifies regular expressions as part of the language; however, regular expressions, such as capture groups and backreferences, are not well supported by symbolic execution engine. ExpoSE present an encoding of JavaScripts regular expressions into a combination of classical regular expressions and SMT, so that the corresponding constraints can be solved by SMT solvers.

8.3.2 Android Applications

There is a growing need for automated testing of Android apps. Android apps are commonly developed in Java; however, the Android platform is different from the "traditional" JDKs on PCs: Android apps are event driven and susceptible to path-divergence due to their reliance on an application development framework.

Different from common software, behaviors from users are much more frequent and important in Android applications. Thus, in order to test these applications automatically, users' behaviors must be simulated accurately. Mirzaei et al. [128] develop a model of Android libraries in Java Pathfinder (JPF) to enable execution of Android apps. Later on, they propose SIG-Droid [129], which leverages the knowledge of Androids ADF specification to automatically extract two models from source code of an app: the Behavior Model (BM), and the Interface Model (IM). These two models are used to generate event sequences to simulate a sequence of user's behaviors.

Anand et al. [130] propose ACTEVE, which applies concolic testing on Android apps. It focuses on tapping events to drive the exploration. Symbolic variables are used for the coordinates of the tapping event, with the frame layout used as the constraints. In the case of tap events, whenever a concrete tap event is input, this instrumentation creates a fresh symbolic tap event and propagates it alongside the concrete event. As the concrete event flows through the SDK and the app, the instrumentation tracks a constraint on the corresponding symbolic event which effectively identifies all concrete events that are handled in the same manner. In order to mitigate path explosion problem, it identifies subsumption between different event sequences and does not extend paths leading to a read-only event.

Jensen et al. [131] propose an approach that, instead of generating test suites with a high coverage, is particularly designed to reach a certain line-of-interest in an Android application. This two-phase technique first discovers and summarizes event handlers of an application by conducting a concolic execution, and then generate event sequences backwardly from the target using these summaries. Thus, an event sequence from the entry of the application toward the given target line is generated automatically and can be used for testing purposes.

8.3.3 Database Management Systems

Traditional symbolic execution approaches are usually not designed to test stored procedures, and consider databases as external and are always facing the challenges introduced by the multilingual nature from applications. To address this problem, Mahmood et al. [132] introduce a dynamic symbolic execution technique for stored procedures in database management systems. It considers database as an internal element for symbolic execution by treating values in database table as symbolic. Data types, expression operations, function calls and database constraints (check, unique, keys, etc.) are extracted and used to build a "plan tree" model, which is later processed using symbolic execution. Then, with the help of SMT solver, it generates test cases that executes different paths and discover input values leading to schema constraint violations or user-defined exceptions.

8.4 Test Optimization

Symbolic execution can generate a large number of test cases; however, it is argued that not all of the test cases are equally important, while some of them can even be considered "unnecessary." Researchers have developed different techniques to optimize the test cases generated by removing unnecessary test cases or prioritizing them.

Nguyen et al. [133] propose "Don't Care Analysis" to reduce the size of test cases generated by symbolic execution. Don't Care Analysis infers a set of don't care symbolic variables: variables that can be assigned to any values, within the context of a specific satisfying assignment, without affecting the overall code coverage. Using don't care analysis, symbolic execution engines can remove assignment statements for don't care variables from the generated test cases, thus make the test cases more succinct while achieving the same code coverage.

Rapos and Dingel [134] leverage fuzzy logic and symbolic execution to prioritize UML-RT test cases, in order to prevent unnecessary generation of tests. It follows the pattern of fuzzy logic control systems to prioritize each test case in a UML-RT test suite, based on natural language rules about testing priority. Wong et al. [135] propose a similar prioritizing technique, Document-Assisted Symbolic Execution (DASE) to enhance the effectiveness of symbolic execution for automatic test generation and bug detection. DASE extracts input constraints from documents and uses them as a "filter" to favor paths that execute the core functionality of the program.

9. SECURITY

9.1 Hybrid Techniques

Fuzzing is a testing approach that consists of feeding large amounts of random inputs to the target program in an attempt to reveal bugs. Recent works attempt to mitigate the weak points of symbolic execution and fuzzing by combining them in a hybrid technique.

Haller et al. [136] uses a combination of fuzzing and taint analysis to guide symbolic execution to areas of the program that may contain buffer overflows. The technique, Dowser, works in five steps: (1) array accesses in loops are collected and ranked according to their complexity; (2) identify through taint analysis which parts of the input (obtained from existing tests or fuzzing) exercise the array accesses; (3) Learn which branches have a higher probability of leading to unique pointer dereferences (*access patterns*) by performing symbolic execution with a small symbolic fragment of the relevant section of the input; (4) Perform symbolic execution with the entire relevant input fragment made symbolic while guiding the exploration toward the branches found in the previous step; (5) use off-the-shelf tools (e.g., AddressSanitizer [137]) with the generated inputs to check the existence of buffer overflows.

Stephens et al. follow the opposite approach with Driller [138]: symbolic execution is used only to help the fuzzing engine to get "unstuck." The underlying fuzzing engine, AFL,[a] keeps track of which basic blocks of the program have been visited so far. Once AFL cannot find inputs that cover new basic blocks after a certain time, Driller performs symbolic execution targeting the frontiers of the explored area of the program so far (called *compartiments* in the work). The generated input is then sent to AFL, which will continue the fuzzing process by generating similar inputs. Driller is one of the components of the Mechanical Phish platform, which achieved third place in the DARPA Cyber Grand Challenge competition.

9.2 Embedded Systems

Applications running in embedded devices are potentially vulnerable to physical attacks. For example, attackers can introduce faults in the program execution by manipulating hardware components (e.g., shining a laser at the memory banks), thus avoiding the execution of security routines. Potet et al. [139] propose Lazart, an approach to evaluate the robustness of applications against fault injection vulnerabilities. Given a target statement, Lazart identifies basic blocks that may impact the reachability of the target. These basic blocks are potential targets for fault injection attacks. Lazart then builds a high-order mutant program [140] that encodes all possible sets of faults. The final step uses concolic execution (through KLEE) to find all paths that

---

[a] http://lcamtuf.coredump.cx/afl/.

lead to the target statement; assertions are used to ensure realistic scenarios (e.g., make sure that only invalid inputs are considered). The number of valid paths found + the number of faults required for each path can be used as a measure of the robustness of the program.

Davidson et al. present FIE [141], a KLEE-based concolic execution tool for 16-bit microprocessor firmwares. Since different microprocessors may have distinct characteristics such as memory layouts and interrupt handlers, FIE allows the user to specify these settings in a modular way. FIE also introduces two optimizations intended to help the exploration get out of deep loops. The first one is *state pruning*, where FIE removes states that match already explored states from the queue. To speed up the matching, FIE stores the diffs of all states for each visited value of the program counter. The second optimization is *memory smudging*: loop counters are replaced by symbolic variables after a certain number of iterations, allowing the analysis to escape the loop at the cost of precision. FIE was evaluated in a set of benchmarks consisting of a corpus of programs taken from github, plus a few well-known firmwares. It was able to reach a high coverage (90%) for small and medium programs (number of executable instructions <500), but it covered a lot less in large programs (<25%).

9.3 Underconstrained Symbolic Execution

Ramos and Engler [142] propose to perform symbolic execution in individual functions of the program (*underconstrained* symbolic execution) in an attempt to avoid the burden of state-space explosion. Although underconstrained symbolic execution is imprecise, the authors claim that it is still powerful enough to be useful in certain scenarios, such as regression testing of patches. To evaluate the technique, the authors developed UC-KLEE, an extension of KLEE for underconstrained symbolic execution. UC-KLEE leverages lazy initialization to handle pointers, and allows the user to specify preconditions to silence false positives encountered by the tool. UC-KLEE was evaluated in two scenarios:

1. checking if patches introduced new crashes by generating test harnesses. If the patched version crashes but the original version don't, an error is reported. Path pruning is used to avoid wasting time in uninteresting paths; for example, paths that lead to an already found error (based on program counter) are pruned.
2. Checking all paths of the function for memory leaks and uninitialized data.

The evaluation was performed on large open-source C programs, such as the Linux Kernel and OpenSSH. Results are positive; UC-KLEE found almost 80 bugs on the subjects, although the false positive rate was considerable.

Jana et al. [143] leverage underconstrained symbolic execution to identify incorrect error handling in C programs. The tool, EPEX, uses an error specification for the API call under test + symbolic execution to look for paths in which the API function may return an error. PEX then checks if the error is handled correctly at that point, i.e., the program either logs the error, returns a valid error value from the caller function, or exits with a nonzero value. To reduce false positives, EPEX does not report an error if all paths ignore the error value. EPEX was evaluated on SSL/TLS libraries, like OpenSSL and GnuTLS, and applications that use them. It was able to find real bugs in most of them, although the false positive rate was higher for the applications due to the complexity and higher number of configuration parameters (i.e., flags to ignore errors).

Yun et al. [144] use a similar idea, *relaxed symbolic execution*, to learn possibly correct API usage patterns directly from the source code. Relaxed symbolic execution is strictly intraprocedural and only unrolls loops a single time, while underconstrained symbolic execution is interprocedural and uses k-bounded loop unrolling. APISan, the presented tool, uses relaxed symbolic execution to extract a few representative paths, which are analyzed to extract *semantic beliefs* about usage patterns API; the probability of a pattern being correct is associated with its frequency in the code. The authors evaluated the effectiveness of APISan in large open-source applications (Linux Kernel, OpenSSL); APISan found 76 unknown bugs (69 confirmed by the developers at the time of writing).

9.4 Ad Hoc Solutions and Enhancements

Wang et al. [145] propose MetaSymploit, a symbolic execution-based approach to automatically generate signatures of attack scripts written in the MetaSploit framework[b] to be used in Intrusion Detection Systems. Given a MetaSploit attack script, MetaSymploit generates a signature in three steps: (1) extract symbolic behavior descriptions of the script, like sequences of api calls and attack payloads, through symbolic execution; (2) Consolidate the information obtained in the previous step into rules that can be used by Intrusion Detection Systems (IDS) to detect the attack. This happens in three stages: extract constant patterns (payload length/offset/content), remove

---

[b] https://www.metasploit.com.

redundancies/benign data (e.g., "Content-Type:text/html"), and derive the context from the sequence of API calls. MetaSymploit seems to be successful in generating signatures for a large amount of real-world MetaSploit scripts; most of the failures can be attributed to limitations in the symbolic execution engine, such as api calls without a model or nontermination due to loops.

Chau et al. [146] propose to test X.509 PKI libraries for noncompliance through the use of symbolic execution. Validating X.509 certificates is a complex task due to the complexity of the specification; applying symbolic execution directly is inviable. The authors propose instead to use *SymCerts*, certificates that mix concrete and symbolic data, to verify the libraries compositionally. This can be done by splitting the certificate fields into independent logical sets that can checked in isolation. For each set, a chain of SymCerts is created with all fields contained in the set made symbolic, and all others made concrete. The chain of SymCerts is used as input for the libraries under test, and the resulting path conditions are used to perform cross-validation. The approach was able to find 48 instances of noncompliance in open-source SSL/TLS libraries.

Păsăreanu et al. [147] propose to combine symbolic execution and Max-SAT solvers to quantify the information leakage of a program after a fixed number of executions. Programs that leak large amounts of information through side-channels, such as execution time or memory consumption, may be vulnerable to attacks. Intuitively, the leakage of a sequence of executions is proportional to the number of distinct *observables* (side-channel values observed); an upper bound can be obtained through well-known information-theoretic measures, such as channel capacity. The proposed approach computes the maximum leakage across k executions in three steps: (1) collect all PCs and their respective observables through symbolic execution. (2) find the k-sequence of executions with most distinct observables through Max-SAT. For each distinct observable sequence, a Max-SAT clause is created containing the disjunction of all PCs that match the sequence. (3) Compute the channel capacity of the sequence. The authors also propose a greedy version of their algorithm that analyses each execution individually; the greedy algorithm is more efficient, but it may not return the actual leakage upper bound.

Xu et al. [148] propose CryptoHunt, a tool that leverages symbolic execution to identify cryptographic functions in obfuscated binaries. CryptoHunt works in three steps: first, the tool extracts a trace of the binary under analysis and attempts to identify possible loop bodies. Next, CryptoHunt encodes the relation between the loop input/output variables as a bit-accurate Boolean formula, which preserves the semantics of the nonobfuscated (original) program. This is done by performing symbolic execution on backward slices starting from the loop output variables. Finally, the tool attempts to prove the equivalence, using SMT solvers, between the formulas extracted from the binary under test and a set of reference formulas for known cryptographic algorithms taken from open-source libraries. Since there are multiple ways of matching the input/output variables for each formula, the authors propose a mapping algorithm based on the matrix mapping problem capable of filtering unfeasible mappings quickly.

Stoenescu et al. [149] present RIVER, a new binary analysis framework. RIVER translates x86 instructions to a intermediate representation that allows reversing (i.e., "undoing") operations. Symbolic execution is implemented on top of this reversible IR, which reduces memory costs by undoing operations instead of backtracking. However, the current implementation of the IL results in a $6\times$ increase in size compared to the original code; the authors claim that it can be reduced to $2\times$ with some optimizations. System calls and external libraries are handled by snapshotting the state before the function call, since there's no way to reverse the side effects in those cases. RIVER is currently used internally at BitDefender; no public releases were available at the time of writing.

Hasabnis and Sekar [150] propose EISSEC, a tool for extracting mappings between a compiler's intermediate representation (IR) and architecture-specific assembly instructions. Modern compilers use architecture-specific sets of rules called *machine descriptions* (MDs) to drive a generic code generator. MDs contain a mix of instruction pattern rules and auxiliary code, which may be many times larger than the rules description; manually extracting IR to assembly mappings from the MD is very costly. EISSEC simplifies this task by computing all solutions to all possible paths on the code generator (and the target MD) through concolic execution. The typical structure of code generators (no complex loops/pointer usage, finite input domain) makes this task feasible. Instead of traditional SAT/SMT solvers, EISSEC uses a CLP solver to check the satisfiability and efficiently enumerate all possible solutions of constraints.

10. PROBABILISTIC SYMBOLIC EXECUTION

Probabilistic symbolic execution (PSE) is an extension of symbolic execution aiming at computing the probability of a specific event to occur during a program execution, assuming the program inputs follow a given probabilistic distribution. The input distribution allows both developer assumptions and data from the real-world to be incorporated in the analysis, thus tailoring the analysis to specific usage profiles. PSE has many potential applications, e.g., it can support debugging by allowing a quantitative ranking of detected program errors [151], for analyzing the control software of an autonomous agent interacting with an uncertain external environment, for computing software reliability or expected execution time [152], or for quantitative information flow analysis [153] for security applications, including preliminary attempts at analyzing encryption routines [154].

PSE has been firstly proposed in [155]. The basic intuition of the authors was to enrich the transitions in a symbolic execution tree with a *count* of the number of solutions, or models, of each path condition. The approach was limited to linear integer constraints, which allow for efficient model counting techniques based on Barvinok's algorithm [156]. The ability to count the number of solutions reaching each node of the symbolic execution tree enabled not only to rank different error paths, but also to steer the symbolic execution driver toward the most unlikely paths, under the intuition these are often less likely to be stressed during quality assurance phases. A set of techniques, including normalization, caching, and relaxation of complex constraints (which entails the computation of sound bounds on the number of approximate solutions, instead of exact counts) have been proposed by the authors to cope with the complexity of a trivial application of model counting. All the input variables in this work are assumed uniformly distributed within a specific interval.

In [157], PSE has been extended to support arbitrary input profiles, nondeterminism, and a more efficient handling of exact counting. This works still focused on linear integer constraints. Input profiles were to be specified as a map from an arbitrary partition of the input domain to the expected probability of each subset in such partition to be observed during execution; elements within the same subset were assumed to be equally probable. For example, the input domain of an integer variable can be split into multiple ranges and each such ranges can be assigned a numerical probability to describe a histogram distribution. This allowed to tailor PSE analyses to specific usage scenarios, formally computing the probability of a target event to occur conditioned to the program being used as specified. To support nondeterminism, the internal nodes of the symbolic execution tree have been classified in *conditional* and *nondeterministic* choices. For conditional nodes, the probability of each successor can be computed using model counting techniques (enhanced to take into account the usage profile). Instead, nondeterministic nodes allowed only to reason in terms of best and worst case scenarios, by selecting either the successor node leading to the largest or the smallest probability of exposing the target event during successive execution steps, respectively. The result of the analysis in presence of nondeterminism is thus no longer a single probability value but a range of probabilities, accompanied by schedules driving the program toward reaching the target event for the maximum and the minimum probabilities in the range, respectively. This analysis is reminiscent of dynamic programming techniques used for the analysis and the synthesis of optimal schedules for Markov decision processes [158], but can be applied directly on the implemented code artifacts instead of requiring abstract modeling.

10.1 Statistical Symbolic Execution

To deal with the scalability issues induced by model counting and nondeterminism, two different approaches have been proposed in [159] and [160], respectively.

In [159], the authors introduced the concept of *statistical symbolic execution*, where the accuracy of the result is traded for scalability by allowing the analysis to only sample a finite number of execution traces and to infer the probability of the target event using statistical methods on those samples. Because an unbiased sampling of the execution paths requires to compute the satisfaction probabilities along the branches traversed by the path, statistical symbolic execution can enhance the statistical estimators with exact information about the parts of the symbolic execution tree traversed during sampling. This distinguishes the approach from classic statistical model checking because the analysis draw conclusions combining a partial exact analysis with a statistical process, guaranteeing a faster converges to the accuracy and confidence goals set by the user. Furthermore, because each path that has already been samples once is exactly analyzed, it can be safely excluded from subsequent sampling rounds. This observation allows to define an iterative sampling procedure that samples new paths only from the parts of the symbolic execution tree that have not been explored yet, increasing the chances of selecting also lower probability paths, which is the main limitation of purely statistical approaches [161].

In [160], the authors propose an approximate solution for the analysis of programs exhibiting nondeterministic behaviors, such as multithreaded programs. To deal with the exponential explosion in the number of possible interleaving, the authors propose a to use a Q-learning [162] algorithm adapted from reinforcement learning. Q-learning aims at establishing a near optimal trade-off between exploration of the possible schedules and exploitation of the information gathered during such process. When the scheduler is presented with a choice between two competing actions, it uses the knowledge gathered before to bias the decision toward the action most likely to lead to the optimal schedules, i.e., the schedule that maximize (or minimize) the probability of satisfying a given property. Similarly to [159], the use of statistical methods allow to handle large programs, but can only provide probabilistic convergence guarantees, i.e., given enough time, the optimal schedule will be found with probability 1.

10.2 Beyond Linear Integer Arithmetics

Probabilistic symbolic execution relies on model counting procedures to compute the probability of satisfying the path condition leading to the occurrence of a target event. While efficient model counting procedures based on Barvinok's algorithm [156] are available for linear integer arithmetics (Latte [163], Barvinok [164]), other domains may require more complex counting procedures if not the (usually unfeasible) enumeration of all models.

Floating-point constraints have been studied in [165, 166] for a direct application to PSE. The authors use Monte Carlo techniques to compute the expected probability of generating a value that satisfy a given constraint from a prescribed input distribution. To increase the precision of the estimation, and to increase the reusability of previous (partial) solutions, the authors introduced a divide-and-conquer strategy where complex constraints are partitioned into independent subconstraints that can be analyzed independently. The results are then composed by propagating both the estimate and a measure of convergence (variance) through the composition rules. Furthermore, interval constraints propagation techniques [167] are used to reduce the uncertainty about the estimate by focusing the Monte Carlo sampling only toward the regions of the input domain that may contain models of a constraints, pruning out the rest of the domain. However, Monte Carlo approaches have usually limited effectiveness when the number of models of a constraint is significantly smaller than the size of the domain, because the randomized sampling is unlikely to hit any such models out of chance. In this case, more complex techniques can be used [154], which rely on bitblasting to reduce the estimation problem to counting the models of a SAT constraint (#SAT).

String constraints and counting procedures for their models have received a significant attention in recent years. In [168], the authors propose a method based on ordinary generating function for the approximate counting of the models of string constraints. The supported constraints include predicates stating a string is matched by a regular expression, equality between strings, substring tests, and a limited set of predicates on the length of a string. The procedure computes approximate counts in the form of intervals that soundly enclose the exact count; the generating functions computing the intervals' bounds are obtained directly from the syntactic structure of the string constraint. In [169], the authors propose an automata-based model counting procedure for string constraints. The procedure is composed of two steps. First the constraint is encoded in an automata that accepts all only the strings satisfying a given constraint. Then, established methods based on generating functions are used to count the number of accepting paths of the automata, up to a maximum string length. This procedure allows to obtain exact counts for linear string constraints. To increase the expressiveness of the constraint language, the authors also support a set of nonlinear string constraints, for which an overapproximating automata is built, in turn producing an upper bound on the number of models.

Mixed theories and other constraints represent the combination of constraints whose satisfaction is to be evaluated over multiple theories. For example, a string may be required to be matched by a certain regular expression, while its length must satisfy a nonlinear integer constraint. In these cases, more general #SMT solvers may be needed. As exact counting in such setting may result in complexity close to the enumeration of all the models of a constraint, approximate solution have been proposed, as in [170], where established hash-based approximate counting techniques from the realm of #SAT [171, 172] are adapted for #SMT problems. This area has the broad potential for supporting model counting for the broader set of constraints currently handled by SMT solvers, though still suffer of limited scalability.

The definition of logics for expressing properties of data structures and the corresponding algorithms for counting the number of their models is also an open problem, with only limited solutions produced so far in literature [173].

11. TOOLS SUPPORT

- Symbolic PathFinder [79] (SPF): Built as an extension of Java Path-Finder, a well-known software model checker, SPF performs a nonstandard interpretation of JVM bytecodes to enable static symbolic execution. SPF supports multithreaded programs by leveraging the underlying model checking framework to explore all possible thread interleavings. The tool is open-source and available for download at https://babelfish.arc.nasa.gov/trac/jpf/wiki/projects/jpf-symbc.

- KLEE [6]: The KLEE LLVM Execution Engine is a symbolic virtual machine for LLVM bytecode, although the main focus are programs written in the C language. KLEE supports a large amount of system calls through a library of models, allowing the software under test to interact with the environment. As one of the most popular open-source symbolic execution tools, many extensions of KLEE are available, such as Cloud9 (distributed symbolic execution, available at http://cloud9.epfl.ch/). KLEE is available for download at https://klee.github.io/.

- JBSE [174]: JBSE is a symbolic virtual machine for JVM bytecodes, in a similar fashion to SPF. The main feature of JBSE is the support for complex, user-friendly assumptions over the heap shape of the program to cull the search space. Users can annotate methods that should be used to validate new objects (*repOK*), specify LICS rules over symbolic references [84] and trigger instrumentation methods when new clauses containing specific references are added to the path condition. The tool is open-source and available at https://github.com/pietrobraione/jbse.

- KeY [175]: KeY is a program verification platform for Java programs. Among other features, KeY contains a symbolic execution engine and a visual symbolic debugger that allows the user to step through the resulting symbolic tree. The tool is open-source and available at https://www.key-project.org/.

- S2E [176]: S2E is a platform for symbolic execution of large systems, such as the combination of binaries + kernel + drivers. Unlike most other symbolic execution tools, S2E performs in vivo analysis through the use of virtualization and dynamic translation of x86/ARM instructions. S2E ameliorates the path explosion problem by interleaving concrete/symbolic execution at different parts of the stack (*selective symbolic execution*) and allowing users to relax the consistency models of the analysis to improve scalability. The tool is open-source (but not free for commercial use) and available at http://s2e.systems/

- SAGE [177]: SAGE is a dynamic symbolic execution tool for x86 binaries utilized internally at Microsoft. SAGE is likely the largest deployment of symbolic execution for test generation, with 3.4 billion constraints generated and solved as of 2014. A commercial service offered by Microsoft that leverages SAGE, named "Security Risk Detection", is available for preview at https://www.microsoft.com/en-us/security-risk-detection.

- Jalangi2: Jalangi2 is a platform for program analysis of Javascript programs. Developers can use Jalangi2 to instrument the source code of their programs, which can then be executed with any out-of-shelf Javascript interpreter (Node.js or a browser). The source code can be found at https://github.com/Samsung/jalangi2.

- Manticore: Manticore is a symbolic execution tool for x86/ARM binaries and Ethereum smart contracts. It supports both out-of-box usage and can be customized through its python interface. The tool is open source and can be found at https://github.com/trailofbits/manticore/.

- FuzzBALL: FuzzBALL is a symbolic execution tool for x86/ARM binaries. FuzzBALL is built on top of the Vine [178] intermediate language/static analysis library. Source code is available at http://bitblaze.cs.berkeley.edu/fuzzball.html

- Angr [179]: Angr is a python platform for program analysis and reverse engineering of x86 binaries, developed for the DARPA Cyber Grand Challenge competition. Among more traditional features, Angr provides tools for automatic exploit generation and patching of binaries and a management interface. The platform is open source and available at http://angr.io/.

- Triton [180]: Triton is a dynamic symbolic execution tool for x86/ARM binaries. Features include a taint propagation engine and a python API for customization. The tool is open source and can be found at https://triton.quarkslab.com/.

- JDart [181]: JDart is a dynamic symbolic execution tool for Java programs, built on top of Java PathFinder. The tool is designed to allow developers to replace existing components, such as search strategies or the bytecode semantics, by custom implementations in an easy way. Other features (provided by extensions) include generation of abstract method summaries and the creation of JUnit test suites. JDart is open source, and can be downloaded at: https://github.com/psycopaths/jdart

- CIVL [182]: CIVL is a framework for the specification and verification of programs written in CIVL-C, an intermediate language similar to C with additional primitives for concurrency. The framework include tools such as a symbolic execution-based model checker (capable of checking safety properties and functional equivalence between two programs) and translators for other languages/APIs (e.g., OpenMP, CUDA, Pthreads). CIVL is open source and can be downloaded at https://vsl.cis.udel.edu/civl/.

12. CONCLUSION

This chapter has reviewed the latest advances in symbolic execution in the last 5 years. Specifically, we have presented recent advances in addressing the main challenges of symbolic execution such as constraint solving, path explosion, concurrency, etc., and we have also reported advances in three application areas of symbolic execution, i.e., testing, security, and probabilistic program analysis. That said, we have only sketched a few important areas of advances here, since there is a lot of work related to this subject and it is impossible to include everything in one article. More work can be found in previous surveys [183–186] for a complete review of symbolic execution, especially for the advances before 5 years ago.

REFERENCES

[1] J.C. King, Symbolic execution and program testing, Commun. ACM 19 (7) (1976) 385–394.
[2] L.A. Clarke, A program testing system, in: Proc. of the 1976 Annual Conference, ACM '76, 1976, pp. 488–491.
[3] D. Kroening, O. Strichman, Decision Procedures: An Algorithmic Point of View, first ed., Springer Publishing Company, Incorporated, 2008. ISBN: 3540741046, 9783540741046.
[4] P. Godefroid, N. Klarlund, K. Sen, DART: directed automated random testing, in: Proceedings of the 2005 ACM SIGPLAN Conference on Programming Language Design and Implementation, PLDI '05, ACM, New York, NY, USA, 2005, ISBN: 1-59593-056-6, pp. 213–223, https://doi.org/10.1145/1065010.1065036.
[5] K. Sen, D. Marinov, G. Agha, CUTE: a concolic unit testing engine for C, in: Proceedings of the 10th European Software Engineering Conference Held Jointly with 13th ACM SIGSOFT International Symposium on Foundations of Software Engineering, ESEC/FSE-13, ACM, New York, NY, USA, 2005, ISBN: 1-59593-014-0, pp. 263–272, https://doi.org/10.1145/1081706.1081750.
[6] C. Cadar, D. Dunbar, D. Engler, KLEE: unassisted and automatic generation of high-coverage tests for complex systems programs, in: Proceedings of the Eighth USENIX Conference on Operating Systems Design and Implementation, OSDI'08, USENIX Association, Berkeley, CA, USA, 2008, pp. 209–224, http://dl.acm.org/citation.cfm?id=1855741.1855756.

[7] N. Tillmann, J. De Halleux, Pex: white box test generation for .NET, in: Proceedings of the Second International Conference on Tests and Proofs, TAP'08, Springer-Verlag, Berlin, Heidelberg, 2008. ISBN: 3-540-79123-X, 978-3-540-79123-2, pp. 134–153. http://dl.acm.org/citation.cfm?id=1792786.1792798.

[8] A. Biere, A. Biere, M. Heule, H. van Maaren, T. Walsh, Handbook of Satisfiability: Volume 185 Frontiers in Artificial Intelligence and Applications, IOS Press, Amsterdam, The Netherlands, 2009. ISBN: 1586039296, 9781586039295.

[9] W. Visser, J. Geldenhuys, M.B. Dwyer, Green: reducing, reusing and recycling Constraints in Program Analysis, in: Proceedings of the ACM SIGSOFT 20th International Symposium on the Foundations of Software Engineering—FSE '12, ACM Press, New York, NY, USA, 2012. ISBN: 9781450316149, p. 1, https://doi.org/10.1145/2393596.2393665. http://dl.acm.org/citation.cfm?doid=2393596.2393665.

[10] X. Jia, C. Ghezzi, S. Ying, Enhancing reuse of constraint solutions to improve symbolic execution, vol. abs/1501.0. in: Proceedings of the 2015 International Symposium on Software Testing and Analysis—ISSTA, ACM Press, New York, NY, USA, 2015. ISBN: 9781450336208, pp. 177–187, https://doi.org/10.1145/2771783.2771806. http://arxiv.org/abs/1501.07174; http://dl.acm.org/citation.cfm?doid=2771783.2771806.

[11] A. Aquino, F.A. Bianchi, M. Chen, G. Denaro, M. Pezzè, V.G. Buffi, Reusing constraint proofs in program analysis, in: International Symposium on Software Testing and Analysis (ISSTA 2015), 2015, ISBN: 9781450336208, pp. 305–315, https://doi.org/10.1145/2771783.2771802.

[12] A. Romano, D. Engler, Expression reduction from programs in a symbolic binary executor, in: Lecture Notes in Computer Science (LNCS) (Including Subseries Lecture Notes in Artificial Intelligence and Lecture Notes in Bioinformatics), ISSN 03029743, vol, 7976, 2013. ISBN: 9783642391750, pp. 301–319, https://doi.org/10.1007/978-3-642-39176-7-19.

[13] J. Lloyd, E. Sherman, Minimizing the size of path conditions using convex polyhedra abstract domain, ACM SIGSOFT Softw. Eng. Notes 40 (1) (2015) 1–5. ISSN: 01635948, https://doi.org/10.1145/2693208.2693244. http://dl.acm.org/citation.cfm?doid=2693208.2693244.

[14] Y. Zhang, Z. Chen, J. Wang, Speculative symbolic execution, in: Proceedings—International Symposium on Software Reliability Engineering, ISSRE, ISSN 10719458, 2012. ISBN: 9780769548883, pp. 101–110, https://doi.org/10.1109/ISSRE.2012.8. arXiv:1205.4951v2.

[15] Y. Zhang, Z. Chen, J. Wang, S2PF speculative symbolic pathFinder, ACM SIGSOFT Softw. Eng. Notes 37 (6) (2012) 1. ISSN: 01635948, https://doi.org/10.1145/2382756.2382792. http://dl.acm.org/citation.cfm?doid=2382756.2382792.

[16] S. Kausler, E. Sherman, User-defined backtracking criteria for symbolic execution, ACM SIGSOFT Softw. Eng. Notes 39 (1) (2014) 1–5. ISSN: 01635948, https://doi.org/10.1145/2557833.2560578. http://dl.acm.org/citation.cfm?doid=2557833.2560578.

[17] N. Bjørner, N. Tillmann, A. Voronkov, Path feasibility analysis for string-manipulating programs, in: S, Kowalewski, A. Philippou (Eds.), Tools and Algorithms for the Construction and Analysis of Systems: 15th International Conference, TACAS 2009, Held as Part of the Joint European Conferences on Theory and Practice of Software, ETAPS 2009, York, UK, March 22–29, 2009. Proceedings, Springer, Berlin, Heidelberg, 2009. ISBN: 978-3-642-00768-2, pp. 307–321, https://doi.org/10.1007/978-3-642-00768-2_27.

[18] A.S. Christensen, A. Møller, M.I. Schwartzbach, Precise analysis of string expressions, in: Proc. 10th International Static Analysis Symposium (SAS), LNCS, vol. 2694, Springer-Verlag, 2003, pp. 1–18. Available from http://www.brics.dk/JSA/.

[19] A. Kiezun, V. Ganesh, P.J. Guo, P. Hooimeijer, M.D. Ernst, HAMPI: a solver for string constraints, in: Proceedings of the 18th International Symposium on Software Testing and Analysis, ISSTA '09, ACM, New York, NY, USA, 2009, ISBN: 978-1-60558-338-9, pp. 105–116, https://doi.org/10.1145/1572272.1572286.

[20] S. Kausler, E. Sherman, Evaluation of string constraint solvers in the context of symbolic execution, in: Proceedings of the 29th ACM/IEEE International Conference on Automated Software Engineering—ASE '14, ACM Press, New York, NY, USA, 2014. ISBN: 9781450330138, pp. 259–270, https://doi.org/10.1145/2642937.2643003. http://dl.acm.org/citation.cfm?doid=2642937.2643003, August.

[21] Y. Zheng, X. Zhang, V. Ganesh, Z3-str: a Z3-based string solver for web application analysis, in: Proceedings of the 2013 9th Joint Meeting on Foundations of Software Engineering, 2013, ISBN: 978-1-4503-2237-9, pp. 114–124, https://doi.org/10.1145/2491411.2491456.

[22] M.T. Trinh, D.H. Chu, J. Jaffar, Progressive reasoning over recursivelydefined strings, in: Lecture Notes in Computer Science (Including Subseries Lecture Notes in Artificial Intelligence and Lecture Notes in Bioinformatics), ISSN 16113349, vol, 9779, 2016. ISBN: 9783319415277, pp. 218–240, https://doi.org/10.1007/978-3-319-41528-4_12, arXiv:1301.4779.

[23] M.-T. Trinh, D.-H. Chu, J. Jaffar, S3: a symbolic string solver for vulnerability detection in web applications, in: Proceedings of the 2014 ACM SIGSAC Conference on Computer and Communications Security, CCS '14, ACM, New York, NY, USA, 2014, ISBN: 978-1-4503-2957-6, pp. 1232–1243, https://doi.org/10.1145/2660267.2660372.

[24] T. Liang, A. Reynolds, C. Tinelli, C. Barrett, M. Deters, A DPLL(T) theory solver for a theory of strings and regular expressions, in: Lecture Notes in Computer Science (LNCS) (Including Subseries Lecture Notes in Artificial Intelligence and Lecture Notes in Bioinformatics), vol, 8559, 2014. ISBN: 9783319088662, pp. 646–662, https://doi.org/10.1007/978-3-319-08867-9_43.

[25] P.A. Abdulla, M.F. Atig, Y.F. Chen, L. Holík, A. Rezine, P. Rümmer, J. Stenman, String constraints for verification, in: Lecture Notes in Computer Science (LNCS) (Including Subseries Lecture Notes in Artificial Intelligence and Lecture Notes in Bioinformatics), vol, 8559, 2014. ISBN: 9783319088662, pp. 150–166, https://doi.org/10.1007/978-3-319-08867-9_10.

[26] P.A. Abdulla, M.F. Atig, Y.F. Chen, L. Holík, A. Rezine, P. Rümmer, J. Stenman, Norn: an SMT solver for string constraints, in: Lecture Notes in Computer Science (Including Subseries Lecture Notes in Artificial Intelligence and Lecture Notes in Bioinformatics), ISSN 16113349, vol, 9206, Springer, Cham, 2015. ISBN: 9783319216898, pp. 462–469, https://doi.org/10.1007/978-3-319-21690-4_29.

[27] M. Borges, M. D'Amorim, S. Anand, D. Bushnell, C.S. Păsăreanu, in: Symbolic execution with interval solving and meta-heuristic search, Proceedings—IEEE Fifth International Conference on Software Testing, Verification and Validation, ICST 2012, ISSN 2159-4848, vol. 1, 2012. ISBN: 9780769546704, pp. 111–120, https://doi.org/10.1109/ICST.2012.91.

[28] P. Dinges, G. Agha, Solving complex path conditions through heuristic search on induced polytopes, in: Proceedings of the 22nd ACM SIGSOFT International Symposium on Foundations of Software Engineering, C, 2014, ISBN: 978-1-4503-3056-5, pp. 425–436, https://doi.org/10.1145/2635868.2635889.

[29] R. Bagnara, M. Carlier, R. Gori, A. Gotlieb, in: Symbolic path-oriented test data generation for floating-point programs, Proceedings—IEEE 6th International Conference on Software Testing, Verification and Validation, ICST 2013, ISSN 2159-4848, 2013. ISBN: 9780769549682, pp. 1–10, https://doi.org/10.1109/ICST.2013.17.

[30] Z. Fu, Z. Su, Xsat: a fast floating-point satisfiability solver, in: Lecture Notes in Computer Science (Including Subseries Lecture Notes in Artificial Intelligence and Lecture Notes in Bioinformatics), ISSN 16113349, vol, 9780, 2016. ISBN: 9783319415390, pp. 187–209, https://doi.org/10.1007/978-3-319-41540-6_11, arXiv:1301.4779.

[31] X. Li, Y. Liang, H. Qian, Y.-Q. Hu, L. Bu, Y. Yu, X. Chen, X. Li, Symbolic execution of complex program driven by machine learning based constraint solving, in: Proceedings of the 31st IEEE/ACM International Conference on Automated Software Engineering—ASE 2016, 2016. ISBN: 9781450338455, pp. 554–559, https://doi.org/10.1145/2970276.2970364. http://dl.acm.org/citation.cfm?doid=2970276.2970364.

[32] A. Tiwari, P. Lincoln, A Nonlinear Real Arithmetic Fragment, Springer-Verlag, Berlin, Heidelberg, 2014, pp. 729–736.

[33] L. Hadarean, K. Bansal, D. Jovanović, C. Barrett, C. Tinelli, A tale of two solvers: eager and lazy approaches to bit-vectors, in: Lecture Notes in Computer Science (LNCS) (Including Subseries Lecture Notes in Artificial Intelligence and Lecture Notes in Bioinformatics), ISSN 16113349, vol. 8559, 2014. ISBN: 9783319088662, pp. 680–695, https://doi.org/10.1007/978-3-319-08867-9_45.

[34] J.C. Reynolds, Separation logic: a logic for shared mutable data structures, in: Proceedings of the 17th Annual IEEE Symposium on Logic in Computer Science, LICS '02, IEEE Computer Society, Washington, DC, USA, 2002. ISBN: 0-7695-1483-9, pp. 55–74, http://dl.acm.org/citation.cfm?id=645683.664578.

[35] R. Piskac, T. Wies, D. Zufferey, Automating separation logic using SMT, in: Lecture Notes in Computer Science (LNCS) (Including Subseries Lecture Notes in Artificial Intelligence and Lecture Notes in Bioinformatics), ISSN 03029743, vol, 8044, 2013. ISBN: 9783642397981, pp. 773–789, https://doi.org/10.1007/978-3-642-39799-8_54.

[36] P. Daca, T.A. Henzinger, A. Kupriyanov, Array folds logic, in: Lecture Notes in Computer Science (Including Subseries Lecture Notes in Artificial Intelligence and Lecture Notes in Bioinformatics), ISSN 16113349, vol. 9780, 2016. ISBN: 9783319415390, pp. 230–248, https://doi.org/10.1007/978-3-319-41540-6_13, arXiv:1603.06850, https://link.springer.com/chapter/10.1007/978-3-319-41540-6_13.

[37] K. Bansal, A. Reynolds, T. King, C. Barrett, T. Wies, Deciding local theory extensions via e-matching, in: Computer Aided Verification: 27th International Conference, CAV 2015, San Francisco, CA, USA, July 18–24, 2015, Proceedings, Part II, 2015, pp. 87–105, https://doi.org/10.1007/978-3-319-21668-3_6.

[38] M. Cristiá, G. Rossi, in: A decision procedure for sets, binary relations and partial functions, Computer Aided Verification: 28th International Conference, CAV 2016, Toronto, ON, Canada, July 17–23, 2016, Proceedings, Part I, ISSN 18651348, vol. 1, 2016. ISBN: 9783319576329, pp. 179–198, https://doi.org/10.1007/978-3-319-57633-6_11, arXiv:9780201398298.

[39] Y. Li, Z. Su, L. Wang, X. Li, Steering symbolic execution to less traveled paths, ACM SIGPLAN Notices 48 (10) (2013) 19–32. ISSN: 03621340, https://doi.org/10.1145/2544173.2509553. http://dl.acm.org/citation.cfm?id=2544173.2509553.

[40] H. Seo, S. Kim, How we get there: a context-guided search strategy in concolic testing, in: Proceedings of the 22nd ACM SIGSOFT International Symposium on Foundations of Software Engineering 2014, ISBN: 978-1-4503-3056-5, pp. 413–424, https://doi.org/10.1145/2635868.2635872.

[41] M. Christakis, P. Müller, V. Wüstholz, in: Guiding dynamic symbolic execution toward unverified program executions, Proceedings of the 38th International Conference on Software Engineering—ICSE '16, ISSN 02705257, 2016. ISBN: 9781450339001, pp. 144–155, https://doi.org/10.1145/2884781.2884843. http://dl.acm.org/citation.cfm?doid=2884781.2884843.

[42] D. Coughlin, B.-Y.E. Chang, A. Diwan, J.G. Siek, Measuring enforcement windows with symbolic trace interpretation: what well-behaved programs say, in: Proc. ISSTA, 2012. ISBN: 9781450314541, pp. 276–286, https://doi.org/10.1145/2338965. 2336786. http://dl.acm.org/citation.cfm?id=04000800.2336786.

[43] Y. Zhang, Z. Chen, J. Wang, W. Dong, Z. Liu, in: Regular property guided dynamic symbolic execution, Proceedings—International Conference on Software Engineering, ISSN 02705257, vol. 1, 2015. ISBN: 9781479919345, pp. 643–653, https://doi.org/10.1109/ICSE.2015.80.

[44] G. Yang, S. Khurshid, S. Person, N. Rungta, in: Property differencing for incremental checking, Proceedings of the 36th International Conference on Software Engineering—ICSE 2014, ISSN 02705257, 2014. ISBN: 9781450327565, pp. 1059–1070, https://doi.org/10.1145/2568225.2568319. http://dl.acm.org/ citation.cfm?doid=2568225.2568319.

[45] G. Yang, S. Person, N. Rungta, S. Khurshid, Directed incremental symbolic execution, ACM Trans, Softw. Eng. Methodol. 24 (1) (2014) 3:1–3:42. ISSN: 1049-331X, https://doi.org/10.1145/2629536.

[46] S. Person, G. Yang, N. Rungta, S. Khurshid, in: Directed incremental symbolic execution, Proceedings of the 32nd ACM SIGPLAN conference on Programming language design and implementation—PLDI '11, ISSN 03621340, 2011. ISBN: 9781450306638. p, p. 504, https://doi.org/10.1145/1993498.1993558. http:// portal.acm.org/citation.cfm?doid=1993498.1993558.

[47] J. Backes, S. Person, N. Rungta, O. Tkachuk, Regression verification using impact summaries, in: Lecture Notes in Computer Science (LNCS) (Including Subseries Lecture Notes in Artificial Intelligence and Lecture Notes in Bioinformatics), ISSN 03029743, vol, 7976, 2013. ISBN: 9783642391750, pp. 99–116, https://doi.org/ 10.1007/978-3-642-39176-7-7.

[48] S. Makhdoom, M.A. Khan, J.H. Siddiqui, Incremental symbolic execution for automated test suite maintenance, in: Proceedings of the 29th ACM/IEEE International Conference on Automated Software Engineering—ASE '14, ACM Press, New York, NY, USA, 2014. ISBN: 9781450330138, pp. 271–276, https://doi.org/ 10.1145/2642937.2642961. http://dl.acm.org/citation.cfm?doid=2642937.2642961.

[49] Q. Yi, Z. Yang, S. Guo, C. Wang, J. Liu, C. Zhao, in: Postconditioned symbolic execution, 2015 IEEE 8th International Conference on Software Testing, Verification and Validation, ICST 2015—Proceedings, ISSN 2159-4848, 2015. ISBN: 9781479971251https://doi.org/10.1109/ICST.2015.7102601.

[50] Y. Li, S.C. Cheung, X. Zhang, Y. Liu, Scaling up symbolic analysis by removing Z-equivalent states, Tsem14 23 (4) (2014) 1–32. ISSN: 1049331X, https://doi.org/ 10.1145/2652484. http://dl.acm.org/citation.cfm?id=2668018.2652484.

[51] G. Yang, C.S. Păsăreanu, S. Khurshid, Memoized symbolic execution, in: Proceedings of the 2012 International Symposium on Software Testing and Analysis—ISSTA 2012, ACM Press, New York, NY, USA, 2012, ISBN: 9781450314541, p. 144. https://dl.acm. org/citation.cfm?id=2336771.

[52] V. Kuznetsov, J. Kinder, S. Bucur, G. Candea, Efficient state merging in symbolic execution, ACM SIGPLAN Notices 47 (6) (2012) 193–204. ISSN: 03621340, https:// doi.org/10.1145/2345156.2254088. http://dl.acm.org/citation.cfm?doid=2345156. 2254088.

[53] D. Scheurer, H. Reiner, A General Lattice Model for Merging Symbolic Execution Branches, vol, 10009, Springer, Cham, 2016. ISBN: 978-3-319-47845-6, pp. 57–73, https://doi.org/10.1007/978-3-319-47846-3.

[54] J. Jaffar, V. Murali, J.a. Navas, Boosting concolic testing via interpolation, in: Proceedings of the 2013 Ninth Joint Meeting on Foundations of Software Engineering—ESEC/FSE 2013, 2013. ISBN: 9781450322379, pp. 53–63, https://doi.org/10.1145/2491411. 2491425. http://dl.acm.org/citation.cfm?id=2491411.2491425.

[55] K. Sen, G. Necula, L. Gong, W. Choi, MultiSE: multi-path symbolic execution using value summaries, in: Joint Meeting on Foundations of Software Engineering, 2015. ISBN: 9781450336758, pp. 842–853, https://doi.org/10.1145/2786805.2786830. http://www.eecs.berkeley.edu/Pubs/TechRpts/2014/EECS-2014-173.html%5Cn http://dl.acm.org/citation.cfm?doid=2786805.2786830.

[56] T. Avgerinos, A. Rebert, S.K. Cha, D. Brumley, Enhancing symbolic execution with veritesting, in: Proceedings of the 36th International Conference on Software Engineering—ICSE, 2014. ISBN: 9781450327565, pp. 1083–1094, https://doi. org/10.1145/2568225.2568293. ISSN 9781450327565. http://dl.acm.org/citation. cfm?doid=2568225.2568293.

[57] J. Strejček, M. Trtík, Abstracting path conditions, in: Proc. ISSTA, 2012. ISBN: 9781450314541, pp. 155–165, https://doi.org/10.1145/2338965.2336772. http:// dl.acm.org/citation.cfm?id=04000800.2336772, arXiv:1112.5671.

[58] R. Kersten, S. Person, N. Rungta, O. Tkachuk, Improving coverage of test cases generated by symbolic pathfinder for programs with loops, ACM SIGSOFT Softw. Eng. Notes 40 (1) (2015) 1–5. ISSN: 01635948, https://doi.org/10.1145/ 2693208.2693243. http://dl.acm.org/citation.cfm?doid=2693208.2693243.

[59] X. Xie, Y. Liu, W. Le, X. Li, H. Chen, S-looper: automatic summarization for multipath string loops, in: Proceedings of the 2015 International Symposium on Software Testing and Analysis—ISSTA 2015, 2015. ISBN: 9781450336208, pp. 188–198, https://doi.org/10.1145/2771783.2771815. http://dl.acm.org/citation. cfm?doid=2771783.2771815.

[60] W. Le, in: Segmented symbolic analysis, Proceedings—International Conference on Software Engineering, ISSN 02705257, 2013. ISBN: 9781467330763, pp. 212–221, https://doi.org/10.1109/ICSE.2013.6606567.

[61] J.H. Siddiqui, S. Khurshid, in: Scaling symbolic execution using ranged analysis, Proceedings of the ACM International Conference on Object Oriented Programming Systems Languages and Applications, ISSN 03621340, 2012. ISBN: 978-1-4503-1561-6, pp. 523–536, https://doi.org/10.1145/2384616.2384654. http://doi.acm. org/10.1145/2384616.2384654.

[62] R. Qiu, Scaling and Certifying Symbolic Execution (PhD dissertation), University of Texas at Austin, 2016.

[63] R. Qiu, S. Khurshid, C.S. Păsăreanu, G. Yang, A synergistic approach for distributed symbolic execution using test ranges, in: ICSE '17—Companion, 2017, pp. 130–132, https://doi.org/10.1109/ICSE-C.2017.116.

[64] Y. Kim, M. Kim, Y.J. Kim, Y. Jang, in: Industrial application of concolic testing approach: a case study on libexif by using CREST-BV and KLEE, Proceedings—International Conference on Software Engineering, ISSN 02705257, 2012. ISBN: 9781467310673, pp. 1143–1152, https://doi.org/10.1109/ICSE.2012.6227105.

[65] Y. Lin, T. Miller, Looking closer at compositional symbolic execution, in: Proceedings of the ASWEC 2015 24th Australasian Software Engineering Conference, ASWEC '15 Vol, II,, ACM, New York, NY, USA, 2015. ISBN: 978-1-4503-3796-0, pp. 138–140, https://doi.org/10.1145/2811681.2817758.

[66] P. Godefroid, Compositional dynamic test generation, SIGPLAN Not. 42 (1) (2007) 47–54. ISSN: 0362-1340, https://doi.org/10.1145/1190215.1190226.

[67] P. Godefroid, N. Klarlund, K. Sen, DART: directed automated random testing, in: Proceedings of the 2005 ACM SIGPLAN Conference on Programming Language Design and Implementation, PLDI '05, ACM, New York, NY, USA, 2005, ISBN: 1-59593-056-6, pp. 213–223, https://doi.org/10.1145/1065010.1065036.

[68] S. Anand, P. Godefroid, N. Tillmann, Demand-driven compositional symbolic execution, in: Proceedings of the Theory and Practice of Software, 14th International Conference on Tools and Algorithms for the Construction and Analysis of Systems, TACAS'08/ ETAPS'08, Springer-Verlag, Berlin, Heidelberg, 2008. ISBN: 3-540-78799-2, 978-3-540-78799-0, pp. 367–381. http://dl.acm.org/citation.cfm?id=1792734.1792771.

[69] P. Godefroid, A.V. Nori, S.K. Rajamani, S.D. Tetali, Compositional may-must program analysis: unleashing the power of alternation, SIGPLAN Not, 45 (1) (2010) 43–56. ISSN: 0362-1340, https://doi.org/10.1145/1707801.1706307.

[70] J.M. Rojas, C.S. Păsăreanu, Compositional Symbolic Execution Through Program Specialization, 2013.

[71] N.D. Jones, C.K. Gomard, P. Sestoft, Partial Evaluation and Automatic Program Generation, Prentice-Hall, Inc., Upper Saddle River, NJ, USA, 1993. ISBN: 0-13-020249-5.

[72] R. Qiu, G. Yang, C.S. Păsăreanu, S. Khurshid, Compositional symbolic execution with memoized replay, in: Proceedings of the 37th International Conference on Software Engineering—Volume 1, IEEE Press, Piscataway, NJ, USA, 2015. ISBN: 978-1-4799-1934-5ICSE '15, pp. 632–642. http://dl.acm.org/citation.cfm?id=2818754.2818832.

[73] Y. Lin, T. Miller, H. Sondergaard, Compositional symbolic execution using fine-grained summaries, in: Proceedings of the 2015 24th Australasian Software Engineering Conference (ASWEC), IEEE Computer Society, Washington, DC, USA, 2015. ISBN: 978-1-4673-9390-4ASWEC'15, pp. 213–222, https://doi.org/10.1109/ASWEC.2015.32.

[74] S. Ognawala, M. Ochoa, A. Pretschner, T. Limmer, MACKE: compositional analysis of low-level vulnerabilities with symbolic execution, in: 2016 31st IEEE/ACM International Conference on Automated Software Engineering (ASE), 2016, pp. 780–785.

[75] I. Pustogarov, T. Ristenpart, V. Shmatikov, Using program analysis to synthesize sensor spoofing attacks, in: Proceedings of the 2017 ACM on Asia Conference on Computer and Communications Security, ASIA CCS'17, ACM, New York, NY, USA, 2017. ISBN: 978-1-4503-4944-4, pp. 757–770, https://doi.org/10.1145/3052973.3053038.

[76] Q.L. Le, J. Sun, W.-N. Chin, Satisfiability modulo heap-based programs, in: Lecture Notes in Computer Science (Including Subseries Lecture Notes in Artificial Intelligence and Lecture Notes in Bioinformatics), vol. 9779, 2016. ISBN: 9783319415277, pp. 382–404, https://doi.org/10.1007/978-3-319-41528-4_21. arXiv:1301.4779.

[77] P. Müller, M. Schwerhoff, A.J. Summers, Automatic verification of iterated separating conjunctions using symbolic execution, in: Lecture Notes in Computer Science (Including Subseries Lecture Notes in Artificial Intelligence and Lecture Notes in Bioinformatics), vol. 9779, 2016. ISBN: 9783319415277, pp. 405–425, https://doi.org/10.1007/978-3-319-41528-4_22, arXiv:1603.00649.

[78] J. Geldenhuys, N. Aguirre, M.F. Frias, W. Visser, Bounded lazy initialization, in: G. Brat, N. Rungta, A. Venet (Eds.), NASA Formal Methods: 5th International Symposium, NFM 2013, Moffett Field, CA, USA, May 14–16, 2013. Proceedings, Springer, Berlin, Heidelberg, 2013. ISBN: 978-3-642-38088-4, pp. 229–243, https://doi.org/10.1007/978-3-642-38088-4_16.

[79] K.S. Luckow, C.S. Păsăreanu, Symbolic pathFinder v7, ACM SIGSOFT Softw. Eng. Notes 39 (1) (2014) 1–5. ISSN: 01635948, https://doi.org/10.1145/2557833.2560571. http://dl.acm.org/citation.cfm?doid=2557833.2560571.

[80] N. Rosner, J. Geldenhuys, N. Aguirre, W. Visser, M. Frias, BLISS: improved symbolic execution by bounded lazy initialization with SAT support, IEEE Trans. Softw. Eng. 41 (7) (2015) 1–1, ISSN: 0098-5589, https://doi.org/10.1109/TSE.2015.2389225. http://ieeexplore.ieee.org/document/7004061/.

[81] J. Belt, Robby, P. Chalin, J. Hatcliff, X. Deng, Efficient symbolic execution of value-based data structures for critical systems, in: NASA Formal Methods: 4th International Symposium, NFM 2012, Norfolk, VA, USA, April 3–5, 2012. Proceedings, Springer, Berlin, Heidelberg, 2012, pp. 295–309, https://doi.org/10.1007/978-3-642-28891-3_29.

[82] B. Carre, J. Garnsworthy, in: SPARK—an annotated Ada subset for safety-critical programming, Proceedings of the Conference on TRI-ADA 1990, Baltimore, MD, USA, 1990, pp. 392–402, https://doi.org/10.1145/255471.255563.

[83] S. Khurshid, C.S. Păsăreanu, W. Visser, Generalized symbolic execution for model checking and testing, in: H. Garavel, J. Hatcliff (Eds.), Tools and Algorithms for the Construction and Analysis of Systems: 9th International Conference, TACAS 2003 Held as Part of the Joint European Conferences on Theory and Practice of Software, ETAPS 2003 Warsaw, Poland, April 7–11, 2003 Proceedings, Springer, Berlin, Heidelberg, 2003. ISBN: 978-3-540-36577-8, pp. 553–568, https://doi.org/10.1007/3-540-36577-X_40.

[84] P. Braione, G. Denaro, M. Pezzè, Enhancing symbolic execution with built-in term rewriting and constrained lazy initialization, in: Proceedings of the 2013 9th Joint Meeting on Foundations of Software Engineering—ESEC/FSE 2013, 2013. ISBN: 9781450322379, p. 411, https://doi.org/10.1145/2491411.2491433. http://dl.acm.org/citation.cfm?doid=2491411.2491433.

[85] P. Braione, G. Denaro, M. Pezzè, Symbolic execution of programs with heap inputs, in: Proceedings of the 2015 10th Joint Meeting on Foundations of Software Engineering, 2015. ISBN: 978-1-4503-3675-8, pp. 602–613, https://doi.org/10.1145/2786805.2786842.

[86] P. Braione, G. Denaro, M. Pezzè, JBSE: a symbolic executor for Java programs with complex heap inputs, in: Proceedings of the 2016 24th ACM SIGSOFT International Symposium on Foundations of Software Engineering—FSE 2016, ACM Press, New York, NY, USA, 2016. ISBN: 9781450342186, pp. 1018–1022, https://doi.org/10.1145/2950290.2983940. http://dl.acm.org/citation.cfm?doid=2950290.2983940.

[87] A. Romano, D.R. Engler, symMMU: symbolically executed runtime libraries for symbolic memory access, in: Proceedings of the 29th ACM/IEEE International Conference on Automated Software Engineering—ASE '14, ACM Press, New York, NY, USA, 2014. ISBN: 9781450330138, pp. 247–258, https://doi.org/10.1145/2642937.2642974. http://dl.acm.org/citation.cfm?doid=2642937.2642974.

[88] A. Fromherz, K.S. Luckow, C.S. Păsăreanu, Symbolic arrays in symbolic pathFinder, ACM SIGSOFT Softw. Eng. Notes 41 (6) (2016) 1–5. ISSN: 01635948, https://doi.org/10.1145/3011286.3011296. http://dl.acm.org/citation.cfm?doid=3011286.3011296.

[89] S. Blackshear, B.-Y.E. Chang, M. Sridharan, Thresher: precise refutations for heap reachability, in: Proceedings of the 34th ACM SIGPLAN Conference on Programming Language Design and Implementation, PLDI '13, ACM, New York, NY, USA, 2013. ISBN: 978-1-4503-2014-6, pp. 275–286, https://doi.org/10.1145/2491956.2462186.

[90] H. Zhu, G. Petri, S. Jagannathan, Poling: SMT aided linearizability proofs, in: Lecture Notes in Computer Science (Including Subseries Lecture Notes in Artificial Intelligence and Lecture Notes in Bioinformatics), ISSN 16113349, vol. 9207, , 2015. ISBN: 9783319216676, pp. 3–19, https://doi.org/10.1007/978-3-319-21668-3 1.

[91] H.V. Nguyen, H.A. Nguyen, T.T. Nguyen, A.T. Nguyen, T.N. Nguyen, Dangling references in multi-configuration and dynamic PHP-based Web applications, in: 2013 28th IEEE/ACM International Conference on Automated Software Engineering, ASE 2013–Proceedings, 2013. ISBN: 9781479902156, pp. 399–409, https://doi.org/10.1109/ASE.2013.6693098.

[92] Y. Smaragdakis, J. Evans, C. Sadowski, J. Yi, C. Flanagan, Sound Predictive Race Detection in Polynomial Time, in: Proceedings of the 39th Annual ACM SIGPLAN-SIGACT Symposium on Principles of Programming Languages, POPL '12, ACM, New York, NY, USA, 2012. ISBN: 978-1-4503-1083-3, pp. 387–400, https://doi.org/10.1145/2103656.2103702.

[93] C. Wang, S. Kundu, M. Ganai, A. Gupta, Symbolic predictive analysis for concurrent programs, in: Proceedings of the Second World Congress on Formal Methods, FM '09, 2009. ISBN: 978-3-642-05088-6, pp. 256–272.

[94] M. Said, C. Wang, Z. Yang, K. Sakallah, Generating data race witnesses by an SMT-based analysis, in: Proceedings of the Third International Conference on NASA Formal Methods, NFM'11, Springer-Verlag, Berlin, Heidelberg, 2011. ISBN: 978-3-642-20397-8, pp. 313–327. http://dl.acm.org/citation.cfm?id=1986308.1986334.

[95] P. Liu, O. Tripp, X. Zhang, IPA: improving predictive analysis with pointer analysis, in: Proceedings of the 25th International Symposium on Software Testing and Analysis, ISSTA 2016, ACM, New York, NY, USA, 2016. ISBN: 978-1-4503-4390-9, pp. 59–69, https://doi.org/10.1145/2931037.2931046.

[96] N. Razavi, F. Ivancic, V. Kahlon, A. Gupta, Concurrent test generation using concolic multi-trace analysis, in: Programming Languages and Systems—10th Asian Symposium, APLAS 2012, Kyoto, Japan, December 11–13, 2012. Proceedings, 2012, pp. 239–255, https://doi.org/10.1007/978-3-642-35182-2_17.

[97] J. Burnim, K. Sen, Heuristics for scalable dynamic test generation, in: Proceedings of the 2008 23rd IEEE/ACM International Conference on Automated Software Engineering, IEEE Computer Society, Washington, DC, USA, 2008. ISBN: 978-1-4244-2187-9ASE '08, pp. 443–446, https://doi.org/10.1109/ASE.2008.69.

[98] A. Farzan, A. Holzer, N. Razavi, H. Veith, Con2Colic testing, in: Proceedings of the 2013 9th Joint Meeting on Foundations of Software Engineering, ESEC/FSE 2013, ACM, New York, NY, USA, 2013. ISBN: 978-1-4503-2237-9, pp. 37–47, https://doi.org/10.1145/2491411.2491453.

[99] P. Deligiannis, A.F. Donaldson, Z. Rakamaric, Fast and precise symbolic analysis of concurrency bugs in device drivers (T), in: Proceedings of the 2015 30th IEEE/ACM International Conference on Automated Software Engineering (ASE), ASE '15, IEEE Computer Society, Washington, DC, USA, 2015. ISBN: 978-1-5090-0025-8, pp. 166–177, https://doi.org/10.1109/ASE.2015.30.

[100] V. Herdt, H.M. Le, D. Große, R. Drechsler, ParCoSS: Efficient Parallelized Compiled Symbolic Simulation, in: Springer International Publishing, Cham, 2016. ISBN: 978-3-319-41540-6, pp. 177–183, https://doi.org/10.1007/978-3-319-41540-6_10.

[101] G. Li, P. Li, G. Sawaya, G. Gopalakrishnan, I. Ghosh, S.P. Rajan, GKLEE: concolic verification and test generation for GPUs, in: Proceedings of the 17th ACM SIGPLAN Symposium on Principles and Practice of Parallel Programming, PPoPP '12, ACM, New York, NY, USA, 2012. ISBN: 978-1-4503-1160-1, pp. 215–224, https://doi.org/10.1145/2145816.2145844.

[102] S. Burckhardt, P. Kothari, M. Musuvathi, S. Nagarakatte, A randomized scheduler with probabilistic guarantees of finding bugs, in: Proceedings of the Fifteenth Edition of ASPLOS on Architectural Support for Programming Languages and Operating Systems, ASPLOS XV, ACM, New York, NY, USA, 2010. ISBN: 978-1-60558-839-1, pp. 167–178, https://doi.org/10.1145/1736020.1736040.

[103] M. Musuvathi, S. Qadeer, Iterative context bounding for systematic testing of multithreaded programs, in: Proceedings of the 28th ACM SIGPLAN Conference on Programming Language Design and Implementation, PLDI '07, ACM, New York, NY, USA, 2007. ISBN: 978-1-59593-633-2, pp. 446–455, https://doi.org/10.1145/1250734.1250785.

[104] K. Kähkönen, O. Saarikivi, K. Heljanko, Using unfoldings in automated testing of multithreaded programs, in: Proceedings of the 27th IEEE/ACM International Conference on Automated Software Engineering, ASE 2012, ACM, New York, NY, USA, 2012. ISBN: 978-1-4503-1204-2, pp. 150–159, https://doi.org/10.1145/2351676.2351698.

[105] K.L. McMillan, Using unfoldings to avoid the state explosion problem in the verification of asynchronous circuits, in: Proceedings of the Fourth International Workshop on Computer Aided Verification, CAV'92, Springer-Verlag, London, UK, 1993. ISBN: 3-540-56496-9, pp. 164–177. http://dl.acm.org/citation.cfm?id=647761.735341.

[106] S. Guo, M. Kusano, C. Wang, Z. Yang, A. Gupta, Assertion guided symbolic execution of multithreaded programs, in: Proceedings of the 2015 10th Joint Meeting on Foundations of Software Engineering, ESEC/FSE 2015, 2015. ISBN: 978-1-4503-3675-8, pp. 854–865.

[107] S. Guo, M. Kusano, C. Wang, Conc-iSE: incremental symbolic execution of concurrent software, in: Proceedings of the 31st IEEE/ACM International Conference on Automated Software Engineering, ASE 2016, ACM, New York, NY, USA, 2016. ISBN: 978-1-4503-3845-5, pp. 531–542, https://doi.org/10.1145/2970276. 2970332.

[108] T. Bergan, L. Ceze, D. Grossman, Input-covering schedules for multithreaded programs, in: Proceedings of the 2013 ACM SIGPLAN International Conference on Object Oriented Programming Systems Languages & Applications, OOPSLA '13, ACM, New York, NY, USA, 2013. ISBN: 978-1-4503-2374-1, pp. 677–692, https://doi.org/10.1145/2509136.2509508.

[109] T. Bergan, D. Grossman, L. Ceze, Symbolic execution of multithreaded programs from arbitrary program contexts, in: Proceedings of the 2014 ACM International Conference on Object Oriented Programming Systems Languages & Applications, OOPSLA '14, ACM, New York, NY, USA, 2014. ISBN: 978-1-4503-2585-1, pp. 491–506, https://doi.org/10.1145/2660193.2660200.

[110] K. Vorobyov, P. Krishnan, Combining static analysis and constraint solving for automatic test case generation, in: Proceedings—IEEE 5th International Conference on Software Testing, Verification and Validation, ICST 2012, 2012. ISBN: 9780769546704, pp. 915–920, https://doi.org/10.1109/ICST.2012.196.

[111] C. Pacheco, M.D. Ernst, Randoop: feedback-directed random testing for Java, in: Companion to the 22nd ACM SIGPLAN Conference on Object-oriented Programming Systems and Applications Companion, OOPSLA'07, ACM, New York, NY, USA, 2007. ISBN: 978-1-59593-865-7, pp. 815–816, https://doi.org/ 10.1145/1297846.1297902.

[112] P. Garg, F. Ivancic, G. Balakrishnan, N. Maeda, A. Gupta, in: Feedbackdirected unit test generation for C/C++ using concolic execution, Proceedings—International Conference on Software Engineering, ISSN 02705257, 2013. ISBN: 9781467330763, pp. 132–141, https://doi.org/10.1109/ICSE.2013.6606559.

[113] S. Bardin, N. Kosmatov, F. Cheynier, in: Efficient leveraging of symbolic execution to advanced coverage criteria, Proceedings—IEEE 7th International Conference on Software Testing, Verification and Validation, ICST 2014, ISSN 2159-4848, vol. 7, 2014. ISBN: 9780769551852, pp. 173–182, https://doi.org/10.1109/ICST.2014.30. arXiv:1308.4045.

[114] D. You, S. Rayadurgam, M. Whalen, M.P.E. Heimdahl, G. Gay, Efficient observability-based test generation by dynamic symbolic execution, in: 2015 IEEE 26th International Symposium on Software Reliability Engineering, ISSRE 2015, 2016. ISBN: 9781509004065, pp. 228–238, https://doi.org/10.1109/ISSRE. 2015.7381816.

[115] M. Whalen, G. Gay, D. You, M.P.E. Heimdahl, M. Staats, in: Observable modified condition/decision coverage, 2013 35th International Conference on Software Engineering (ICSE), ISSN 0270-5257, 2013, pp. 102–111, https://doi.org/10.1109/ ICSE.2013.6606556.

[116] K. Jamrozik, G. Fraser, N. Tillmann, J. De Halleux, Augmented dynamic symbolic execution, in: Proceedings of the 27th IEEE/ACM International Conference on Automated Software Engineering—ASE 2012, 2012. ISBN: 9781450312042, p. 254, https://doi.org/10.1145/2351676.2351716. http://dl.acm.org/citation.cfm? doid=2351676.2351716.

[117] D. Qi, H.D.T. Nguyen, A. Roychoudhury, in: Path exploration based on symbolic output, Proceedings of the 19th ACM SIGSOFT Symposium and the 13th European Conference on Foundations of Software Engineering SE—ESEC/FSE '11, ISSN 1049-331X, vol. 22, 2011. ISBN: 978-1-4503-0443-6, pp. 278–288, https://doi.org/10.1145/2025113.2025152. https://doi.org/10.1145/2025113.2025152%5Cn http://portaciteulike-article-id:9935732%5Cn.

[118] P. Dan Marinescu, C. Cadar, in: Make test-zesti: a symbolic execution solution for improving regression testing, Proceedings—International Conference on Software Engineering, ISSN 02705257, 2012. ISBN: 9781467310673, pp. 716–726, https://doi.org/10.1109/ICSE.2012.6227146.

[119] C. Cadar, H. Palikareva, Shadow symbolic execution for better testing of evolving software, in: ICSE, ISSN 02705257, 2014. ISBN: 978-1-4503-2768-8, pp. 432–435, https://doi.org/10.1145/2591062.2591104. http://dl.acm.org/citation.cfm?doid= 2591062.2591104%5Cnhttp://doi.acm.org/10.1145/2591062.

[120] W. Le, S.D. Pattison, in: Patch verification via multiversion interprocedural control flow graphs, Proceedings of the 36th International Conference on Software Engineering—ICSE 2014, ISSN 02705257, 2014. ISBN: 9781450327565, pp. 1047–1058, https://doi.org/10.1145/2568225.2568304. http://dl.acm.org/citation. cfm?doid=2568225.2568304.

[121] H. Yoshida, S. Tokumoto, M.R. Prasad, I. Ghosh, T. Uehara, FSX: fine-grained incremental unit test generation for C/C++ programs, in: Proceedings of the 25th International Symposium on Software Testing and Analysis—ISSTA 2016, ACM Press, New York, NY, USA, 2016. ISBN: 9781450343909, pp. 106–117, https://doi.org/10.1145/2931037.2931055.

[122] R.E. Bryant, Symbolic Boolean manipulation with ordered binary-decision diagrams, ACM Comput. Surv. 24 (3) (1992) 293–318. ISSN: 0360-0300, https://doi.org/10.1145/136035.136043.

[123] T. Su, Z. Fu, G. Pu, J. He, Z. Su, in: Combining symbolic execution and model checking for data flow testing, Proceedings—International Conference on Software Engineering, ISSN 02705257, vol. 1, 2015. ISBN: 9781479919345, pp. 654–665, https://doi.org/10.1109/ICSE.2015.81.

[124] P. Zhang, S. Elbaum, M.B. Dwyer, Compositional load test generation for software pipelines, in: Proceedings of the 2012 International Symposium on Software Testing and Analysis–ISSTA 2012, 2012. ISBN: 9781450314541, p. 89, https://doi.org/10.1145/2338965.2336764. http://dl.acm.org/citation.cfm?doid=2338965.2336764.

[125] G. Li, E. Andreasen, I. Ghosh, SymJS: automatic symbolic testing of JavaScript web applications, in: Proceedings of the 22nd ACM SIGSOFT International Symposium on Foundations of Software Engineering—FSE 2014, 2014. ISBN: 9781450330565, pp. 449–459, https://doi.org/10.1145/2635868.2635913. http://www.cs.utah.edu/ligd/publications/SymJS-FSE14.pdf%5Cnpapers3://publication/uuid/E49C07EA-502A-4F49-A818-BC9DFA1EF9BA%5Cnhttp://dl.acm.org/citation.cfm?id=2635913.

[126] M. Dhok, M.K. Ramanathan, N. Sinha, Type-aware concolic testing of JavaScript programs, in: Proceedings of the 38th International Conference on Software Engineering—ICSE '16, 2016. ISBN: 9781450339001, ISSN 9781450321389, pp. 168–179, https://doi.org/10.1145/2884781.2884859. arXiv:1508.06655v1, http://dl.acm.org/citation.cfm?doid=2884781.2884859.

[127] B. Loring, D. Mitchell, J. Kinder, ExpoSE: practical symbolic execution of standalone javascript, in: Proceedings of the 24th ACM SIGSOFT International SPIN Symposium on Model Checking of Software, SPIN 2017, ACM, New York, NY, USA, 2017. ISBN: 978-1-4503-5077-8, pp. 196–199, https://doi.org/10.1145/3092282.3092295.

[128] N. Mirzaei, S. Malek, C.S. Păsăreanu, N. Esfahani, R. Mahmood, Testing android apps through symbolic execution, ACM SIGSOFT Softw. Eng. Notes 37 (6) (2012) 1. ISSN: 01635948, https://doi.org/10.1145/2382756.2382798. http://dl. acm.org/citation.cfm?doid=2382756.2382798.

[129] N. Mirzaei, H. Bagheri, R. Mahmood, S. Malek, SIG-Droid: automated system input generation for android applications, in: 2015 IEEE 26th International Symposium on Software Reliability Engineering, ISSRE 2015, 2016. ISBN: 9781509004065, pp. 461–471, https://doi.org/10.1109/ISSRE.2015.7381839.

[130] S. Anand, M. Naik, M.J. Harrold, H. Yang, Automated concolic testing of smartphone apps, in: Proceedings of the ACM SIGSOFT 20th International Symposium on the Foundations of Software Engineering—FSE '12, 2012. ISBN: 9781450316149, p. 1, https://doi.org/10.1145/2393596.2393666. http://dl.acm.org/citation.cfm? doid=2393596.2393666.

[131] C.S. Jensen, M.R. Prasad, A. Møller, Automated testing with targeted event sequence generation, in: Proceedings of the 2013 International Symposium on Software Testing and Analysis—ISSTA 2013, ACM Press, New York, NY, USA, 2013. ISBN: 9781450321594, p. 67, https://doi.org/10.1145/2483760.2483777. http://dl.acm. org/citation.cfm?doid=2483760.2483777.

[132] M.S. Mahmood, M. Abdul Ghafoor, J.H. Siddiqui, Symbolic execution of stored procedures in database management systems, in: Proceedings of the 31st IEEE/ACM International Conference on Automated Software Engineering—ASE 2016, 2016. ISBN: 9781450338455, pp. 519–530, https://doi.org/10.1145/2970276.2970318. http://dl.acm.org/citation.cfm?doid=2970276.2970318.

[133] C. Nguyen, H. Yoshida, M. Prasad, I. Ghosh, K. Sen, in: Generating succinct test cases using don't care analysis, 2015 IEEE 8th International Conference on Software Testing, Verification and Validation, ICST 2015–Proceedings, ISSN 2159-4848, 2015. ISBN: 9781479971251, https://doi.org/10.1109/ICST.2015.7102590.

[134] E.J. Rapos, J. Dingel, in: Using fuzzy logic and symbolic execution to prioritize UML-RT test cases, 2015 IEEE 8th International Conference on Software Testing, Verification and Validation (ICST), ISSN 2159-4848, 2015. ISBN: 978-1-4799-7125-1, pp. 1–10, https://doi.org/10.1109/ICST.2015.7102610. http://ieeexplore. ieee.org/document/7102610/.

[135] E. Wong, L. Zhang, S. Wang, T. Liu, L. Tan, in: DASE: documentassisted symbolic execution for improving automated software testing, Proceedings—International Conference on Software Engineering, ISSN 02705257, vol. 1, 2015. ISBN: 9781479919345, pp. 620–631, https://doi.org/10.1109/ICSE.2015.78.

[136] I. Haller, A. Slowinska, M. Neugschwandtner, H. Bos, Dowser: a guided fuzzer to find buffer overflow vulnerabilities, in: SEC'13 Proceedings of the 22nd USENIX Conference on Security, 2013. ISBN: 9781931971034, pp. 49–64.

[137] K. Serebryany, D. Bruening, A. Potapenko, D. Vyukov, AddressSanitizer: a fast address sanity checker, in: Proceedings of the 2012 USENIX Conference on Annual Technical Conference, USENIX ATC'12, USENIX Association, Berkeley, CA, USA, 2012, p. 28. http://dl.acm.org/citation.cfm?id=2342821.2342849.28.

[138] N. Stephens, J. Grosen, C. Salls, A. Dutcher, R. Wang, J. Corbetta, Y. Shoshitaishvili, C. Kruegel, G. Vigna, Driller: augmenting fuzzing through selective symbolic execution, in: NDSS, 2016. ISBN: 189156241X, pp. 21–24, https://doi.org/10.14722/ndss.2016.23368. February.

[139] M.L. Potet, L. Mounier, M. Puys, L. Dureuil, in: Lazart: a symbolic approach for evaluation the robustness of secured codes against control flow injections, Proceedings—IEEE Seventh International Conference on Software Testing, Verification and Validation, ICST 2014, ISSN 2159-4848, 2014. ISBN: 9780769551852, pp. 213–222, https://doi.org/10.1109/ICST.2014.34.

[140] Y. Jia, M. Harman, Higher Order Mutation Testing, Inf. Softw. Technol. 51 (10) (2009) 1379–1393. ISSN: 0950-5849, https://doi.org/10.1016/j.infsof.2009.04.016.

[141] D. Davidson, B. Moench, S. Jha, T. Ristenpart, FIE on firmware: finding vulnerabilities in embedded systems using symbolic execution finding vulnerabilities in embedded systems using symbolic execution, in: Proceedings of the 22nd USENIX Security Symposium, 2013. ISBN: 9781931971034, pp. 463–478. https://www.usenix.org/conference/usenixsecurity13/technical-sessions/paper/davidson.

[142] D.A. Ramos, D. Engler, Under-constrained symbolic execution: correctness checking for real code, in: USENIX Security Symposium, 2015. ISBN: 9781931971232, pp. 49–64. https://www.usenix.org/conference/usenixsecurity15/technical-sessions/presentation/ramos.

[143] S. Jana, Y. Kang, S. Roth, B. Ray, Automatically detecting error handling bugs using error specifications, in: USENIX Security, 2016. ISBN: 978-1-931971-32-4, pp. 345–362. https://www.usenix.org/conference/usenixsecurity16/technical-sessions/presentation/jana.

[144] I. Yun, C. Min, X. Si, Y. Jang, T. Kim, M. Naik, APISan: sanitizing API usages through semantic cross-checking, in: USENIX Security Symposium 2016. ISBN: 9781931971324, pp. 363–378.

[145] R. Wang, P. Ning, T. Xie, Q. Chen, MetaSymploit: day-one defense against script-based attacks with security-enhanced symbolic analysis, in: Proceedings of the USENIX Security Symposium 2013. ISBN: 9781931971034.

[146] S.Y. Chau, O. Chowdhury, E. Hoque, H. Ge, A. Kate, C. Nita-Rotaru, N. Li, SymCerts: Practical Symbolic Execution for Exposing Noncompliance in X.509 Certificate Validation Implementations, in: 2017 IEEE Symposium on Security and Privacy (SP), 2017, pp. 503–520, https://doi.org/10.1109/SP.2017.40.

[147] C.S. Păsăreanu, Q.S. Phan, P. Malacaria, Multi-run side-channel analysis using symbolic execution and Max-SMT, in: 2016 IEEE 29th Computer Security Foundations Symposium (CSF), 2016, pp. 387–400, https://doi.org/10.1109/CSF.2016.34.

[148] D. Xu, J. Ming, D. Wu, in: Cryptographic Function Detection in Obfuscated Binaries via Bit-Precise Symbolic Loop Mapping, Proceedings—IEEE Symposium on Security and Privacy, ISSN 10816011, 2017. ISBN: 9781509055326, pp. 921–937, https://doi.org/10.1109/SP.2017.56.

[149] T. Stoenescu, A.S. B, S. Predut, F. Ipate, RIVER: A Binary Analysis Framework Using Symbolic Execution and Reversible x86 Instructions, vol. 9995, 2016. ISBN: 978-3-319-48988-9, pp. 779–785, https://doi.org/10.1007/978-3-319-48989-6.

[150] N. Hasabnis, R. Sekar, Extracting instruction semantics via symbolic execution of code generators, in: Proceedings of the 2016 24th ACM SIGSOFT International Symposium on Foundations of Software Engineering—FSE 2016, 2016. ISBN: 9781450342186, pp. 301–313, https://doi.org/10.1145/2950290.2950335. http://dl.acm.org/citation.cfm?id=2950290.2950335.

[151] A. Orso, G. Rothermel, Software testing: a research travelogue (2000-2014), in: Proceedings of the on Future of Software Engineering, FOSE 2014, ACM, 2014. ISBN: 978-1-4503-2865-4, pp. 117–132, https://doi.org/10.1145/2593882.2593885.

[152] B. Chen, Y. Liu, W. Le, Generating performance distributions via probabilistic symbolic execution, in: 2016 IEEE/ACM 38th International Conference on Software Engineering (ICSE), 2016, pp. 49–60, https://doi.org/10.1145/2884781.2884794.

[153] Q.-S. Phan, P. Malacaria, C.S. Păsăreanu, M. D'Amorim, Quantifying information leaks using reliability analysis, in: Proceedings of the 2014 International SPIN Symposium on Model Checking of Software—SPIN 2014, 2014. ISBN: 9781450324526, pp. 105–108, https://doi.org/10.1145/2632362.2632367. http://doi.acm.org/10.1145/2632362.2632367%5Cnhttp://dl.acm.org/citation.cfm?doid=2632362.2632367.

[154] M. Borges, Q.-S. Phan, A. Filieri, C.S. Păsăreanu, Model-counting approaches for nonlinear numerical constraints, in: C. Barrett, M. Davies, T. Kahsai (Eds.), NASA Formal Methods: Ninth International Symposium, NFM 2017, Moffett Field, CA, USA, May 16–18, 2017, Proceedings, Springer International Publishing, Cham, 2017. ISBN: 978-3-319-57288-8, pp. 131–138, https://doi.org/10.1007/978-3-319-57288-8_9.

[155] J. Geldenhuys, M.B. Dwyer, W. Visser, Probabilistic symbolic execution, in: Proceedings of the 2012 International Symposium on Software Testing and Analysis—ISSTA 2012, 2012. ISBN: 9781450314541, p. 166, https://doi.org/10.1145/2338965.2336773. http://dl.acm.org/citation.cfm?id=2338965.2336773.

[156] A. Barvinok, J.E. Pommersheim, An algorithmic theory of lattice points, in: New Perspectives in Algebraic Combinatorics, vol. 38, Cambridge University Press, Cambridge, 1999, p. 91.

[157] A. Filieri, C.S. Păsăreanu, W. Visser, in: Reliability analysis in symbolic PathFinder, Proceedings—International Conference on Software Engineering, ISSN 02705257, 2013. ISBN: 9781467330763, pp. 622–631, https://doi.org/10.1109/ICSE.2013.6606608.

[158] C. Baier, J.-P. Katoen, K.G. Larsen, Principles of Model Checking, MIT Press, 2008.

[159] A. Filieri, C.S. Păsăreanu, W. Visser, J. Geldenhuys, Statistical symbolic execution with informed sampling, in: Proceedings of the 22nd ACM SIGSOFT International Symposium on Foundations of Software Engineering, 2014. ISBN: 978-1-4503-3056-5, pp. 437–448, https://doi.org/10.1145/2635868.2635899.

[160] K. Luckow, C.S. Păsăreanu, M.B. Dwyer, A. Filieri, W. Visser, Exact and approximate probabilistic symbolic execution for nondeterministic programs, in: Proceedings of the 29th ACM/IEEE International Conference on Automated Software Engineering—ASE '14, 2014. ISBN: 9781450330138, pp. 575–586, https://doi.org/10.1145/2642937.2643011. http://dl.acm.org/citation.cfm?doid=2642937.2643011.

[161] G. Rubino, B. Tuffin, Rare Event Simulation Using Monte Carlo Methods, John Wiley & Sons, 2009.

[162] L.P. Kaelbling, M.L. Littman, A.W. Moore, Reinforcement learning: a survey, J. Artif. Intell. Res. 4 (1996) 237–285, https://doi.org/10.1613/jair.301.

[163] V. Baldoni, N. Berline, J.A. De Loera, B. Dutra, M. Köppe, S. Moreinis, G. Pinto, M. Vergne, J. Wu, A User's Guide for LattE Integrale v1.7.2, 2014.

[164] S. Verdoolaege, R. Seghir, K. Beyls, V. Loechner, M. Bruynooghe, Counting integer points in parametric polytopes using Barvinok's rational functions, Algorithmica 48 (1) (2007) 37–66. ISSN: 0178-4617.

[165] M. Borges, A. Filieri, M. d'Amorim, C.S. Păsăreanu, W. Visser, Compositional solution space quantification for probabilistic software analysis, in: Proceedings of the 35th ACM SIGPLAN Conference on Programming Language Design and Implementation, PLDI '14, ACM, New York, NY, USA, 2014. ISBN: 978-1-4503-2784-8, pp. 123–132, https://doi.org/10.1145/2594291.2594329.

[166] M. Borges, A. Filieri, M. D'Amorim, C. Păsăreanu, Iterative distribution-aware sampling for probabilistic symbolic execution, in: 2015 10th Joint Meeting of the European Software Engineering Conference and the ACM SIGSOFT Symposium on the Foundations of Software Engineering, ESEC/FSE 2015—Proceedings, 2015. ISBN: 9781450336758, pp. 866–877, https://doi.org/10.1145/2786805.2786832.

[167] L. Granvilliers, F. Benhamou, Algorithm 852: realpaver: an interval solver using constraint satisfaction techniques, ACM Trans. Math. Softw. 32 (1) (2006) 138–156. ISSN: 0098-3500, https://doi.org/10.1145/1132973.1132980.

[168] L. Luu, S. Shinde, P. Saxena, B. Demsky, A Model Counter for Constraints over Unbounded Strings, in: Proceedings of the 35th ACM SIGPLAN Conference on Programming Language Design and Implementation, , PLDI'14, ACM, New York, NY, USA, 2014. ISBN: 978-1-4503-2784-8, pp. 565–576, https://doi.org/10.1145/2594291.2594331.

[169] A. Aydin, L. Bang, T. Bultan, Automata-based model counting for string constraints, in: D. Kroening, C.S. Păsăreanu (Eds.), Computer Aided Verification: 27th International Conference, CAV 2015, San Francisco, CA, USA, July 18–24, 2015, Proceedings, Part I, Springer International Publishing, Cham, 2015. ISBN: 978-3-319-21690-4, pp. 255–272, https://doi.org/10.1007/978-3-319-21690-4_15.

[170] D. Chistikov, R. Dimitrova, R. Majumdar, Approximate counting in SMT and value estimation for probabilistic programs, Acta Inform. 54 (8) (2017) 729–764. ISSN: 1432-0525, https://doi.org/10.1007/s00236-017-0297-2.

[171] S. Chakraborty, K.S. Meel, M.Y. Vardi, A scalable approximate model counter, in: C. Schulte (Ed.), Proceedings of the 19th International Conference on Principles and Practice of Constraint Programming, Springer, Berlin, Heidelberg, 2013. ISBN: 978-3-642-40627-0, pp. 200–216, https://doi.org/10.1007/978-3-642-40627-0_18.

[172] C.P. Gomes, A. Sabharwal, B. Selman, Model counting, in: Handbook of Satisfiability, Frontiers in Artificial Intelligence and Applications, IOS Press, 2009, pp. 633–654, https://doi.org/10.3233/978-1-58603-929-5-633.

[173] A. Filieri, M.F. Frias, C.S. Păsăreanu, W. Visser, Model counting for complex data structures, in: Lecture Notes in Computer Science (Including Subseries Lecture Notes in Artificial Intelligence and Lecture Notes in Bioinformatics), ISSN 16113349, vol. 9232, 2015. ISBN: 9783319234038, pp. 222–241, https://doi.org/10.1007/978-3-319-23404-5_15.

[174] P. Braione, G. Denaro, M. Pezzè, Symbolic execution of programs with heap inputs, in: Proceedings of the 2015 10th Joint Meeting on Foundations of Software Engineering, 2015. ISBN: 978-1-4503-3675-8, pp. 602–613, https://doi.org/10.1145/2786805.2786842.

[175] W. Ahrendt, B. Beckert, R. Bubel, R. Hähnle, P.H. Schmitt, M. Ulbrich (Eds.), Deductive Software Verification—the Key Book—From Theory to Practice, Lecture Notes in Computer Science, vol. 10001, Springer, 2016. ISBN: 978-3-319-49811-9, https://doi.org/10.1007/978-3-319-49812-6.

[176] V. Chipounov, V. Kuznetsov, G. Candea, The S2E platform: design, implementation, and applications, ACM Trans. Comput. Syst. 30 (1) (2012) 2:1–2:49, ISSN: 0734-2071, https://doi.org/10.1145/2110356.2110358.

[177] E. Bounimova, P. Godefroid, D. Molnar, Billions and billions of constraints: whitebox fuzz testing in production, in: Proceedings of the 2013 International Conference on Software Engineering, ICSE '13, IEEE Press, Piscataway, NJ, USA, 2013. ISBN: 978-1-4673-3076-3, pp. 122–131, http://dl.acm.org/citation.cfm?id=2486788.2486805.

[178] D. Song, D. Brumley, H. Yin, J. Caballero, I. Jager, M.G. Kang, Z. Liang, J. Newsome, P. Poosankam, P. Saxena, BitBlaze: a new approach to computer security via binary analysis, in: R. Sekar, A.K. Pujari (Eds.), Information Systems Security: 4th International Conference, ICISS 2008, Hyderabad, India, December 16–20, 2008. Proceedings, Springer, Berlin, Heidelberg, 2008. ISBN: 978-3-540-89862-7, pp. 1–25, https://doi.org/10.1007/978-3-540-89862-7_1.

[179] Y. Shoshitaishvili, R. Wang, C. Salls, N. Stephens, M. Polino, A. Dutcher, J. Grosen, S. Feng, C. Hauser, C. Kruegel, G. Vigna, SoK: (State of) the art of war: offensive techniques in binary analysis, in: IEEE Symposium on Security and Privacy, 2016.

[180] F. Saudel, J. Salwan, Triton: a dynamic symbolic execution framework, in: Symposium sur la sécurité des technologies de l'information et des communications, SSTIC, France, Rennes, June 3–5, 2015, SSTIC, 2015, pp. 31–54.

[181] K. Luckow, M. Dimjašević, D. Giannakopoulou, F. Howar, M. Isberner, T. Kahsai, Z. Rakamarić, V. Raman, JDart: a dynamic symbolic analysis framework, in: M. Chechik, J.-F. Raskin (Eds.), Lecture Notes in Computer Science, Proceedings of the 22nd International Conference on Tools and Algorithms for the Construction and Analysis of Systems (TACAS), vol. 9636, Springer, 2016, pp. 442–459.

[182] S.F. Siegel, M.B. Dwyer, G. Gopalakrishnan, Z. Luo, Z. Rakamaric, R. Thakur, M. Zheng, T.K. Zirkel, CIVL: the concurrency intermediate verification language, Department of Computer and Information Sciences, University of Delaware, 2014. Tech. Rep. UD-CIS-2014/001.

[183] C.S. Păsăreanu, W. Visser, A survey of new trends in symbolic execution for software testing and analysis, Int, J. Softw. Tools Technol. Transf. 11 (4) (2009) 339–353. ISSN: 1433-2779, https://doi.org/10.1007/s10009-009-0118-1.

[184] C. Cadar, K. Sen, Symbolic execution for software testing: three decades later, Commun. ACM 56 (2) (2013) 82–90. ISSN: 0001-0782, https://doi.org/10.1145/2408776.2408795.

[185] T. Chen, X.-S. Zhang, S.-Z. Guo, H.-Y. Li, Y. Wu, State of the art: dynamic symbolic execution for automated test generation, Future Gener. Comput. Syst. 29 (7) (2013) 1758–1773. ISSN: 0167-739X.

[186] C. Cadar, P. Godefroid, S. Khurshid, C.S. Păsăreanu, K. Sen, N. Tillmann, W. Visser, Symbolic execution for software testing in practice: preliminary assessment, in: Proceedings of the 33rd International Conference on Software Engineering, ICSE 2011, Waikiki, Honolulu, HI, USA, May 21–28, 2011, 2011, pp. 1066–1071, https://doi.org/10.1145/1985793.1985995.

ABOUT THE AUTHORS

Guowei Yang is an Assistant Professor in the Department of Computer Science at Texas State University. He received the PhD degree from the University of Texas at Austin in 2013. His research addresses various elements of how to enhance software reliability and dependability, including software verification and testing, software maintenance and evolution, program analysis, and formal methods. https://cs.txstate.edu/~g_y10.

Antonio Filieri is a Lecturer (Assistant Professor) at Imperial College London, UK. His main research interests are in the application of mathematical methods for Software Engineering, in particular, Probability, Statistics, Logic, and Control theory. His recent work includes exact and approximate methods for probabilistic symbolic execution, incremental verification, quantitative software modeling and verification at runtime, and control-theoretical software adaptation. https://antonio.filieri.name.

Mateus Borges is a PhD Candidate at the Department of Computing, Imperial College London, UK. His research interests span the areas of Software Engineering and Program Analysis, with a focus on improving the accuracy, scalability, and effectiveness of testing, debugging, and verification techniques. https://www.mateusborges.com.

Donato Clun is a PhD Candidate at the Department of Computing, Imperial College London, UK. His research focuses on symbolic execution, grammar-driven fuzzing, input format recovery, and automated debugging.

Junye Wen is a PhD student in the Department of Computer Science at Texas State University. His research interests focus on symbolic execution, property checking, parallel and incremental algorithms for software analysis. https://userweb.cs.txstate.edu/~j_w236.

CHAPTER SIX

Symbolic Execution and Recent Applications to Worst-Case Execution, Load Testing, and Security Analysis

Corina S. Păsăreanu*, Rody Kersten†, Kasper Luckow‡, Quoc-Sang Phan§

*NASA Ames and Carnegie Mellon University, Mountain View, CA, United States
†Synopsys, Inc., San Francisco, CA, United States
‡Amazon Web Services, San Francisco, CA, United States
§Fujitsu Laboratories of America, Sunnyvale, CA, United States

Contents

1. Introduction	290
2. Symbolic Execution	290
2.1 Complex Heap Data Structures	291
3. Tools and Scalability Challenges	293
3.1 Challenges	295
4. Applications	296
4.1 Worst-Case Execution Time (WCET) Analysis	296
4.2 Performance Testing	298
4.3 Security Analysis	300
4.4 Symbolic Execution and Fuzzing	302
5. Conclusion	305
References	305
About the Authors	313

Abstract

Symbolic execution is a systematic program analysis technique which executes programs on symbolic inputs, representing multiple concrete inputs, and represents the program behavior using mathematical constraints over the symbolic inputs. Solving the constraints with off-the-shelf solvers yields inputs that exercise different program paths. Typical applications of the technique include test input generation and error detection. In this chapter we review symbolic execution and associated tools, and describe some of the main challenges in applying symbolic execution in practice: handling of programs with complex inputs, coping with path explosion, and ameliorating the cost of constraint solving. We also survey promising applications of the technique that go beyond checking functional properties of programs. These include finding worst-case execution time in programs, load testing and security analysis, via combinations of symbolic execution with fuzzing.

1. INTRODUCTION

As computer systems become more pervasive and complex, it has become increasingly important to develop techniques and tools that effectively ensure software dependability. Symbolic execution [1] is a systematic program analysis technique which explores multiple program behaviors at once, by collecting and solving symbolic path conditions collected over program paths. Symbolic execution can be used for finding bugs in software, where it checks for runtime errors or assertion violations during execution and it generates test inputs that trigger those errors.

Nowadays there are many symbolic execution tools available [2–7] which have found numerous vulnerabilities and other interesting bugs in software. Much of the success of symbolic execution in recent years is due to significant advances in constraint solving and decision procedures [8, 9] as well as to the availability of increasingly cheap computational power and cloud computing platforms [4, 5], allowing to scale the technique to large applications.

In this chapter we review symbolic execution and associated tools, and we describe the main challenges in applying symbolic execution in practice: handling of programs with complex inputs, coping with path explosion, and ameliorating the cost of constraint solving. We also survey some applications of the technique that go beyond checking functional properties of programs. These include finding worst-case execution time in programs, load testing and security analysis, via combinations of symbolic execution with fuzzing. These applications are perhaps less studied in the literature but we believe they hold much promise for the future. We conclude with directions for future work.

2. SYMBOLIC EXECUTION

Symbolic execution [1] is a program analysis technique that executes a program on symbolic, instead of concrete, input values and computes the effects of the program as *functions* in terms of these symbolic inputs. The result of symbolically executing a program is a set of symbolic paths, each with a path condition PC, which is a conjunction of constraints over the symbolic inputs that characterizes all the inputs that follow that path. All the PCs are disjoint.

When executing a branching instruction with condition c, symbolic execution systematically explores both branches and updates the path condition accordingly: $PC \leftarrow PC \land c$ for the *then* branch and $PC \leftarrow PC \land \neg c$ for the *else* branch. The feasibility of the path conditions is checked using off-the-shelf constraint solvers such as Z3 [8]. If a path condition is found to be unsatisfiable, symbolic execution stops analyzing that path (since that path is not feasible). For the feasible paths, the models returned by the constraint solver can be used as test inputs that execute these paths. To deal with loops and recursion, typically a bound is put on the exploration depth.

Several tools implement "classic" symbolic execution which is essentially a static analysis technique, as it analyzes a program without running it; in Symbolic PathFinder, the program is actually "run," but this is done inside the *custom* JVM of the Java pathFinder tool. Dynamic symbolic execution techniques, on the other hand, collect symbolic constraints at *run time* during concrete executions. Examples of such dynamic techniques are implemented in DART (Directed Automated Random Testing) [10] and Klee [2].

Dynamic test generation as first proposed by Korel [11], consists of running the program starting with some random inputs, gathering the symbolic constraints on inputs at conditional statements, using a constraint solver to generate new test inputs and repeating the process until a specific program path or statement is reached. DART performs a similar dynamic test generation, where the process is repeated to attempt to cover *all* feasible program paths, and it detects crashes, assert violations, runtime errors, etc. during execution.

2.1 Complex Heap Data Structures

Invented in the 1970s, traditional symbolic execution has been proposed for programs with a fixed number of numerical inputs. However, modern programming languages such as C++ and Java contain a variety of data structures, e.g., linked lists or trees, that might dynamically allocate objects at run time. A naive approach to this problem is to impose a priori bounds on the inputs. For example, for a program that takes a linked list as input, one needs to initialize it with k list nodes, and each one can be symbolic.

However, k have to be defined beforehand. A pessimistically large k leads to path explosion problem, and small k (incorrectly) reduces the search space of symbolic execution. Moreover, it is not straightforward to describe the bounds for data structures such as tree.

To address the problem above, Khurshid et al. [12] introduced the *lazy initialization* algorithm, which has become the state-of-the-art way of handling heap data structures. This algorithm works as follows:

1. When a symbolic input is of reference type, i.e., linked list, execute the program without initializing it.
2. When an uninitialized symbolic variable is dereferenced, exhaustively enumerate all possible concrete objects that it can reference to (i) null; (ii) new object; (iii) previously initialized objects of the same type (i.e., it is an alias)

In the second step, symbolic execution case splits on each of possible choices, which leads to rapid path explosion. Therefore, there have been multiple efforts on improving the enumeration of this step.

Deng et al. proposed the *lazier* algorithm [13], which delays case splitting in lazy initialization by grouping together choices in (ii) and (iii) (nonnull choices), into a symbolic variable. Case splitting on nonnull variables occurs later when they are accessed. The same authors then introduced a more enhanced algorithm, called *lazy#* [14], with group together all choices in (i), (ii), and (iii) in the same manner.

Symbolic initialization [15] uses a guarded value set to capture all choices in (i), (ii), and (iii) in the same symbolic heap. This completely avoids case splitting in symbolic paths when initializing a symbolic variable of reference type. This, however, comes with the cost of solving constraints with greater complexity, since case splitting is actually delegated to the SMT solver.

Geldenhuys et al. [16] took a different approach, instead of delaying case splitting, the authors aim to reduce the number of choices in (iii) by considering only nonisomorphic structures. This is done via precomputed tight field bounds. Computing those bounds is very expensive, but the authors argue that they can be reused to test different methods in the program.

The lazy initialization-based approaches have been adapted to take into account the shape of the input. For example, when the input is designed to be singly linked list, the choices in (iii) should be restricted to avoid configurations of a circular linked list and so on. To address this problem preconditions are used in e.g., [12, 17] to constraint heap inputs. This is implemented as using an API, verify.ignoreIf, to tell symbolic execution to stop exploring when a method *pre()* representing preconditions returns false. This approach delegates testing preconditions to the users. For example, to impose the constraint that the input is a binary tree, such a method *pre()* needs to implement a depth-first search to detect cycles.

Braione et al. [18] introduced *Heap EXploration Logic* (HEX) as a specification language for lazy initialization. When symbolic execution enumerates the choices in (i), (ii), and (iii), the HEX verifier checks those choices against a specification, written in HEX, and prune off invalid states. HEX lacks numerical operators, and thus it cannot express numerical properties of the data structures such as the size of a linked list. The HEX language is not expressive enough to describe shapes of data structures either. Users have to provide methods, called *triggers*, to check properties that cannot be checked by HEX.

Pham et al. [19] use separation logic [20, 21] with inductive definitions to describe the symbolic heap and the shape of the input data structures. When an uninitialized symbolic variable is dereferenced and if it is defined by an inductive predicate, they unfold it to capture the footprints, i.e., the resources it accesses. This unfolding process updates the heap configuration, and a SAT solver for separation logic [22] is then used to check if the updated heap configuration is satisfiable.

3. TOOLS AND SCALABILITY CHALLENGES

Because of its capability of finding subtle bugs, and its applications in a widespread of domains, symbolic execution has been developed on several platforms, for different programming languages. The following table contains a (likely incomplete) list of symbolic executors.

Language	Tool	Link
Java (bytecode)	Symbolic PathFinder [3]	https://babelfish.arc.nasa.gov/trac/jpf/wiki/projects/jpf-symbc
	Java StarFinder [19]	https://github.com/star-finder/jpf-star
	jCUTE	https://github.com/osl/jcute
	janala2	https://github.com/ksen007/janala2
	jDART [23]	https://github.com/psycopaths/jdart
	JBSE [24]	https://github.com/pietrobraione/jbse
	KeY	http://www.key-project.org/

Continued

—cont'd Language	Tool	Link
X86(-64) binaries	Project Springfield	https://www.microsoft.com/en-us/springfield/
	SAGE [5]	—
	Mayhem [6]	http://forallsecure.com/mayhem.html
	Miasm (many different binaries)	https://github.com/cea-sec/miasm
	BAP (Also ARM) [25]	https://github.com/BinaryAnalysisPlatform/bap/
	S^{2}E (Also ARM) [26]	http://s2e.epfl.ch
	Angr	http://angr.io/
	Pathgrind	https://github.com/codelion/pathgrind
	pysymemu	https://github.com/feliam/pysymemu/
	Triton	http://triton.quarkslab.com
C/C++	CUTE [27]	—
	CREST	http://www.burn.im/crest/
	DART [10]	—
	KLOVER [28]	—
	EXE [29]	—
	Otter	https://bitbucket.org/khooyp/otter/overview
LLVM	KLEE [2]	http://klee.github.io/
	Cloud9	http://cloud9.epfl.ch/
	Kite	http://www.cs.ubc.ca/labs/isd/Projects/Kite/
.NET	Pex	http://research.microsoft.com/en-us/projects/pex/
JavaScript	SymJS	http://www.cs.utah.edu/~ligd/publications/SymJS-FSE14.pdf
	Jalangi2	https://github.com/Samsung/jalangi2
	Kudzu [30]	—
Dalvik bytecode	SymDroid [31]	—

—cont'd Language	Tool	Link
Python	PyExZ3 [32]	https://github.com/thomasjball/PyExZ3
VineIL	BitBlaze [33]	http://bitblaze.cs.berkeley.edu
	FuzzBALL [34]	http://bitblaze.cs.berkeley.edu/fuzzball.html
Boogie	Symbooglix [35]	https://github.com/symbooglix/symbooglix
CIVL language	CIVL	http://vsl.cis.udel.edu/civl/
Ruby	Rubyx	http://www.cs.umd.edu/~avik/papers/ssarorwa.pdf

3.1 Challenges

There are two main challenges in scaling up symbolic execution: there are too many paths to explore and the path conditions are too difficult to solve. Addressing these two challenges are active areas of research.

3.1.1 Path Explosion

Recall that symbolic execution explores symbolic paths of the program, which form a (symbolic execution) tree. Each path of the tree is independent of the others, thus there have been multiple efforts on parallelizing symbolic execution [4, 36–39], and distributing the exploration process to multiple workers. This idea is very promising thanks to recent advances in cloud services; however, balanced distribution of workload among workers remains a big challenge, as the depth and breadth of the symbolic execution tree are not known in advance.

Other approaches to the path explosion problem reducing the number of paths using state or path merging and also compositional techniques [40–45]. Some of these approaches use disjunction or set to represent (symbolic) values of the resultant merged states. Thus the reduction of the number of path comes with the cost of solving constraints with greater complexity. Related techniques use different forms of abstraction to reduce the number of paths in looping programs [46, 47].

Other techniques use sampling or different search heuristics [48] to try to hit the bug faster using sampling. The idea is that most symbolic execution engine employs depth-first search, which systematically searches the symbolic execution tree from one side to the other. When the execution tree is too big for exhaustive search, depth-first search may always get stuck in the beginning parts of the tree, and thus sampling can increase the chances of hitting the bugs.

3.1.2 Improving Constraint Solving

As constraint solving is expensive, an intuitive idea is to cache the result, and look up the cache before invoking the solver. KLEE [2] exploits the fact that when a program has independent branches, the path condition will be comprised of independent constraints. Therefore, decomposing the constraints into multiple independent subsets, and caching results for those subsets increases the possibility of cache hit.

Green [49] took one step further, normalizing the constraints and then saving them, together with their results, offline to a database. In this way, constraint solving can be reused across programs, analyses, and solvers. Green was implemented for linear integer constraints, while Cashew [50], built on top of Green, extends the Green approach to string constraints.

In a different context, since symbolic execution often has to be run several times on the same program, e.g., first to check the error, then to verify the program after fixing the error, memoized symbolic execution [51] uses a trie to store the whole symbolic execution tree in the first run, then reuses the summaries from the trie in the following runs.

Another active research area is to extend symbolic execution to programs with complex constraints, such as nonlinear numerical constraints [52] or combination of string constraints and numeric constraints [53, 54].

4. APPLICATIONS

4.1 Worst-Case Execution Time (WCET) Analysis

Symbolic execution has been used in several works related to real-time systems. Real-time systems are characterized by having timing requirements in addition to functional requirements. As an example, systems operating in the safety-critical domain often have hard temporal requirements on responding to stimuli from the environment, such as an airbag that must be deployed within a specific time frame upon collision.

An important aspect of real-time systems is the Worst-Case Execution Time (WCET) of the (real-time) tasks constituting the system. In hard real-time systems (i.e., systems where deadline violations cannot be tolerated), it is often insufficient to rely on measuring execution times of the tasks with various inputs. Symbolic execution has been extensively used in the context of WCET analysis [55–63].

Generally, symbolic execution is used in the field of WCET analysis as the *high-level* analysis that restricts focus to obtaining information about feasibility of program paths. This information is subsequently used in combination with a *low-level* analysis that gathers platform specific information, including behavior of processor-specific features such as caching, pipelines, etc. In this combination, only the feasible paths as determined by symbolic execution are used, allowing higher precision of the analysis result.

An early work that takes this approach is that of [64] that performs timing analysis of SPARK Ada code. The work by [57] uses a similar approach by using cycle-level symbolic execution to integrate path and timing analysis for obtaining tight WCET estimates. Using this technique, the authors were able to perform a perfect WCET estimation for six out of seven test subjects. They also showed that this combination can improve WCET estimation by a factor of twenty when comparing it with a more conservative method that does not prune infeasible paths, but only relies on the structure of the program.

The work of [58] uses symbolic execution to prune infeasible paths in straight-line code (i.e., no loops or recursion)—a commonly found approach to embedded systems development. Imposing such restrictions on the control-flow, guarantees termination for symbolic execution. As with the previous work, the motivation for symbolic execution is to check for feasibility of paths.

The WCET analysis tool, r–TuBound, uses selective symbolic execution [63]. It uses a selective approach to avoid the high computational costs of exhaustive analysis. The symbolic execution is only invoked when the information obtained is limited and when other analysis techniques incorporated in r–TuBound fail.

The work of [60] presents an approach for modeling the real-time tasks of a real-time system written in a variant of the Safety-Critical Java profile. The tool extracts the real-time (periodic and sporadic) tasks and symbolically executes them using Symbolic PathFinder [3]. The paths obtained, are translated into a Network of Timed Automata—the modeling formalism of the UPPAAL model checker—and combined with models of the scheduler. The complete NTA can be used for reasoning about temporal properties that can be expressed in the Timed Computation Tree Logic variant that UPPAAL supports. This includes the *schedulability* of the tasks, i.e., under all different task schedules (taking into account task interactions and sporadically firing tasks), will the system never violate a task deadline? In addition, the tool also supports querying WCETs as well as Best Case Execution Times (BCETs) and response times of the tasks.

4.2 Performance Testing

Symbolic execution effectively enumerates all paths through a program, up to a user-specified bound. It can therefore be used to find performance bottlenecks, e.g., paths that exhibit a large cost with respect to time, memory, power or energy consumption, and so on. By finding a solution for the corresponding path condition, an actual input that triggers this behavior can be generated.

4.2.1 Load Testing

In *load testing*, a system is analyzed with its behavior under peak loads. Typically, one increases the size of the test input to increase the load on the system. In many cases, however, it is possible to increase the load by carefully choosing input values rather than by increasing the input size. Moreover, when simply increasing the input size, certain program behaviors may remain undetected. Larger but similarly shaped input may execute the same behavior more often, yet miss other potentially costly behaviors.

Directed incremental symbolic execution is applied by Zhang et al. to automatically generate load tests in [65]. Their approach is *directed* by a cost model, in the sense that it favors more costly paths. It is *incremental* in that it works in phases. It is implemented in a modified version of SYMBOLIC PATHFINDER.

The user specifies two parameters: the number of test cases to generate and the depth of each phase of symbolic execution. Each phase starts with exhaustive exploration up to the user-specified depth, either from program entry or from a set of locations resulting from the previous phase. Next, the most costly paths are scheduled for further exploration. The number of paths that will be explored further is exactly the number of requested test cases. To increase diversity among paths selected for further exploration, paths are first clustered and in case these do not satisfy a *diversity measure*, further exhaustive exploration of all paths is performed.

Evaluation shows that load tests generated with directed incremental symbolic execution can incite bigger loads, often at smaller input sizes, than human written or randomly generated tests. The approach is also shown to scale up to input sizes of 100 MB.

4.2.2 Finding Performance Bugs

In [66], Burnim et al. apply symbolic execution to find what they call *performance bugs*. Such a bug is said to exist when the complexity of the implementation does not match the theoretical complexity of the implemented algorithm. Their algorithm is called WISE. It uses a clever trick to increase scalability of the analysis, based on the observation that worst-case program behaviors at small input sizes are often good indicators of the worst-case program behavior at larger input sizes.

In a first step, exhaustive exploration at small input sizes is used to construct a *worst-case generator*. Such a generator specifies which paths are likely to lead to the worst case and which are not. For a conditional b in the program, if in the paths leading to the worst case at small input, the same decision (true, false) is always made, it is conjectured that this decision will lead to the worst-case at greater input sizes as well. The generator can then be used at greater input sizes to prune paths that are not likely to lead to the worst case. The paper provides a theoretical guarantee that there is an input size N that is large enough to capture all program behaviors and that, therefore, the generator resulting from exhaustive exploration up to size N accurately predicts the worst-case behavior at any input size $M > N$.

The WISE algorithm is extended in [67] in a tool called SPF-WCA. SPF-WCA generates *guidance policies* which, similarly to worst-case generators, dictate which paths to follow during symbolic execution to discover likely worst-case behaviors. However, the policies are made more expressive by taking into account the history of decisions for each conditional. This means that even though both *true* and *false* are seen on the worst-case paths, the generator can still make a suggestion by looking at patterns in the history of decisions. Precision is also improved by making the policies context-aware, in the sense that only decisions within the same calling context can affect the generator for a conditional. Furthermore, the algorithm in [67] is extended to infer the complexity at any input size, by fitting a function to the results for increasing input sizes. Costs are obtained for input sizes $1\ldots N$, then functions corresponding to common complexity classes are fitted against the results. The application of this work is finding performance related security bugs: if the actual complexity of an algorithm does not match the theoretical complexity, then an adversary can potentially deny service to benign users by sending input that triggers the worst-case complexity. Such inputs can be found by solving the path condition of worst-case paths.

4.3 Security Analysis

4.3.1 Automated Exploit Generation

Automated exploit generation as proposed in [68] uses symbolic execution
to find vulnerabilities and to generate working exploits for them. The
exploits can redirect control flow to execute injected shell-code, perform
a return-to-libc attack, and so on. With the goal being discovering some particular types of exploitable bugs, symbolic execution is used with heuristics
based on domain knowledge about different types of bugs. For example,
buffer overflow can only occur when an input is copied to a buffer with
smaller size, thus the approach uses a light-weight analysis to determine
the minimum length k to overwrite any buffer in the program. Performing
symbolic execution with the precondition that the input should be at least k
significantly prunes off uninteresting input space. Moreover, buffer overflow often occurs at the end of loops, so symbolic execution is customized
to give higher priority to the paths that fully exhaust the loop.

4.3.2 Noninterference Testing

Undesired flows of information between different sensitivity levels can seri-
ously compromise the security of a system. In a security context, a program
can be viewed as a communication channel where information is transmitted
from a source H to a sink O. When H contains confidential information and
O can be observed by public users, information flow from H to O is not d-
esirable. Traditional information flow analysis considers source and sink as
variables of the program: H is an input with sensitive data (e.g., a user
password), and O is the program output. Absence of information flow means
the variable O is not interfered by the variable H, which can be formalized as
a *noninterference* policy [69, 70].

A prominent approach to checking noninterference involves self-
composition [71, 72], which checks the following Hoare triple on the composition of program P:

$$\{L = L_1\}P; P_1\{O = O_1\}$$

Here L is the public input of the program P, and P_1 is a copy of P where L and
O are renamed to L_1 and O_1, respectively. The program P satisfies noninterference if when executing the sequential composition of P and P_1 with
the precondition $L = L_1$, after the execution, the postcondition $O = O_1$ holds.

Symbolic execution was used for checking self-composition as described
in [73]. This approach assumes that the program P can be fully explored to

obtain the set of all symbolic paths, and uses path manipulation to avoid the cost of executing the self-composed program. This work is later generalized in [74], which releases the assumption, and handles recursions and unbounded loops using user-defined loop invariants and method contracts.

Balliu et al. [75] took a different approach and formalized noninterference using an epistemic logic. Formulae in this logic are then checked using an algorithm based on symbolic execution, implemented on top of SYMBOLIC PATHFINDER.

4.3.3 Quantitative Information Flow Analysis

Noninterference is often overly pessimistic and in practice unachievable. To illustrate consider a password checking program whose public output, which rejects or accepts a user-provided input, depends on the value of the password. Such a program does not satisfy noninterference and it leaks a small amount of information, i.e., if the input matches the secret password or not. The program will eventually leak the whole password if the adversary is given enough attempts. However, with a strong password the amount of leaked information is too small, and the program is considered to be secure.

To address the limitation above, quantitative methods for information flow [76, 77] have been developed, which, instead of enforcing zero interference, measure interference. We use the two terms "*interference*" and "*information flow*" interchangeably, since 0 is interfered by H if there is information flow from H to 0. Thus programs with "*small*" interference can be accepted as secure. Information leakage is measured using information theory metrics [78] such as Shannon entropy, Rényi's min-entropy and channel capacity.

To compute channel capacity, i.e., maximum amount of information leakage, symbolic quantitative information flow (SQIF) [79, 80] adds conditions to test every bit of the output 0, and uses symbolic execution to explore all possible values of 0. Using bitvector solvers, SQIF can perform quantitative information flow analysis over programs with nonlinear constraints.

Instead of using symbolic execution to enumerate values of the output 0, and thus compute channel capacity, QILURA [81] uses symbolic execution to partition the input space, and counts (using an off-the shelf model counting tool) the blocks in the partition that lead to leakage of information. By delegating the counting process to Latte [82], a model counter for systems of linear integer inequalities, QILURA achieves significant improvement in performance compared to SQIF. However, by using Latte, it can only analyze programs with linear constraints, and by counting the input, it only returns an upper bound on channel capacity.

4.3.4 Side-Channel Analysis

Side channels allow an attacker to infer information about a secret by observing nonfunctional characteristics of a program, such as execution time or memory consumed. Recall that a program can be viewed as a communication channel where information is transmitted from a source H to a sink O. For side-channel analysis, the sink O is not necessary an output variable but rather a nonfunctional characteristic of program execution, such as running time, power consumption, number of memory accessed or packets transmitted over a network. Side-channel attacks [83–86] have been used successfully to uncover secret information in a variety of applications, including web applications and cryptographic systems.

In previous work [54, 87, 88], we have studied the use of symbolic execution for side channel analysis. Different from SQIF and QILURA, in this line of work we compute Shannon entropy of the leakage and we tackle the problem of multi-run attacks, that is we consider scenarios when an adversary can execute the program multiple times with different and *gradually* uncover a secret. Solving this problem is difficult, since quantifying leakage for a weak or random single-runattack could not provide a guarantee for all possible attacks, and thus we need to synthesize optimal attacks.

4.4 Symbolic Execution and Fuzzing

An idea that has been shown to be particularly promising in recent years is the combination of symbolic execution with other testing techniques that are less expensive, but also are limited in their ability of achieving a high coverage of program paths. Symbolic execution is then invoked on demand, to increase coverage.

Particularly promising is the combination of symbolic execution and fuzzing. Fuzzing is an automated testing technique that has been used successfully to discover security vulnerabilities and other bugs in software [89, 90]. In its simplest, black-box, form, a program is run on randomly generated or mutated inputs, in search of cases where the program crashes or hangs. More advanced techniques may take input formatting into account, e.g., in the form of grammars, or leverage program instrumentation or program analysis to gather information about the program paths exercised by the inputs, in order to increase coverage.

Fuzzing has shown to be very effective at finding security vulnerabilities in practice. For instance, the popular fuzzing tool AFL [91] was instrumental in finding several of the Stagefright vulnerabilities in Android, as well as numerous bugs in (security-critical) applications and libraries such as BASH,

BIND, OPENSSL, OPENSSH, GNUTLS, GNUPG, PHP, APACHE, IJG JPEG, LIBJPEG-TURBO, and many more. The work in [92] presents KELINCI, an AFL-based fuzzer for Java that found similar vulnerabilities in Apache Commons Imaging and OpenJDK 9.

Fuzzing has its limitations. As inputs are tested randomly, every input value has the same probability of getting tested and code coverage is generally low. Consider, for example, the code in Listing 1. This function has a bug when the value of its input is exactly 1234. The chance of randomly testing this input is only 1 in 2^{32}.

Listing 1 Function that is problematic for fuzzing

```
void ex(int x) {
    if (x == 1234)
        abort();
}
```

This is exactly the type of problem that symbolic execution is good at. It will easily find and solve the constraint $x = 1234$ leading to new behavior. The techniques described in this sections leverage these complementary strengths of fuzz testing and symbolic execution.

4.4.1 EvoSuite

In [93] Galeotti et al. observed that if there is a change in fitness after a mutation on a primitive value, then the variable this value is assigned to is important. Thus they use dynamic symbolic execution with this variable being symbolic, to derive new values for it. On the other hand, if there is no change in fitness after a mutation or the changes in fitness are not related to a primitive value, then their adaptive algorithm does not apply dynamic symbolic execution. This approach is embodied in the EvoSuite tool.

4.4.2 SAGE and Project Springfield

A promising approach that combines symbolic execution with fuzzing is implemented in the SAGE tool which has been continued with PROJECT SPRINGFIELD. SAGE (Scalable Automated Guided Execution) [5] extends DART with a directed search algorithm. Instead of negating only the final condition of a complete symbolic execution, this *generational search* negates all conditions on the path (in conjunction with the path condition for the path leading up to them). This results in a large number of new test inputs, instead of just one. SAGE is used extensively at Microsoft and has been very successful at finding security-related bugs. Out of all bugs discovered in

Windows 7, approximately one in three was found using SAGE. This is notable, as it was the last tool applied, so none of these bugs were found by other tools [5]. Microsoft is currently in the process of making SAGE available to the public as a cloud service under the name PROJECT SPRINGFIELD.[a]

4.4.3 Driller

DRILLER [7] is another promising tool that combines the AFL fuzzer with the ANGR symbolic execution engine. AFL is a security-oriented gray-box fuzzer that employs compile-time instrumentation and genetic algorithms to automatically discover test cases that trigger new internal states in C programs, improving the functional coverage for the fuzzed code. DRILLER is based on the idea that software consists of different *compartments*. Within a compartment, decisions are fairly uniformly distributed and, as such, fuzzing works very well. Jumps between compartments, however, may be less trivial for a fuzzer to detect. For instance, an application may expect a certain file header that is essentially a magic number as in Listing 1. To ensure progress, DRILLER invokes the symbolic execution engine whenever the fuzzer appears to be "stuck." It symbolically traces the program for all inputs that AFL found "interesting," then finds decisions that have unexplored branches and invokes a solver to generate inputs that drive execution down that branch. As this is expected to help crossover to new program compartments, fuzzing continues from these generated inputs. DRILLER was evaluated on 126 applications released in the qualifying event of the DARPA Cyber Grand Challenge. It identified the same number of vulnerabilities, in a similar time-frame, as the tool that performed best at the event.

4.4.4 Mayhem

MAYHEM [6] is a symbolic execution engine that aims to find security vulnerabilities in binaries. It has a strong focus on the ubiquitous buffer overflow, and other memory-related vulnerabilities. MAYHEM augments path constraints with additional, security-related information such as if a user can load their own code into memory. If such an augmented path condition is satisfiable, then the program is vulnerable. To be able to capture such security-related properties, MAYHEM uses an index-based memory model to allow using symbolic values to point to memory locations.

---

[a] https://www.microsoft.com/en-us/security-risk-detection/.

It also uses a combination of dynamic symbolic execution and traditional symbolic execution, which is referred to as *hybrid* symbolic execution. A *Concrete Executor Client* (CEC) explores paths concretely. However, it does keep track of which inputs are considered symbolic and performs a dynamic taint analysis. When a basic block is reached that contains tainted instructions, it is passed to the *Symbolic Executor Server* (SES) that is running in parallel. After symbolic execution, the SES instructs the CEC on a particular path to execute. When memory is strained, MAYHEM threads can store their state to be efficiently restarted later.

The tool has been combined with a fuzzer (MURPHY) and in 2016, it won the DARPA Cyber Grand Challenge, in which seven autonomous computer systems competed live in a search for security vulnerabilities.[b]

5. CONCLUSION

In this chapter we reviewed symbolic execution techniques and tools and we described recent applications, including finding worst-case execution time in programs, load testing and security analysis, via combinations of symbolic execution with fuzzing. There are other promising directions for symbolic execution, among them the extension of symbolic execution to probabilistic reasoning [94, 95], with applications to reliability analysis and quantitative information flow (which we described briefly in this chapter). An in-depth review of those techniques is left for the future.

Symbolic execution is increasingly used not only in academic settings but also in industry, e.g., in Microsoft, NASA, IBM, and Fujitsu, and even at the Pentagon [96]. Many symbolic execution engines have been built targeting different programming languages and architectures. This trend is expected to intensify in the future. Symbolic execution in a distributed setting, leveraging cloud technology, such as Cloud9 [4], SAGE [5], and MergePoint [43], is expected to further extend the applicability of the technique in practice.

REFERENCES

[1] J.C. King, Symbolic execution and program testing, Commun. ACM 19 (7) (1976) 385–394. ISSN: 0001-0782, https://doi.org/10.1145/360248.360252.

[2] C. Cadar, D. Dunbar, D. Engler, KLEE: unassisted and automatic generation of high-coverage tests for complex systems programs, in: Proceedings of the 8th USENIX Conference on Operating Systems Design and Implementation, OSDI'08, USENIX Association, Berkeley, CA, USA, 2008, pp. 209–224, http://dl.acm.org/citation.cfm?id=1855741.1855756.

---

[b] https://www.darpa.mil/news-events/2016-08-04.

[3] C.S. Păsăreanu, W. Visser, D. Bushnell, J. Geldenhuys, P. Mehlitz, N. Rungta, Symbolic pathfinder: integrating symbolic execution with model checking for Java bytecode analysis, Autom. Softw. Eng. (2013) 1–35, ISSN: 0928-8910, https://doi.org/10.1007/s10515-013-0122-2.

[4] L. Ciortea, C. Zamfir, S. Bucur, V. Chipounov, G. Candea, Cloud9: a software testing service, SIGOPS Oper. Syst. Rev. 43 (4) (2010) 5–10, ISSN: 0163-5980, https://doi.org/10.1145/1713254.1713257.

[5] P. Godefroid, M.Y. Levin, D. Molnar, SAGE: whitebox fuzzing for security testing, Queue 10 (1) (2012) 20:20–20:27. ISSN: 1542-7730, https://doi.org/10.1145/2090147.2094081.

[6] S.K. Cha, T. Avgerinos, A. Rebert, D. Brumley, Unleashing mayhem on binary code, in: Proceedings of the 2012 IEEE Symposium on Security and Privacy, SP '12, IEEE Computer Society, Washington, DC, USA, 2012, ISBN: 978-0-7695-4681-0, pp 380–394,https://doi.org/10.1109/SP.2012.31.

[7] N. Stephens, J. Grosen, C. Salls, A. Dutcher, R. Wang, J. Corbetta, Y. Shoshitaishvili, C. Kruegel, G. Vigna, Driller: Augmenting Fuzzing Through Selective Symbolic Execution, in: 23nd Annual Network and Distributed System Security Symposium, NDSS 2016, San Diego, California, USA, February 21–24, 2016, 2016, http://www.internetsociety.org/sites/default/files/blogs-media/driller-augmenting-fuzzing-through-selective-symbolic-execution.pdf.

[8] L. De Moura, N. Bjørner, Z3: an efficient SMT solver, in: Proceedings of the 14th International Conference on Tools and Algorithms for the Construction and Analysis of Systems, TACAS'08, Springer-Verlag, Berlin, Heidelberg, 2008, pp. 337–340. ISBN: 3-540-78799-2, 978-3-540-78799-0, http://dl.acm.org/citation.cfm?id=1792734.1792766.

[9] A. Aydin, L. Bang, T. Bultan, Automata-Based Model Counting for String Constraints, Springer International Publishing, Cham, 2015, pp. 255–272, https://doi.org/10.1007/978-3-319-21690-4_15.

[10] P. Godefroid, N. Klarlund, K. Sen, DART: directed automated random testing, PLDI '05, ACM, 2005, ISBN: 1-59593-056-6, pp 213–223, https://doi.org/10.1145/1065010.1065036.

[11] B. Korel, A dynamic approach of test data generation, in: Proceedings. Conference on Software Maintenance 1990, 1990, pp. 311–317, https://doi.org/10.1109/ICSM.1990.131379.

[12] S. Khurshid, C.S. Păsăreanu, W. Visser, Generalized symbolic execution for model checking and testing, in: Proceedings of the 9th International Conference on Tools and Algorithms for the Construction and Analysis of Systems, TACAS'03, Springer-Verlag, Berlin, Heidelberg, 2003, ISBN: 3-540-00898-5, pp 553–568, http://dl.acm.org/citation.cfm?id=1765871.1765924.

[13] X. Deng, J. Lee, Robby, Bogor/Kiasan: a K-bounded symbolic execution for checking strong heap properties of open systems, in: Proceedings of the 21st IEEE/ACM International Conference on Automated Software Engineering, ASE '06, IEEE Computer Society, Washington, DC, USA, 2006, ISBN: 0-7695-2579-2, pp. 157–166, https://doi.org/10.1109/ASE.2006.26.

[14] X. Deng, Robby, J. Hatcliff, Towards a case-optimal symbolic execution algorithm for analyzing strong properties of object-oriented programs, in: Proceedings of the Fifth IEEE International Conference on Software Engineering and Formal Methods SEFM '07, IEEE Computer Society, Washington, DC, USA, 2007, ISBN: 0-7695-2884-8, pp 273–282, https://doi.org/10.1109/SEFM.2007.43.

[15] B. Hillery, E. Mercer, N. Rungta, S. Person, Exact heap summaries for symbolic execution, in: Proceedings of the 17th International Conference on Verification, Model Checking, and Abstract Interpretation, vol. 9583, VMCAI 2016, Springer-Verlag, New York, NY, USA, 2016, ISBN: 978-3-662-49121-8, pp 206–225, https://doi.org/10.1007/978-3-662-49122-5_10.

[16] J. Geldenhuys, N. Aguirre, M.F. Frias, W. Visser, Bounded lazy initialization, in: G. Brat, N. Rungta, A. Venet (Eds.), NASA Formal Methods, 5th International Symposium, NFM 2013, Moffett Field, CA, USA, May 14–16, 2013. Proceedings, Springer, Berlin, Heidelberg, 2013, ISBN: 978-3-642-38088-4, pp. 229–243, https://doi.org/10.1007/978-3-642-38088-4_16.

[17] W. Visser, C.S. Păsăreanu, S. Khurshid, Test input generation with Java PathFinder, in: Proceedings of the 2004 ACM SIGSOFT International Symposium on Software Testing and Analysis, ISSTA '04, ACM, New York, NY, USA, 2004, ISBN: 1-58113-820-2, pp. 97–107, https://doi.org/10.1145/1007512.1007526.

[18] P. Braione, G. Denaro, M. Pezzè, Symbolic execution of programs with heap inputs, in: Proceedings of the 2015 10th Joint Meeting on Foundations of Software Engineering, ESEC/FSE 2015, ACM, New York, NY, USA, 2015, ISBN: 978-1-4503-3675-8, pp. 602–613, https://doi.org/10.1145/2786805.2786842.

[19] L.H. Pham, Q.L. Le, Q.-S. Phan, J. Sun, S. Qin, Enhancing symbolic execution of heap-based programs with separation logic for test input generation, CoRR abs/1712.06025 (2017), https://arxiv.org/abs/1712.06025.

[20] S.S. Ishtiaq, P.W. O'Hearn, BI as an assertion language for mutable data structures, in: Proceedings of the 28th ACM SIGPLAN-SIGACT Symposium on Principles of Programming Languages, POPL'01, ACM, New York, NY, USA, 2001, ISBN: 1-58113-336-7, pp. 14–26, https://doi.org/10.1145/360204.375719.

[21] J. Reynolds, Separation logic: a logic for shared mutable data structures, in: LICS 2002, pp. 55–74.

[22] Q.L. Le, J. Sun, W.-N. Chin, Satisfiability Modulo Heap-Based Programs, in: Springer International Publishing, Cham, 2016, ISBN: 978-3-319-41528-4, pp. 382–404, https://doi.org/10.1007/978-3-319-41528-4_21.

[23] K. Luckow, M. Dimjašević, D. Giannakopoulou, F. Howar, M. Isberner, T. Kahsai, Z. Rakamarić, V. Raman, JDart: a dynamic symbolic analysis framework, in: M. Chechik, J.-F. Raskin (Eds.), Lecture Notes in Computer Science, Proceedings of the 22nd International Conference on Tools and Algorithms for the Construction and Analysis of Systems (TACAS), vol. 9636, Springer, 2016, pp. 442–459.

[24] P. Braione, G. Denaro, M. Pezzè, JBSE: a symbolic executor for java programs with complex heap inputs. in: Proceedings of the 2016 24th ACM SIGSOFT International Symposium on Foundations of Software Engineering, FSE 2016, ACM, New York, NY, USA, 2016, ISBN: 978-1-4503-4218-6, pp. 1018–1022, https://doi.org/10.1145/2950290.2983940.

[25] D. Brumley, I. Jager, T. Avgerinos, E.J. Schwartz, BAP: a binary analysis platform, Springer, Berlin, Heidelberg, 2011, ISBN: 978-3-642-22110-1, pp. 463–469, https://doi.org/10.1007/978-3-642-22110-1_37.

[26] V. Chipounov, V. Kuznetsov, G. Candea, S2E: a platform for in-vivo multi-path analysis of software systems, in: Proceedings of the Sixteenth International Conference on Architectural Support for Programming Languages and Operating Systems, ASPLOS XVI, ACM, New York, NY, USA, 2011, ISBN: 978-1-4503-0266-1, pp. 265–278, https://doi.org/10.1145/1950365.1950396.

[27] K. Sen, D. Marinov, G. Agha, CUTE: a concolic unit testing engine for C, in: Proceedings of the 13th ACM SIGSOFT International Symposium on Foundations of Software Engineering, ESEC/FSE-13, ACM, New York, NY, USA, 2005, ISBN: 1-59593-014-0, pp. 263–272, https://doi.org/10.1145/1081706.1081750.

[28] G. Li, I. Ghosh, S.P. Rajan, KLOVER: a symbolic execution and automatic test generation tool for C++ programs, in: Proceedings of the 23rd International Conference on Computer Aided Verification, CAV'11, Springer-Verlag, Berlin, Heidelberg, 2011, ISBN: 978-3-642-22109-5, pp. 609–615. http://dl.acm.org/citation.cfm?id=2032305.2032354.

[29] C. Cadar, V. Ganesh, P.M. Pawlowski, D.L. Dill, D.R. Engler, EXE: automatically generating inputs of death, in: Proceedings of the 13th ACM Conference on Computer and Communications Security, CCS '06, ACM, New York, NY, USA, 2006, ISBN: 1-59593-518-5, pp. 322–335, https://doi.org/10.1145/1180405.1180445.

[30] P. Saxena, D. Akhawe, S. Hanna, F. Mao, S. McCamant, D. Song, A symbolic execution framework for Javascript, in: Proceedings of the 2010 IEEE Symposium on Security and Privacy, SP '10, IEEE Computer Society, Washington, DC, USA, 2010, ISBN: 978-0-7695-4035-1, pp. 513–528, https://doi.org/10.1109/SP.2010.38.

[31] J. Jeon, K.K. Micinski, J.S. Foster, SymDroid: symbolic execution for Dalvik bytecode, 2012. Technical report.

[32] T. Ball, J. Daniel, Deconstructing dynamic symbolic execution, in: Proceedings of the 2014 Marktoberdorf Summer School on Dependable Software Systems Engineering. The 2014 Marktober Summer School on Deop, IOS Press, 2015, https://www.microsoft.com/en-us/research/publication/deconstructing-dynamic-symbolic-execution/.

[33] D. Song, D. Brumley, H. Yin, J. Caballero, I. Jager, M.G. Kang, Z. Liang, J. Newsome, P. Poosankam, P. Saxena, BitBlaze: a new approach to computer security via binary analysis, Springer, Berlin, Heidelberg, 2008, ISBN: 978-3-540-89862-7, pp. 1–25, https://doi.org/10.1007/978-3-540-89862-7_1.

[34] D. Babić, L. Martignoni, S. McCamant, D. Song, Statically-directed dynamic automated test generation, in: Proceedings of the 2011 International Symposium on Software Testing and Analysis, ISSTA '11, ACM, New York, NY, USA, 2011, ISBN: 978-1-4503-0562-4, pp. 12–22, https://doi.org/10.1145/2001420.2001423.

[35] D. Liew, C. Cadar, A.F. Donaldson, Symbooglix: a symbolic execution engine for boogie programs, in: 2016 IEEE International Conference on Software Testing, Verification and Validation (ICST), 2016, pp. 45–56, https://doi.org/10.1109/ICST.2016.11.

[36] A. King, Distributed Parallel Symbolic Execution, Kansas State University, (Master thesis), 2009.

[37] J.H. Siddiqui, S. Khurshid, ParSym: parallel symbolic execution, in: ICSTE, vol. 1, 2010, pp. V1-405-V1-409, https://doi.org/10.1109/ICSTE.2010.5608866.

[38] M. Staats, C. Păsăreanu, Parallel symbolic execution for structural test generation. in: ISSTA '10, ACM, New York, NY, USA, 2010, ISBN: 978-1-60558-823-0, pp. 183–194, https://doi.org/10.1145/1831708.1831732.

[39] E.L. Gunter, D. Peled, Unit checking: symbolic model checking for a unit of code, in: Verification: Theory and Practice, Essays Dedicated to Zohar Manna on the Occasion of his 64th Birthday, 2003, pp. 548–567, https://doi.org/10.1007/978-3-540-39910-0_24.

[40] P. Godefroid, Compositional dynamic test generation, in: Proceedings of the 34th Annual ACM SIGPLAN-SIGACT Symposium on Principles of Programming Languages, POPL '07, ACM, New York, NY, USA, 2007, ISBN: 1-59593-575-4, pp. 47–54, https://doi.org/10.1145/1190216.1190226.

[41] S. Anand, P. Godefroid, N. Tillmann, Demand-driven compositional symbolic execution, in: Proceedings of the Theory and Practice of Software, 14th International Conference on Tools and Algorithms for the Construction and Analysis of Systems, TACAS'08/ETAPS'08, Springer-Verlag, Berlin, Heidelberg, 2008, pp. 367–381. ISBN 3-540-78799-2, 978-3-540-78799-0. http://dl.acm.org/citation.cfm?id=1792734.1792771.

[42] V. Kuznetsov, J. Kinder, S. Bucur, G. Candea, Efficient state merging in symbolic execution, in: Proceedings of the 33rd ACM SIGPLAN Conference on Programming Language Design and Implementation, PLDI '12, ACM, New York, NY, USA, 2012, ISBN: 978-1-4503-1205-9, pp. 193–204, https://doi.org/10.1145/2254064.2254088.

[43] T. Avgerinos, A. Rebert, S.K. Cha, D. Brumley, Enhancing symbolic execution with veritesting, in: Proceedings of the 36th International Conference on Software Engineering, ICSE 2014, ACM, New York, NY, USA, 2014, ISBN: 978-1-4503-2756-5, pp. 1083–1094, https://doi.org/10.1145/2568225.2568293.

[44] K. Sen, G. Necula, L. Gong, W. Choi, MultiSE: multi-path symbolic execution using value summaries, in: Proceedings of the 2015 10th Joint Meeting on Foundations of Software Engineering, ESEC/FSE 2015, ACM, New York, NY, USA, 2015, ISBN: 978-1-4503-3675-8, pp. 842–853, https://doi.org/10.1145/2786805.2786830.

[45] R. Qiu, G. Yang, C.S. Păsăreanu, S. Khurshid, Compositional symbolic execution with memoized replay. in: 37th IEEE/ACM International Conference on Software Engineering, ICSE 2015, Florence, Italy, May 16–24, 2015, Volume 1, 2015, pp. 632–642, https://doi.org/10.1109/ICSE.2015.79.

[46] J. Jaffar, V. Murali, J.A. Navas, Boosting concolic testing via interpolation, in: Joint Meeting of the European Software Engineering Conference and the ACM SIGSOFT Symposium on the Foundations of Software Engineering, ESEC/FSE'13, Saint Petersburg, Russian Federation, August 18–26, 2013, 2013, pp. 48–58, https://doi.org/10.1145/2491411.2491425.

[47] P. Godefroid, D. Luchaup, Automatic partial loop summarization in dynamic test generation, in: Proceedings of the 20th International Symposium on Software Testing and Analysis, ISSTA 2011, Toronto, ON, Canada, July 17–21, 2011, 2011, pp. 23–33, https://doi.org/10.1145/2001420.2001424.

[48] A. Filieri, C.S. Păsăreanu, W. Visser, J. Geldenhuys, Statistical symbolic execution with informed sampling, in: Proceedings of the 22nd ACM SIGSOFT International Symposium on Foundations of Software Engineering, FSE 2014, ACM, New York, NY, USA, 2014, ISBN: 978-1-4503-3056-5, pp. 437–448, https://doi.org/10.1145/2635868.2635899.

[49] W. Visser, J. Geldenhuys, M.B. Dwyer, Green: reducing, reusing and recycling constraints in program analysis, in: Proceedings of the ACM SIGSOFT 20th International Symposium on the Foundations of Software Engineering, FSE '12, ACM, New York, NY, USA, 2012, ISBN: 978-1-4503-1614-9, pp. 58:1–58:11, https://doi.org/10.1145/2393596.2393665.

[50] T. Brennan, N. Tsiskaridze, N. Rosner, A. Aydin, T. Bultan, Constraint normalization and parameterized caching for quantitative program analysis, in: Proceedings of the 2017 11th Joint Meeting on Foundations of Software Engineering, ESEC/FSE 2017, ACM, New York, NY, USA, 2017, ISBN: 978-1-4503-5105-8, pp. 535–546, https://doi.org/10.1145/3106237.3106303.

[51] G. Yang, C.S. Păsăreanu, S. Khurshid, Memoized symbolic execution. in: Proceedings of the 2012 International Symposium on Software Testing and Analysis, ISSTA 2012, ACM, New York, NY, USA, 2012, ISBN: 978-1-4503-1454-1, pp. 144–154, https://doi.org/10.1145/2338965.2336771.

[52] M. Souza, M. Borges, M. d'Amorim, C.S. Păsăreanu, CORAL: solving complex constraints for symbolic pathfinder, in: Proceedings of the Third International Conference on NASA Formal Methods, NFM'11, Springer-Verlag, Berlin, Heidelberg, 2011, ISBN: 978-3-642-20397-8, pp. 359–374. http://dl.acm.org/citation.cfm?id=1986308.1986337.

[53] Y. Zheng, X. Zhang, V. Ganesh, Z3-str: a Z3-based string solver for web application analysis, in: Proceedings of the 2013, Ninth Joint Meeting on Foundations of Software Engineering, ESEC/FSE 2013, ACM, New York, NY, USA, 2013, ISBN: 978-1-4503-2237-9, pp. 114–124, https://doi.org/10.1145/2491411.2491456.

[54] L. Bang, A. Aydin, Q.-S. Phan, C.S. Păsăreanu, T. Bultan, String analysis for side channels with segmented oracles, in: Proceedings of the 2016 24th ACM SIGSOFT International Symposium on Foundations of Software Engineering, FSE 2016, ACM, New York, NY, USA, 2016, ISBN: 978-1-4503-4218-6, pp. 193–204, https://doi.org/10.1145/2950290.2950362.

[55] B. Benhamamouch, B. Monsuez, F. Védrine, Computing WCET using symbolic execution, in: Proceedings of the Second International Conference on Verification and Evaluation of Computer and Communication Systems, VECoS'08, British Computer Society, Swinton, UK, 2008, pp. 128–139, http://dl.acm.org/citation.cfm?id=2227461.2227475.

[56] D. Kebbal, P. Sainrat, Combining symbolic execution and path enumeration in worst-case execution time analysis, in: F. Mueller (Ed.), OpenAccess Series in Informatics (OASIcs), 6th International Workshop on Worst-Case Execution Time Analysis (WCET'06), ISSN 2190-6807, vol. 4, Schloss Dagstuhl-Leibniz-Zentrum fuer Informatik, Dagstuhl, Germany, 2006, ISBN: 978-3-939897-03-3, https://doi.org/10.4230/OASIcs.WCET.2006.675, http://drops.dagstuhl.de/opus/volltexte/2006/675.

[57] T. Lundqvist, P. Stenström, An integrated path and timing analysis method based on cycle-level symbolic execution, Real-Time Syst. 17 (2–3) (1999) 183–207. ISSN: 0922-6443, https://doi.org/10.1023/A:1008138407139.

[58] F. Stappert, P. Altenbernd, Complete worst-case execution time analysis of straight-line hard real-time programs, J. Syst. Archit. 46 (4) (2000) 339–355, https://doi.org/10.1016/S1383-7621(99)00010-7.

[59] J. Knoop, L. Kovács, J. Zwirchmayr, WCET Squeezing: On-demand Feasibility Refinement for Proven Precise WCET-bounds, in: Proceedings of the 21st International Conference on Real-Time Networks and Systems, RTNS '13, ACM, New York, NY, USA, 2013, ISBN: 978-1-4503-2058-0, pp. 161–170, https://doi.org/10.1145/2516821.2516847.

[60] K.S. Luckow, C.S. Păsăreanu, B. Thomsen, Symbolic execution and timed automata model checking for timing analysis of java real-time systems, EURASIP J. Embed. Syst. 2015 (1) (2015) 2. ISSN: 1687-3963, https://doi.org/10.1186/s13639-015-0020-8.

[61] D. Kebbal, P. Sainrat, Combining symbolic execution and path enumeration in worst-case execution time analysis, in: F. Mueller (Ed.), OpenAccess Series in Informatics (OASIcs), 6th International Workshop on Worst-Case Execution Time Analysis (WCET'06), ISSN 2190-6807, vol. 4, Schloss Dagstuhl-Leibniz-Zentrum fuer Informatik, Dagstuhl, Germany, 2006 ISBN: 978-3-939897-03-3, https://doi.org/10.4230/OASIcs.WCET.2006.675, http://drops.dagstuhl.de/opus/volltexte/2006/675.

[62] R. Bodík, R. Gupta, M.L. Soffa, Refining data flow information using infeasible paths, SIGSOFT Softw. Eng. Notes 22 (6) (1997) 361–377. ISSN: 0163-5948, https://doi.org/10.1145/267896.267921.

[63] A. Biere, J. Knoop, L. Kovács, J. Zwirchmayr, The auspicious couple: symbolic execution and WCET analysis, in: C. Maiza (Ed.), OpenAccess Series in Informatics (OASIcs), 13th International Workshop on Worst-Case Execution Time Analysis, ISSN 2190-6807, vol. 30, Schloss Dagstuhl-Leibniz-Zentrum fuer Informatik, Dagstuhl, Germany, 2013, ISBN: 978-3-939897-54-5, pp. 53–63, https://doi.org/10.4230/OASIcs.WCET.2013.53. http://drops.dagstuhl.de/opus/volltexte/2013/4122.

[64] R. Chapman, A. Burns, A. Wellings, Integrated program proof and worst-case timing analysis of SPARK Ada., in: Proc. ACM SIGPLAN Workshop on Languages, Compilers and Tools for Real-Time Systems (LCT-RTS'94), ACM Press, 1994.

[65] P. Zhang, S. Elbaum, M.B. Dwyer, Automatic generation of load tests, in: Proceedings of the 2011 26th IEEE/ACM International Conference on Automated Software Engineering, ASE '11, IEEE Computer Society, Washington, DC, USA, 2011, ISBN: 978-1-4577-1638-6, pp. 43–52, https://doi.org/10.1109/ASE.2011.6100093.

[66] J. Burnim, S. Juvekar, K. Sen, WISE: automated test generation for worst-case complexity, in: 2009 IEEE 31st International Conference on Software Engineering, ISSN: 0270-5257, 2009, pp. 463–473, https://doi.org/10.1109/ICSE.2009.5070545.

[67] K. Luckow, R. Kersten, C.S. Păsăreanu, Symbolic complexity analysis using context-preserving histories, in: Proceedings of the 10th IEEE International Conference on Software Testing, Verification and Validation (ICST 2017), 2017 (to appear).

[68] T. Avgerinos, S.K. Cha, A. Rebert, E.J. Schwartz, M. Woo, D. Brumley, Automatic exploit generation, Commun. ACM 57 (2) (2014) 74–84. ISSN: 0001-0782, https://doi.org/10.1145/2560217.2560219.

[69] E.S. Cohen, Information transmission in sequential programs, in: R.A. DeMillo, D.P. Dobkin, A.K. Jones, R.J. Lipton (Eds.), Foundations of Secure Computation, Academic Press, 1978, pp. 297–335.

[70] J.A. Goguen, J. Meseguer, Security policies and security models, in: IEEE Symposium on Security and Privacy, 1982, pp. 11–20.

[71] A. Darvas, R. Hähnle, D. Sands, A theorem proving approach to analysis of secure information flow, in: Proceedings of the Second International conference on Security in Pervasive Computing, SPC'05, Springer-Verlag, Berlin, Heidelberg, 2005, pp. ISBN 3-25521-4, 978-540-25521-5,193–209, https://doi.org/10.1007/978-3-540-32004-3_20.

[72] G. Barthe, P.R. D'Argenio, T. Rezk, Secure information flow by self-composition, in: Proceedings of the 17th IEEE workshop on Computer Security Foundations, CSFW '04, IEEE Computer Society, Washington, DC, USA, 2004, ISBN: 0-7695-2169-X, https://doi.org/10.1109/CSFW.2004.17.

[73] Q.-S. Phan, Self-composition by symbolic execution, in: OpenAccess Series in Informatics (OASIcs), 2013 Imperial College Computing Student Workshop, ISSN 2190-6807, vol. 35, Schloss Dagstuhl-Leibniz-Zentrum fuer Informatik, Dagstuhl, Germany, 2013, ISBN: 978-3-939897-63-7, pp. 95–102, http://drops.dagstuhl.de/opus/volltexte/2013/4277.

[74] Q.H. Do, R. Bubel, R. Hähnle, Exploit generation for information flow leaks in object-oriented programs, in: ICT Systems Security and Privacy Protection: 30th IFIP TC 11 International Conference, SEC 2015, Hamburg, Germany, May 26–28, 2015, Proceedings Springer International Publishing, Cham, 2015, ISBN: 978-3-319-18467-8, pp. 401–415, https://doi.org/10.1007/978-3-319-18467-8_27.

[75] M. Balliu, M. Dam, G.L. Guernic, ENCoVer: symbolic exploration for information flow security, in: Proceedings of the 2012 IEEE 25th Computer Security Foundations Symposium, CSF '12, IEEE Computer Society, Washington, DC, USA, 2012, ISBN: 978-0-7695-4718-3, pp. 30–44, https://doi.org/10.1109/CSF.2012.24.

[76] D. Clark, S. Hunt, P. Malacaria, A static analysis for quantifying information flow in a simple imperative language, J. Comput. Secur. 15 (3) (2007) 321–371. ISSN: 0926-227X, http://dl.acm.org/citation.cfm?id=1370628.1370629.

[77] P. Malacaria, Assessing security threats of looping constructs, in: Proceedings of the 34th Annual ACM SIGPLAN-SIGACT Symposium on Principles of Programming Languages, POPL '07, ACM, New York, NY, USA, 2007, ISBN: 1-59593-575-4, pp. 225–235, https://doi.org/10.1145/1190216.1190251.

[78] T.M. Cover, J.A. Thomas, Elements of Information Theory, Wiley-Interscience, New York, NY, USA, 1991, ISBN: 0-471-06259-6.

[79] Q.-S. Phan, P. Malacaria, O. Tkachuk, C.S. Păsăreanu, Symbolic quantitative information flow, SIGSOFT Softw. Eng. Notes 37 (6) (2012) 1–5, ISSN: 0163-5948, https://doi.org/10.1145/2382756.2382791.

[80] Q.-S. Phan, P. Malacaria, Abstract model counting: a novel approach for quantification of information leaks, in: Proceedings of the Ninth ACM Symposium on Information, Computer and Communications Security, ASIA CCS '14 ACM, New York, NY, USA, 2014, ISBN: 978-1-4503-2800-5, pp 283–292, https://doi.org/10.1145/2590296.2590328.

[81] Q.-S. Phan, P. Malacaria, C.S. Păsăreanu, M. d'Amorim, Quantifying information leaks using reliability analysis, in: Proceedings of the 2014 International SPIN Symposium on Model Checking of Software, SPIN 2014, ACM, New York, NY, USA, 2014, ISBN: 978-1-4503-2452-6, pp 105–108, https://doi.org/10.1145/2632362.2632367.

[82] LattE, http://www.math.ucdavis.edu/latte/.

[83] P.C. Kocher, Timing attacks on implementations of diffie-hellman, RSA, DSS, and other systems, in: Proceedings of the 16th Annual International Cryptology Conference on Advances in Cryptology, CRYPTO '96, Springer-Verlag, London, UK, 1996, ISBN: 3-540-61512-1, pp 104–113, http://dl.acm.org/citation.cfm?id=646761.706156.

[84] J. Kelsey, Compression and information leakage of plaintext, in: Revised Papers from the 9th International Workshop on Fast Software Encryption, FSE '02, Springer-Verlag, London, UK, 2002, ISBN: 3-540-44009-7, pp. 263–276, http://dl.acm.org/citation.cfm?id=647937.741226.

[85] D. Brumley, D. Boneh, Remote timing attacks are practical, in: Proceedings of the 12th Conference on USENIX Security Symposium, Vol. 12, SSYM'03, USENIX Association, Berkeley, CA, 2003, p. 1. http://dl.acm.org/citation.cfm?id=1251353.1251354.

[86] S. Chen, R. Wang, X. Wang, K. Zhang, Side-channel leaks in web applications: a reality today, a challenge tomorrow, Proceedings of the 2010 IEEE Symposium on Security and Privacy, SP '10, IEEE Computer Society, Washington, DC, USA, 2010, ISBN: 978-0-7695-4035-1, pp. 191–206, https://doi.org/10.1109/SP.2010.20.

[87] C.S. Păsăreanu, Q.-S. Phan, P. Malacaria, Multi-run side-channel analysis using symbolic execution and max-SMT, in: Proceedings of the 2016 IEEE 29th Computer Security Foundations Symposium, CSF '16, IEEE Computer Society, Washington, DC, USA, 2016, pp. 387–400, https://doi.org/10.1109/CSF.2016.34.

[88] Q.-S. Phan, L. Bang, C.S. Păsăreanu, P. Malacaria, T. Bultan, Synthesis of adaptive side-channel attacks, in: 2017 IEEE 30th Computer Security Foundations Symposium (CSF), CSF '17, IEEE Computer Society, Washington, DC, USA, 2017.

[89] B.P. Miller, L. Fredriksen, B. So, An Empirical Study of the Reliability of UNIX Utilities. Commun. ACM 33 (12) (1990) 32–44. ISSN: 0001-0782, https://doi.org/10.1145/96267.96279.

[90] M. Sutton, A. Greene, P. Amini, Fuzzing: Brute Force Vulnerability Discovery, Pearson Education, 2007.

[91] M. Zalewski, American Fuzzy Lop (AFL), 2017, http://lcamtuf.coredump.cx/afl/ (accessed August 11, 2017).

[92] R. Kersten, K. Luckow, C.S. Păsăreanu, POSTER: AFL-based Fuzzing for Java with Kelinci, Proceedings of the 2017 ACM SIGSAC Conference on Computer and Communications Security, CCS '17, ACM, New York, NY, USA, 2017, ISBN: 978-1-4503-4946-8, pp 2511–2513, https://doi.org/10.1145/3133956.3138820.

[93] J.P. Galeotti, G. Fraser, A. Arcuri, Improving search-based test suite generation with dynamic symbolic execution, in: 2013 IEEE 24th International Symposium on Software Reliability Engineering (ISSRE), ISSN 1071-9458, 2013, pp. 360–369, https://doi.org/10.1109/ISSRE.2013.6698889.

[94] M. Borges, A. Filieri, M. d'Amorim, C.S. Păsăreanu, W. Visser, Compositional solution space quantification for probabilistic software analysis, ACM SIGPLAN Conference on Programming Language Design and Implementation, PLDI '14, Edinburgh, United Kingdom—June 9–11, 2014, 2014, pp. 123–132, https://doi.org/10.1145/2594291.2594329.

[95] A. Filieri, C.S. Păsăreanu, W. Visser, Reliability analysis in symbolic pathfinder, in: 35th International Conference on Software Engineering, ICSE '13, San Francisco, CA, USA, May 18–26, 2013, 2013, pp. 622–631, https://doi.org/10.1109/ICSE.2013.6606608.

[96] https://www.cyberscoop.com/mayhem-darpa-cyber-grand-challenge-dod-voltron/.

ABOUT THE AUTHORS

Corina S. Păsăreanu is an associate research professor with CyLab at Carnegie Mellon University, working at the Silicon Valley campus with NASA Ames Research Center. She is an ACM Distinguished Scientist, known for her influential research on software model checking, symbolic execution and assume-guarantee compositional verification, using abstraction and learning-based methods. She is the recipient of an ACM SIGSOFT Distinguished Paper Award in 2002, the ICSE 2010 Most Influential Paper Award, the 2010 ACM SIGSOFT Impact Paper Award, the ISSTA 2018 Retrospective Impact Paper Award, the ASE 2018 Most Influential Paper Award, and the ESEC/FSE 2018 Test of Time Award. More information is available at her website: https://ti.arc.nasa.gov/profile/pcorina/.

Rody Kersten is a senior software engineer at Synopsys, in the Core Analysis team for the Coverity Static Application Security Testing product. He has received an MSc (2010) and PhD (2015) in Computer Science from Radboud University Nijmegen in The Netherlands. He is a former postdoctoral researcher at Carnegie Mellon University, and a former assistant professor at Open University of the Netherlands. His research interests include static analysis, formal verification, symbolic execution, and fuzz testing, with a focus on software resource consumption (time, memory, energy). As a software engineer, he has contributed to a variety of software analysis applications, including the industry-leading Static Application Security Testing product Coverity.

Kasper Søe Luckow is a software development engineer in the Automated Reasoning Group at Amazon Web Services. He received a BSc degree (2009) and MSc Cum Laude degree (2011) in Software Engineering from Aalborg University and a PhD degree (2014) in Computer Science from the same university. From 2014 to 2017 he was a postdoctoral researcher at Carnegie Mellon University. His research interests include program analysis, verification, and testing.

Quoc-Sang Phan is currently a member of research staff at Fujitsu Laboratories of America. He received his PhD degree in Computer Science from Queen Mary University of London in 2015. From 2015 to 2017, he was a postdoctoral researcher at Carnegie Mellon University. His research interests include symbolic execution and fuzzing techniques for software security.

CHAPTER SEVEN

Experiences With Replicable Experiments and Replication Kits for Software Engineering Research

Steffen Herbold, Fabian Trautsch, Patrick Harms, Verena Herbold, Jens Grabowski
University of Goettingen, Göttingen, Germany

Contents

1. Introduction   316
2. What Is Replication   318
3. Replication Kits   319
4. Experience Reports   320
    4.1 Software Defect Prediction   320
    4.2 Analysis of Test Types Characteristics   325
    4.3 Developer Contribution Behavior   327
    4.4 User-Oriented Usability Studies   327
5. Discussion   330
    5.1 Contents of Replication Kits   330
    5.2 Publishing of Replication Kits   332
    5.3 A Checklist for Replication Kits   333
    5.4 Platforms for Replicable Research   333
    5.5 The Role of Replication Kits in Review Processes   337
6. Conclusion   339
References   340
About the Authors   342

Abstract

Replications and replicable research are currently gaining traction in the software engineering research community. Our research group made an effort in the recent years to make our own research accessible for other researchers, through the provision of replication kits that allow rerunning our experiments. Within this chapter, we present our experiences with replication kits. We first had to learn which contents are required, how to structure them, how to document them, and also how to best share them with other researchers. While this sounds very straightforward, there are many small potential

Advances in Computers, Volume 113
ISSN 0065-2458
https://doi.org/10.1016/bs.adcom.2018.10.003

mistakes, which may have a strong negative impact on the usefulness and long-term availability of replication kits. We derive best practices for the content and the sharing of replication kits based on our experiences. Moreover, we outline how platforms for replicable research may further help our community, especially with problems related to the external validity of results. Finally, we discuss the lack of integration of replication kits into most current review processes at conferences and journals. We also give one example for a review process into which replication kits were well-integrated. Altogether, this chapter demonstrates that making research replicable is a challenging task and there is a long road ahead until our community has a generally accepted and enforced standard of replicability.

1. INTRODUCTION

While replication was a neglected topic for software engineering research for a long time, it is currently gaining traction. Siegmund et al. [1] not only found that replication studies are underrepresented in the field of software engineering, they also determined that most researchers in the community agree with this statement and that more replications are required. The reasons for this are manifold: in a paper from 2016, Trautsch et al. [2] point out that the non–availability of implementations is one of the points that is problematic, if a study should be replicated. But, as Robles emphasizes in his paper from 2010 [3], only one fifth of the 171 papers he reviewed published "the complete set of tools and scripts to fully reproduce the study" [3]. This lack increases the burden on researchers who want to conduct replications.

If we do not address the problems with the replication of studies, our community runs into the same situation at it is currently the case in, e.-g., medicine or psychology where a replication often fails [4, 5]. The emerging importance of this problem is highlighted by the results of a survey, which was carried out by the high-impact journal Nature from 2016 [6] in which 1576 researchers from different disciplines took part. The results show that "more than 70% of researchers have tried and failed to reproduce another scientist's experiments, and more than half have failed to reproduce their own experiments" [6]. One of the reasons for this crisis is that researchers did not share their raw data and/or details on the collection method. We as software engineering researchers do our experiments with different subjects than in, e.g., medicine. Nevertheless, if we run into the same situation as the mentioned disciplines, the resulting problem will be the same: a lot of invalid research will be done. Hence, findings of such studies do not contribute to our existing body of knowledge, as they are not reproducible and, therefore, they can not be validated.

Another overlooked aspect of replicable research is that they are important for novice researchers. Novice researchers, e.g., at the beginning of their PhD studies do not know the relevant tools and techniques which are standard in their field of research, yet. In case existing research provides the means for easy replication, novice researchers can use this as a starting point, which would lower the entry barrier into a new field of research. One way to ease this (at least partially) is replicating research as a first step. You can see and try out how others have approached the problem. Additionally, issues that may occur in the implementation of an own approach or in the following analysis of the data can be detected earlier. Furthermore, a replication of earlier work should be done anyhow, as a comparison of one's own work with related work should be done by, e.g., applying both approaches to the same data.

However, a novice researcher is not the only one that benefits from replicating research in her early stage of research. The replication has value for the whole community. A conceptual replication that, e.g., uses more or other data raises the external validity of an approach or a conclusion made in a paper. Moreover, it can even detect problems within a paper that might not be obvious at first.

While it is easy to ask for more replications and demand that researchers make sure that their work is easy to replicate, practical insights on this are not available. In recent years, we conducted multiple replication studies. Moreover, we made sure that replicability of all of our research improves. Within this chapter, we want to describe our experiences. We will not discuss the contents of our replication work, but rather which problems we faced, how we solved them, and how we learned over time to make our own research easy to replicate. Our goal behind this is twofold. One the one hand, we want to motivate more researchers to conduct replications, which can be exciting research projects with a big impact. On the other hand, we want to show which steps must be taken to make research easy to replicate, which problems may occur, and how they can be resolved.

The remainder of this chapter is structured as follows. First, we clarify our terminology. In Section 2, we define the terminology regarding replications. Afterwards, we define what a replication kit is in Section 3. The main part of this chapter is our experience report provided in Section 4. We describe how we created replication kits for research on software defect prediction, analysis of test types characteristics, developer contribution behavior modeling, and usability. In Section 5, we discuss our experiences and derive best practices for replication kits, discuss how platforms for replicable research may help our community and consider the role of replication kits in review processes. Finally, we conclude the chapter in Section 6.

2. WHAT IS REPLICATION

While most researchers agree that replication is a key element to ensure validity of results, many use this term differently as replication can have different forms. Most importantly, people often believe replication is the same as with being able to execute the experiment again. While this is a major component of one type of replication, just the rerun of an experiment is not a replication. Moreover, replication does not mean that exactly the same experiment must be conducted again. Shull et al. [7] provide a good summary of what replication means in the context of software engineering research. Based on behavioral research [8] and previous definitions for the software engineering domain [9], Shull et al. differentiate between *exact* and *conceptual* replications.

Definition 1 (Exact Replication). "An exact replication is one in which the procedures of an experiment are followed as closely as possible to determine whether the same results can be obtained" [7].

Definition 2 (Conceptual Replication). "A conceptual replication is one in which the same research question or hypothesis is evaluated by using a different experimental procedure, i.e. many or all of the variables [...] are changed" [7].

If we dissect the definitions by Shull et al., we see that a replication consists of three elements of replications: experiment procedures, research questions, and results. What all replications have in common is that the research question remains the same. Thus, the underlying goal of replications is to extend the existing knowledge regarding a research question. This additional knowledge can be regarding the internal validity of results, the external validity of results, or both. The internal validity may be improved, e.g., with more insights into how results were produced through an exact replication. The external validity may be improved, e.g., with demonstrating that the results remain the same even if different data or statistical procedures are used in a conceptual validation. Exact and conceptual replications may both also be negative. For exact replications, this may be due to a bug in an implementation that was fixed for the replication. A negative

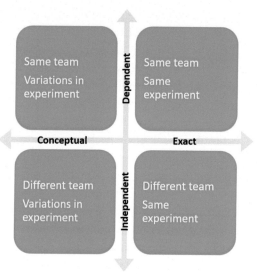

Fig. 1 Summary of the different types of replications according to Shull et al. [7].

conceptual replication may also be due to mistakes in the original study, but also due to different results on other data.

Shull et al. also identified two subtypes of exact and conceptual replications. They refer to them as dependent replications and independent replications, which are related to the goal of the replication. If the goal of the replication is to show that a given effect is robust, the replication should ideally be *independent*, i.e., performed by a different set of researchers that use their own experimental design. As such, independent replications are a subtype of the conceptual replications. If the goal of a replication is to consider additional factors under which research results hold, it is useful to apply only small variations to the experiment procedure, e.g., slightly different data. Shull et al. call this type of replication dependent replications. Such replications are often performed by the same set of researchers. Moreover, since the experimental procedures are followed as closely as possible with minor exceptions, this is subtype of the exact replications. Fig. 1 summarizes the four types of replications identified by Shull et al.

3. REPLICATION KITS

Shull et al. [7] identified *lab packages* [9] as a valuable resource for replication. They describe lab packages as "an exhaustive documentation of the original experiment [...] with relevant details excluded from the report. [...] The lab package is a simple way to transmit all necessary details of the experiment in order to be able to perform an exact replication" [7]. Shull et al. viewed lab packages purely in the view of replications. Without lab packages, exact replications are nearly impossible. For conceptual replications, lab packages can still be very valuable, as they allow to better analyze how and where modifications to the original experimental procedure may be performed. The drawback of lab packages is that they may propagate problems and bias of the original study to other works [10].

Lab packages and their definitions stem from behavioral sciences. As such, they mostly consider experiments with human subjects and the protocols required for such experiments. However, a lot of recent empirical software engineering research does not evaluate human subjects, but rather source code or tools. As such, the implementations that perform the analysis are often the focus of the investigation and not just a means to an end. Simplified, often we evaluate and compare tools to each other. As a consequence, a large part of "protocol" of experiments is nothing else but the source code and binaries of prototypes that were developed to conduct the research, e.g., to evaluate how good a certain test method or mutation operator is.

Due to this, we do use the term *replication kit* instead of lab package in this paper. A replication kit should directly enable the (semi)automated exact replication of an experiment that is reported. Typically, a hyperlink to the replication kit is provided in a publication. Other researchers can visit the replication kit, which supports and easier replication of prior work.

4. EXPERIENCE REPORTS

Now that we have outlined what replication is and, especially, what replication kits are, we want to present our experiences both with replication of prior work, as well as then generation of replication kits. Through our experience reports, we hope to be able to help other researchers to create more and better replication kits and avoid problems we encountered while performing our research.

4.1 Software Defect Prediction

Most of our experience regarding replication kits comes from defect prediction research. We started working on the topic of defect prediction in 2013 [11]. While our descriptions in the paper were sufficient to reimplement our approach and perform a replication, we did not provide a replication kit.

At that time, we did not see any replication kits in the state of the art and did not yet realize how problematic this is.

While we continued to work on our research, we found contracting studies on the potential positive impact of local models in the literature [12–14]. When we evaluated the publications in detail, we found that they used different and relatively small data sets. Therefore, we decided to conduct a replication study on local models for cross-project defect prediction [15]. Our goal was to increase the external validity and resolve the contradictions of the prior studies. However, none of the publications provided replication kits, which we could use to replicate their work on other data. Therefore, we had to reimplement the previously considered approaches from scratch based on the descriptions in the publications. As an additional consequence, we could not validate the results our implementations produced against the results of the original implementations, which means that they may be different as we may have misinterpreted the descriptions of the approaches in the publications. These problems made us aware of the importance of replication kits. If the authors would have provided replications kits together with the publications, our work would have been less effort and at the same time have a better construct validity. We concluded that the conceptual replication we performed was not sufficient, but we needed to provide a replication kit.

This was our first replication kit and we were not aware of any best practices regarding how to create replication kits. For our first submission of the article, we created an archive with the binary for the experiment execution, copies of the data from public repositories we used, execution scripts for Windows and Linux, and a text file with a description of the contents of the replication kit, software requirements, and a short guide on how to execute the experiments. Based on feedback from our reviewers, we later also included the source code for the experiment execution and build scripts in the replication kit.

Our replication kit played another important role in the review process. Since the results of our study were mostly negative, one reviewer rightly questioned if this may be due to our choice of classifier. Even before this question arose, we had performed an experiment to determine that best classifier to make sure that our results are not negatively impacted due to our choice of the classification model. However, we had not reported on the experiment for classifier selection in the paper, because our focus was on local vs global models and not which classifiers perform best. We also did not want to change this focus, by adding more of the same results, just with different classifiers. The replication kit provided a convenient solution for this problem. We included results with additional classifiers in the replication kit, and simply stated in the paper what our internal experiments revealed: there are no big differences in conclusions and our chosen classifier is the best. This way, we could keep the focus of the paper narrow, but were also able to address this potential threat to the validity of our results.

One of our key results from our replication of local models was that a relatively naïve approach is actually quite good. This prompted us to work on a large conceptual replication of 24 proposed approaches from the cross-project defect prediction state of the art [16]. Again, we could not use prior implementations, because they were in general not available. Additionally, some descriptions in the papers were imprecise. This actually meant that we could not replicate all publications as is, but had to make assumptions on how the approaches may have been implemented for the original work. These assumptions are documented as part of the replication for transparency. One publication had to be excluded, because key components of the approach were not described. A replication kit could have mitigated this. The implementation effort to replicate the 24 publications required us to write over 35,000 lines of code (including documentation). In case implementations or replication kits would have been available, the effort would probably have been much lower, because we would have been able to reuse code from the original work. Moreover, same as in our previous replication, we could not validate our implementations against the original work, which poses a major threat to the construct validity.

In general, our approach for the replication kit for this study was the same as before. However, we directly knew that we wanted to include all source code. Additionally, we realized that we should also include the statistical analysis in the replication kit, because of the central nature with regards to the replication. Since the scope of our experiments was very large,[a] we could not include all visualizations of the data in the publication. We used the opportunity to include more visualizations in the replication kit to resolve this problem. Thus, we did not only use the replication kit to include more information to satisfy reviewer requests. Instead, we proactively used the replication kit to provide more data about our results to the community.

We requested a double-blind review for the publication, which the editors of the IEEE Transaction on Software Engineering agreed to. However, this complicated the usage of the replication kit during the submission, because the replication kit had also to be blinded. This posed a problem

---

[a] The results of 67,326 are included in the replication kit.

for our code to run the experiments, for which we used our own tool called CrossPare [17]. Because we had published CrossPare already, including this in the replication kit would have made the authors obvious and broken the double blindness. Including a blinded version of the tool was also not an option, due to several reasons.

1. We would have had to remove all license statements, which would be legally problematic. This would have forced us to distribute code that whose license is owned by our employer, i.e., the University of Goettingen. We would have had to at least clarify this with the legal department, to ensure that the confidentiality of the review process allows for a distribution of code without proper licensing.

2. Because our package names use the prefix de.ugoe.cs, we would have had to modify them. While this is in principle a simple rename refactoring, we would have broken the internal logic of CrossPare this way, because it relied on lazy loading of experiment configurations at runtime through reflection.

3. We would have had to modify hundreds of Javadoc comments, because they contain the author names.

If only the Javadoc comments would have been the problem, this would have been feasible. However, changing the package names would have broken the program and required more changes to the internal logic—thereby actually introducing threats to the construct validity of our replication kit. Still, they may have been feasible, albeit it would be a software engineering bad practice and therefore to some degree ironic for a publication in the IEEE Transaction on Software Engineering. However, the license problem was an obvious blocker. We cannot simply abandon the Intellectual Property Rights (IPR) of the university for a review process, even if the software is licensed as open source.

This decision was actually not well received by one of our reviewers, who would have liked to look at the source code. We think this is an interesting aspect that warrants further attention. Most papers still do not provide replication kits. In that case, reviewers usually just accept that and move on. However, if a replication kit with additional information is provided, that includes parts of the source code, etc., but excludes parts of the complete information required to perform the replication, the replication kit may actually be viewed in a negative light. So instead of being a bonus, this actually becomes a problem. This is a somewhat paradox situation, mainly due to the fact that replication kits are still optional and there is no agreement in the community regarding their importance, what they should contain, and most importantly, how they fit into a double-blind review processes.

The above experiences describe our learning curve with replications in general as well as the required contents of replication kits. However, where a replication kit is hosted and how it is referenced in a publication is equally important. The best replication kit is worthless, if it becomes unavailable. Thus, researchers have to make sure that their links in publications remain available. That this is actually a problem for software engineering research was confirmed by Leitner [18].

For the replication kit of our work on local vs global models [15], we decided to upload the replication kit to a locally hosted web space and used a PID handle to reference the replication kit in the paper [19]. Our idea behind this approach was that the local Web space is a fragile location, while the handle is permanently available and the URL it points to can be updated, in case a new location is needed. Still, the choice of a local Web space for hosting was bad. Our link to the replication kit could have been broken, e.g., the author leaving the institution and thereby losing the local Web space. While the Permanent IDentifier (PID) allows updating the URL, this still would have required conscious effort on the author's side—which likely would just have been forgotten.

Once we became aware of this problem, we looked for a better solution. This way, we discovered Zenodo [20], an open sharing platform hosted by the CERN [21] that supports reliable long-term archiving of digital research artifacts. Moreover, the uploaded data get assigned a Digital Object Identifier (DOI) which can be used for reliable citation of the artifacts. Another advantage of Zenodo is that it works well together with GitHub [22]. With only a few clicks, a repository hosted on GitHub can be backed up on Zenodo. Thus, our solution to host and share a replication kit was the following:

- We use GitHub to upload the replication kit. This enables convenient access, as well as later updates to the replication kit, e.g., with more additional information and visualizations.
- The documentation of the replication kit is provided as README.md file on the top level of the repository. This means that the documentation of the replication is directly available in a well formatted and easy to read format for researchers, without even the need to download the replication kit.
- We use Zenodo to archive the repository and cite the DOI created by Zenodo in the publication. This ensure long-term availability of the data and prevents link rot.

We moved our replication kit for the local vs global models to Github and archived it on Zenodo [23]. We then updated the link of the handle. Thus, we now ensured in retrospect the long-term availability of the replication kit. For our replication kit of the larger benchmark [16], we directly followed this practice and made it available via GitHub and Zenodo [24] and cited the DOI in the publication. By now, this is our standard practice for defect prediction publications and was also used for another publication on defect prediction in the automotive industry [25, 26].

4.2 Analysis of Test Types Characteristics

We were able to profit from the experience that was gathered during our work on software defect prediction, as we were starting with the research on test type characteristics. Concretely, we are working on differences between unit and integration tests, regarding both their adoption in practice as well as their defect revelation capabilities. As it was explained in Section 1, a replication can help especially novice researchers in many ways. Hence, as a novice researcher started with this topic, we tried to replicate work in this field, but failed miserably. The research that we do on test type character-istics is not a large dedicated research field like defect prediction, but it is a field that is directly in between software testing an mining software repos-itories (MSR). In our research we want to compare the influence of test types on software quality. This includes research questions like "Do unit tests detect more defects than integration tests?" or "Do we really need unit tests or are integration tests sufficient?". These are interesting questions, as the results can be used to create new development models or they can give hints on which types of testing might be favorable. Moreover, this kind of research provides empirical data to discussions that arise in this field (e.g., [27]), where empirical data were not available before.

We failed in replicating work in this field, because of several reasons. One major problem was that none of the papers that we looked at provided a replication kit. Furthermore, most papers described their implementation (e.g., how they separate test types from each other) not in a sufficient way and artifacts (e.g., raw data, processed data) were missing. This did not only raise the barrier for us to start our research, as we had a high overhead in creating and validating implementations which were already done before, but it also raised a couple of deeper problems. Due to the lack of available implementations, we were not able to validate our own implementations against the original work introducing threats to the construct validity of our work. Furthermore, as some descriptions were not sufficient for an exact reimplementation of the prior work, we needed to reconstruct some parts of the implementations based on our assumptions, introducing additional threats to the internal validity of our studies. Moreover, we were not able to test our approaches against the original data or compare the outcome of other approaches with our own implementations, as the raw or processed data were not available. This would have been a solution which we could have used to mitigate the introduced problems to the construction and internal validity. This kind of experience can be very demotivating, especially for novice researchers, as they need to create implementations which do not contribute new knowledge to the body of software engineering research. Moreover, the researchers do not get any real outcome out of it, as the creation of implementations that are already existing is often not directly contributing to answering their own research questions.

To resolve this problem for future research, we provided a replication kit [28] together with our first paper [29] on this topic. The replication kit is hosted locally, as at this time we were not aware of the long-term archiving option via Zenodo, which is explained in Section 4.1.

Before we were able to create our replication kit, we needed to make some decisions on what to include in it. While it was given that we needed to include our data collection and analysis implementations, we did not know what data we should include. This question was raised, because this kind of replication kit was different from the defect prediction replication kits discussed before, as we did not use "standard" data sets but collected our own data sets from projects. Hence, we needed to decide on which data need to be included: do we include our processed data only? Do we include our raw and processed data? Do we include the project repositories of all analyzed projects? After some discussions, it was clear that we definitely needed to include the raw data and the processed data. But, it was tough to make a decision regarding the project repositories. On the one hand, we have collected our data from the source code of them and therefore they should be included. Especially, as a replication of our work may be prevented if the projects are deleted from GitHub or their history changed (e.g., via git rebase [30]). On the other hand, the inclusion of the project repositories would bloat the replication kit extremely. In the end, we decided against copies of the repositories we analyzed in the replication kit. Our reasons where twofold. First, it is unlikely that the past of the projects is actually changed, because such operations are usually used for maintenance of merge commits. Second, the Software Heritage project [31] archives GitHub and even in case of removal of repositories from GitHub, they should still be available in the archives. Thus, we created a replication kit that included our implementations that we have used to collect and analyze the data, a dump of our database in which the data were stored, and a README file explaining the contents of the kit and how to execute the analysis.

4.3 Developer Contribution Behavior

Another area we are working on is the analysis of contribution behavior of developers [32], where we also prepared a replication kit for other researchers. In general, this is very similar to our experiences with defect prediction (see Section 4.1). Because this was before we were aware of the long-term archiving option with a citable DOI offered by Zenodo, we locally hosted the replication kit [33] and later uploaded a copy of the replication kit to GitHub and used Zenodo to create an archive [34].

However, there was one major difference: we created models for developer behavior for single developers, i.e., we analyzed personal data. While the models were afterwards aggregated, we still require knowledge which data belong to which developer for the replication kit. For this, we used the Email addresses of the developers. While this is compliant with European data privacy laws [35], we may not simple share and archive this personal data publicly ourselves. Therefore, we had to anonymize the developer identities prior to the publication of the replication kit. Because we already had a preprocessing script that merged different developer identities, this was a simple step. Instead of merging them into one of the real identities (e.g., merge jon.doe@company.com with jdoe@company.com into only jon.doe@company.com), we simple merged them into the symbolic names dev1, dev2, etc. This way, each developer still has a unique identity within our data, but there is no way to infer the real identity using only our published data. Of course it is still possible to infer the real identities by replicating our data collection and looking at the real identities in the public repositories of the project. However, since we do not share the private data ourselves, we are not a controller of the data and, thus, compliant with the European law.

4.4 User-Oriented Usability Studies

A further area of our research focuses on automated usability engineering of software. Here, we developed an approach that identifies usability issues based on recordings of software usage. The approach mainly consists of three steps. First, we record the usage of the software by intercepting the software internal events that are fired due to user interactions. For websites, this means, e.g., to intercept click events using JavaScript. In the second step, we use the recorded data to generate usage models. These models represent the average system usage. Based on the recorded data and these models, we then detect patterns of user behavior that indicate usability issues in the third step.

To evaluate the results of our approach, we usually perform a separate study with the same analyzed software, in which we apply traditional means of usability engineering. This comprises to ask potential users of a software to use it while we are observing their interactions. This requires sitting next to the users and taking notes of the issues the users have. In addition, we use questionnaires and subsequent interviews to gain information about detailed issues the users had. Subsuming all the data we produce in our studies, we usually have

1. data from applying our approach
   (a) the software we analyzed with respect to usability
   (b) the original recordings of the usage of this software and copies with adaptation, e.g., after cleanup
   (c) our tooling (including source code) to generate usage models and analyze the data
   (d) scripts, parameters, and log files for the execution of our tooling
   (e) statistical data extracted from the tooling execution and scripts for their analysis
   (f) other results from the data processing, i.e., output of our tooling (e.g., task models)
2. data for performing counter evaluations
   (a) notes of traditional usability evaluations
   (b) screen casts (videos) of the usage of the software
   (c) questionnaires or notes from user interviews
   (d) statistical data resulting from questionnaires or notes as well as scripts for their processing and visualization

The first replication kit that we intended to provide for one of our works could have been used to replicate a PhD thesis [36]. With 20 GB, already the uncompressed recorded usage data were rather large. Furthermore, an analysis of the recordings showed that they contained personal data. This was caused by recording also the text that was entered into text fields and because the users entered personal data into text fields where it was not supposed to be entered. Hence, a simple publication of the data was not possible. In addition, we had to follow some institution internal regulations for data archiving as this was considered to be processing personal data what in Germany is only allowed with certain restrictions. Therefore, we did not publish this data set, but archived it without public access. The data in the replication kit contained only the usage recordings, our tooling without source code, scripts for executing our tooling, statistical data from the tooling execution and other results, e.g., the task models. We did not include data for performing the counter evaluations as we were not aware of their value. The replication kit and its internal structure is described using metadata descriptions in the archive system.

The second replication kit [37] was created for a case study, in which we performed an analysis of a Virtual Reality. This replication kit contains only data for performing counter evaluations, as well as the analyzed Virtual Reality. The reason is that in this study, we did not apply our automated approach for usability evaluation but only performed an analysis of the Virtual Reality itself. The data covered screen casts of the software usage, questionnaires, statistical data from these questionnaires and scripts for their processing. Further data collected during the video screen analysis were included in the statistical data, as well. Due to the videos, this replication kit was also rather large with more than 5 GB. For this data set, the videos were also an issue for user privacy as, initially, they contained the audio stream of what our study participants were saying during the study. Hence, we had to adapt the videos accordingly.

The experience from the creation of the first two replication kits helped us in a third case study and its corresponding replication kit. For example, we directly ensured not to record personal data as it is, anyway, not required for our analysis. The third replication kit [38] is for a journal paper that is currently under review. The journal explicitly asks for replication kits, which is not often the case in our research area. Unfortunately, the maximum allowed size for replication kits for the journal is 100 MB, but our replication kit had a size of 2.5 GB. As a consequence, we decided not to provide a replication kit to the journal, but to upload our replication kit to a separate public long-term archive and to include a corresponding link in the paper. This replication kit contains almost all of the above mentioned elements. We did not include the source code of our tooling but provided a reference to its open source publication and mentioned the applied version. We also did not provide log files for applying our tooling as these were of no value in this study. Finally, we removed the audio stream from the screen recordings as done in the other case study.

5. DISCUSSION

Within this section, we discuss our experience with replications. We summarize the best practices regarding the contents required for replication kits, as well as how to publish them. Additionally, we discuss how replication kits may be made obsolete by platforms on which researchers automatically share their data, implementations, and results, without the specific need for replication kits. Finally, we discuss problems with the current review processes at conferences and journals with respect to replication kits.

5.1 Contents of Replication Kits

Fig. 2 summarizes the required contents for replication kits based on our experiences. Thus, we recommend that replication kits should contain the following artifacts.

1. All source code and/or binaries of tools used to execute experiments.
2. All underlying data on which the experiments were executed.
3. All source code and/or binaries of tools used to evaluate the experiment.
4. Exhaustive documentation to enable the use of the above.

The first two artifacts of a replication kit are required to enable exact replications. The first artifact is the source code or binaries required to execute the experiments conducted. Source code is the better option, due to multiple reasons.

- Source code allows deeper insights into the approach and the conducted experiments. Thus, the source code can be used to clarify potential ambiguities or misunderstandings that may be contained in the textual description contained in the publication.

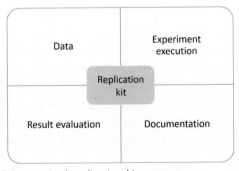

Fig. 2 Overview of the required replication kit contents.

- In case the goal is a dependent replication or an advancement of the approach, source code allows the modification of the original study.

However, it is not always possible to share the source code, e.g., due to proprietary reasons. In this case, an executable binary is a valid replacement for the source code in a replication kit. In case the compilation of the source code is difficult, both binaries and source code may also be provided.

The second artifact is the data that is used for the experiments. Ideally, there is a dedicated folder in which all data are stored. The data may be further separated, e.g., into raw data (e.g., source code samples or answer sheets) and preprocessed data that is used for input of the analysis (e.g., metrics extracted from the source code or a spreadsheet that contains the data from all answer sheets). However, in practice this is often very difficult to achieve. For example, in case source code is analyzed, sharing this code may not be possible or allowed due to licensing reasons. In such a case, all data extracted from the source code should be included and the tools used to obtain this data. In case the source code is available in open source repositories and cannot be shared, e.g., due to size constraints (if hundreds or thousands of projects are analyzed), links to the repositories should be provided to mitigate this. In case tools are used for the preprocessing of data, these tools should also be provided. In case data were manually preprocessed, a description of the performed tasks should be provided. Moreover, in case data were collected through a survey, interviews, or an observation technique, protocols on how these were conducted are required to allow the replication of the data collection. In case tools were used to run experiments on the raw data, e.g., test generation, defect prediction, or something similar, the results of these executions should also be provided as part of the data. These results may only consist of the generated data, which is then used for result evaluation, but may also contain log files of executions.

The first two artifacts are not sufficient for the replication of conclusions, because the evaluation of the data is missing. For this, the third artifact is required, i.e., source code or binaries for the evaluation of results. Such evaluations can be very diverse, e.g., visualizations, summary tables, or statistical tests.

The fourth and final artifact is the (often neglected) documentation. Without guidance on how to compile and execute code, including a documentation of the environment that is required, replication is often only very hard to achieve. Moreover, in case manual steps are required, e.g., to copy data from one tool to another, it cannot be expected that other researchers know this without a proper documentation. Thus, while technically not required for replications, documentation is in practice nearly as important as the tools and the data itself.

In addition to the four components that all replication kits should contain, there is a fifth component which can be very valuable.

5. Additional results and visualizations that did not fit into the publication due to space restrictions.

Because of page limits we usually have at conference, but also sometimes in journals, we often shorten or condense information, or can only fit the detailed results of parts of our research within the paper. With replication kits, we do not have such restrictions. You can add additional figures, spreadsheets with more data or results, or similarly expand on what you can present within the paper. Moreover, since the replication kit does not have to be printed, you can even include things that are impossible to include in a paper otherwise, e.g., interactive visualizations. Basically, replication kits provide a convenient means for an appendix without page restrictions.

5.2 Publishing of Replication Kits

Based on our experience, we can clearly recommend the publishing of replication kits using Zenodo [20] and then citing the replication kit with the generated DOI. Hosting on other systems that provide the same long-term archiving features including a DOI or a similar PID would also be acceptable. Same as others, we can only strongly suggest to not host replication kits on normal Web space of your institutions, due to the possibility of link rot [18]. Moreover, we found that GitHub provides a convenient way for uploading and maintaining replication kits. Due to the integration with Zenodo, it is easy to create a long-term archive of a GitHub repository. The advantage of using GitHub in comparison to direct upload of Zenodo is that it is easier to update the replication kit, e.g., due to a major revision of a journal article. Zenodo can assign a new DOI for all releases of the replication kit on GitHub, as well as a DOI that always points to the latest release of the replication kit.

However, we are also aware of two problems for which we cannot suggest general guidelines. The first is the handling of private data (e.g., names, email addresses) in replication kits. Publication of such data is often prohibited by data privacy laws. Thus, authors must always make sure that the data comply with the data privacy laws and, if required, remove the privacy related parts from the data, e.g., through anonymization, pseudonymization, or a similar technique. It might also be a solution to remove such parts from a replication kit (e.g., answer sheets for a survey) and only make them available upon request with a separate privacy agreement.

The second problem are very large replication kits, e.g., due to videos, or large amounts of data that were collected automatically. Zenodo allows archives with up to 50 gigabytes without restrictions. Larger data sets may be possible upon request, but we have no experience with this so far. While this is sufficient for nearly all replication kits, we actually face the problem that we are currently using a data base that already contains several terabytes of data and is still growing. A replication kit with this data base will be a completely new challenge for us, for which we do not have a solution yet.

5.3 A Checklist for Replication Kits

In order to support other researchers, we created a checklist for replication kits. This checklist summarizes our experiences which we discussed above regarding the contents and publishing of replication kits. Overall, our checklist consists of 26 points, which demonstrates that the creation of replication kits is no trivial task. The checklist is depicted in Table 1. The compartments in the tables contain the checklist items regarding separate concerns, i.e., contents regarding the data, novel software engineering approaches, the result evaluation, documentation, additional contents, publishing of the replication kit, and other issues, respectively.

5.4 Platforms for Replicable Research

While we think that the consistent provision of replication kits for research results would already be a major improvement for the state of practice of software engineering research, we also identified problems that would not be resolved that way, e.g., the often small size of data sets that are used or the heavy reuse of data, which can lead to problems like inadvertent p-hacking. If data are reused which are later shown to be problematic, the validity of the results of many papers may be compromised. We discussed these and other problems for the validity of research results for software repository mining studies in our recent publications [28, 39].

However, our work is not restricted to the analysis of the problem. We also suggest a potential solution through a commonly used platform. Instead of stand-alone replication kits, research would be performed and shared directly in a common platform, that hosts all data, implementations of analyses, and results. In case more data are added, analyses can automatically be

Table 1 Checklist for Replication Kits

Contents regarding data that is used
  1. Are the source code and compilation scripts for developed data collection tools included?
  2. Are binaries of all tools required for the data collection available?
  3. Are scripts for the execution of the data collection included?
  4. Are the software artifacts from which data are collected included? Alternatively, are links and exact information regarding the analyzed version provided?
  5. Are for all surveys/interviews/observation techniques the protocols of how they were conducted available?
  6. Are all raw results from surveys/interviews/observations included?
  7. Is the source code for data preprocessing included?
  8. Are manually performed preprocessing steps described?
  9. Does the data contain any private information which may not be published?

Contents regarding novel software engineering approaches
  10. Are the source code and compilation scripts for developed tools for a novel software engineering approach included?
  11. Are binaries of all tools required for a novel software engineering approach included?
  12. Are additional source code, compilation scripts, and binaries of approaches used for comparison included?
  13. Are scripts for the execution of the software engineering approaches on the provided data that replicate the results reported in a publication, included?

Contents regarding results and evaluations
  14. Are the complete raw results of all experiments included?
  15. Are the source code and compilation scripts for the results evaluation (e.g., visualizations, statistical tests) included?
  16. Are binaries of all tools required for the evaluation included?
  17. Are scripts for the execution of the result evaluation included?

Concents regarding documentation
  18. Is the required environment (e.g., operating system, software versions of dependencies) fully described?
  19. Is all provided source code documented?
  20. Does the documentation describe the structure of the replication kit, i.e., which contents can be found where?
  21. Is a description of how the results from the publication can be replicated provided?

Optional contents
  22. Are additional relevant result and visualizations included?
  23. Is all additionally provided information described?

Publishing of the replication kit
  22. Is the replication kit referenced using a DOI or similar unique identifier?
  23. Is the long-term availability of the replication kit assured?

Table 1 Checklist for Replication Kits—cont'd

Other issues

24. Was the completeness of the replication kit and the documentation validated by a colleague in a clean environment?

25. Is all legally relevant information (e.g., copyrights, license statements) provided?

26. In case any of the questions above is answered with no, are reasons for the exclusion given?

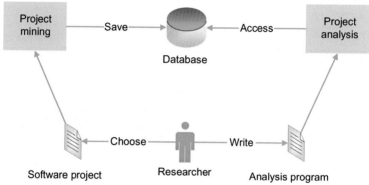

Fig. 3 Concept for a platform for replicable research.

executed against the new data, to generate new and more results. This way, problems with replicability are minimized. In our work, we provided a preliminary analysis of such a platform for (MSR) studies. Fig. 3 shows the concept of our proposed solution. In general, the goal is that researchers can select projects and provide analytic programs, the desired analysis is then executed automatically on the platform. Thus, the platform takes care of the extract–transform–load (ETL) of the data for the chosen project and stores this data in a database, i.e., the database in the center of Fig. 3. The analytic program then accesses this database to perform the desired analysis. The results of the analysis are again, stored in the database. Researchers can then access their results by retrieving the relevant information from the database.

We implemented this concept as a prototype called SmartSHARK. Fig. 4 shows the components of our prototype. From the researcher's point of view, SmartSHARK is an extensible Web application. The Web application is the central location to access the results of experiments that were

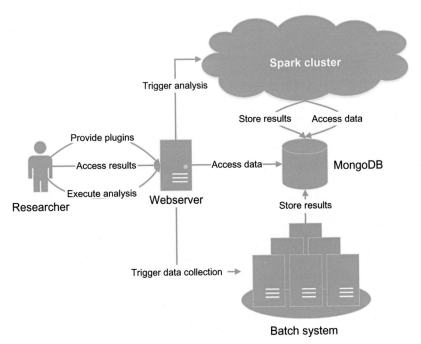

Fig. 4 Structure of the SmartSHARK prototype.

performed on the platform. Through the provision of plugins that are loaded into the platform, researchers can define methods for the ETL of data and define analytics programs. Thus, the plugins allow the definition of new experiments that are executed on the platform. Our prototypes allows two types of plugins: plugins that are executed on a batch system and plugins that define Apache Spark [40] jobs. Plugins that run on the batch system are for the execution independently parallel running task. Such tasks include, e.g., the collection of metadata from software repositories like version control systems, issue tracking systems, and mailing lists. The collected metadata may be relatively straightforward in form of software metrics, but also more complex like the results of test executions or the performance of test generation approaches, as measured with mutation testing. Another advantage of the batch system is that this allows an easy way to integrate existing tools that can be executed on the command line. Plugins that run on the Spark cluster can perform complex analysis on the data stored in the central database, e.g., through machine learning, statistical tests, or similar. The prototype uses MongoDB [41] as the central database. The advantage of MongoDB is that the document storage supports the storage of diverse and potentially nested data. Moreover, MongoDB provides a grid-based file storage.

In our work, we showed that different tasks can be carried out using our prototype, including defect prediction, the analysis of developer social networks, and effort analysis. However, our prototype is still in its early stages. For example, we cannot yet compare different approaches on a similar topic, e.g., multiple defect prediction approaches with each other. Still, we believe that the concept of directly sharing results on a platform with common interfaces can be a game changer for software engineering research, if this is extended with direct comparisons of related approaches. While we believe that no such general platform for all topics of empirical software engineering research is possible, we are confident that this can be solved for specific problems. For example, for test generation through search-based software testing, regular challenges are held to compare different approaches [42]. These challenges demonstrate that platforms for automated comparisons are possible and can help the community. Similarly, we believe that this will be possible for other research topics, e.g., defect prediction. However, even if such platforms are created and maintained by researchers, we still must accept them as a community and use them to compare our research, which is a vital success factor that may be very hard to achieve.

5.5 The Role of Replication Kits in Review Processes

Finally, we want to comment on the current state of practice regarding replication kits in review processes. Currently, replication kits are usually seen as a bonus, i.e., their presence is not required and they are, in our experience, often only acknowledged by the reviewers at technical tracks and journals, and not actually validated by executing them or checking them otherwise. To our mind, this is problematic due to two reasons. Most importantly, we think that we as a community should stop seeing replication kits as a bonus, and instead see them as a normal and mandatory part of publications, same as, e.g., discussing threats to the validity. Enforcing this change starts with the reviewers. Unless a good reason for the absence of a replication kit in an empirical paper is given, this should count as a strong negative aspect during the evaluation of a publication. To our mind, not providing a replication kit is similar to not discussing threats to the validity—and this is a common reason for rejections and revisions.

As a corollary, the evaluation of the contents of a replication kit must also become a part of the review process. Otherwise, this would be the same as only checking if a publication contains a section called "Threats to validity" without actually reading the section. We are aware that this is additional work for reviewers. However, replication kits provide a valuable source to evaluate the validity of the presented results. For example, the replication kit allows reviewers to check if the description of the approach is consistent with the implementation. Thus, the construct validity can be evaluated by the reviewers instead of just acknowledging that the implementation may be broken as a threat to validity. While a full review of source code, documentations, etc., would require extensive effort, spot checking a replication kit or just checking if it is actually running is feasible for reviewers. To further remove the additional burden on reviewers, there could be a single reviewer that is designated to review the replication kit.

Another unresolved problem with replication kits are double-blind reviews, i.e., venues that require authors to hide their identity. Blinding replication kits means removing all identifying information from the source code, scripts, etc. While this is feasible for short programs, it cannot easily be done for larger software products. For example, we are using a tool called AutoQUEST [43] with over 150,000 lines of code, of which over 48,000 lines of code contain comments [44]. These comments are usually Javadoc comments and also contain identifying information about the authors. Moreover, each file in the project has an open source license statement referencing the University of Goettingen as a header. Both could probably be blinded using string replacement tools. However, the legal implications of removing licenses are not clear to us. We are neither sure if this would be allowed for us, nor are we sure what it would mean for the license of the source code, if we were to distribute it without license. Then there are the technical problems of blinding. All our Java code is located in packages which are named de.ugoe.cs.*, i.e., even the package names identify our institution. Thus, we would have to change them, too. This would break the software, as we sometimes use reflection, i.e., loading of classes using their string identification at runtime. Thus, we would also need to adopt those strings as well, and hope that everything would be running again. This would be feasible, but from a software engineering point of view an extremely defect prone bad practice. Unfortunately, the tool would still not be blind, because we also require a locally hosted Maven repository for third party artifacts. This could never be blinded, without breaking the build process.

We asked PC chairs at three recent conferences that use double-blind review regarding these anonymization issues and got the following answer:

we should either blind the code or not submit the source code at all as part of the review process. There was a separate artifact evaluation process for one of the conferences, only for accepted papers, after the technical review process. This response highlights our prior point: the actual evaluation of the artifacts is not part of the review process, and replication kits are not seen as something that should always be provided, yet.

To the best of our knowledge, this situation applies to all software engineering venues that use double-blind review, with one exception, i.e., the International Symposium on Software Testing and Analysis 2016 (ISSTA'16) [45]. In that year, the program committee chairs adopted a double-blind review procedure that incorporated the unblinded review of artifacts already as part of the technical evaluation of papers. While their approach was more work for the program committee, we hope that more future conference adopt similar procedures and, thereby, acknowledge artifacts, especially replication kits, as what they should be: a vital part of our research.

6. CONCLUSION

Within this chapter, we presented our experiences with replicable software engineering research. Our experiences show that replicable research requires dedicated effort. The provision of good replication kits is more than just the publishing of a link to some source code within a publication. Our experiences show that there are many nooks and crannies which researchers must be aware of in order to make their replication kits usable for other researchers. Potential problems include aspects related to the content, e.g., the source code, the documentation, or the raw data, but may also be related to the hosting of the replication kits to prevent link rot. Based on our experiences, we provided best practices for the contents and publishing of replication kits. Our hope is that this chapter contributes to advancing the state of practice of software engineering research and that more and more researchers publish replication kits together with their research.

However, we do not think that replication kits solve all problems with replications in software engineering research, e.g., the general lack of replications or the heavy reuse of data. In our future work, we will continue to investigate how platforms designed for replicable research may help the community and improve the external validity of results. Furthermore, the spread of replication kits has not yet altered review processes at conferences and journals significantly. Currently, reviewers see replication kits as a benefit, and not something mandatory. Moreover, the review of replication kits is often not part of the technical evaluation of a paper. As a consequence, replication kits are usually either not reviewed or reviewed independently of the technical evaluation of papers as part of a separate artifact evaluation. A shift in mentality is required such that there are real incentives to always provide replication kits and such that replication kits and their correctness become a natural part of review processes. Moreover, we believe that replication kits provide challenges for double-blind reviews, which are not yet adequately resolved. Thus, we think that our research community still has a long road ahead until we have a generally accepted and enforced standard of replicability.

REFERENCES

[1] J. Siegmund, N. Siegmund, S. Apel, Views on internal and external validity in empirical software engineering, in: IEEE/ACM 37th IEEE International Conference on Software Engineering (2015), vol. 1, IEEE, 2015, pp. 9–19.
[2] F. Trautsch, Replication Kit, 2016, https://user.informatik.uni-goettingen.de/ftrauts/replicationkit01.tar.gz.
[3] G. Robles, Replicating MSR: a study of the potential replicability of papers published in the mining software repositories proceedings, in: Proceedings of the Seventh IEEE Working Conference on Mining Software Repositories, IEEE, 2010, pp. 171–180.
[4] J.P.A. Ioannidis, Why most published research findings are false, PLoS Med. 2 (8) (2005) e124.
[5] Open Science Collaboration, et al. Estimating the reproducibility of psychological science, Science 349 (6251) (2015) aac4716.
[6] M. Baker, 1,500 scientists lift the lid on reproducibility, 2016, http://www.nature.com/news/1-500-scientists-lift-the-lid-on-reproducibility-1.19970?WT.mc_id=SFB_NNEWS_1508_RHBox (accessed 26.11.2018).
[7] F.J. Shull, J.C. Carver, S. Vegas, N. Juristo, The role of replications in empirical software engineering, Empir. Softw. Eng. 13 (2) (2008) 211–218.
[8] P.C. Cozby, S. Bates, Methods in Behavioral Research, McGraw Hill, 2011.
[9] A. Brooks, M. Roper, M. Wood, J. Daly, J. Miller, Replication's role in software engineering, in: F. Shull, J. Singer, D.I.K. Sjøberg (Eds.), Guide to Advanced Empirical Software Engineering, Springer London, London, 2008. ISBN: 978-1-84800-044-5, pp. 365–379, https://doi.org/10.1007/978-1-84800-044-5_14.
[10] J. Miller, Applying meta-analytical procedures to software engineering experiments, J. Syst. Softw. 54 (1) (2000) 29–39. ISSN: 0164-1212, https://doi.org/10.1016/S0164-1212(00)00024-8.
[11] S. Herbold, Training data selection for cross-project defect prediction, in: Proceedings of Ninth International Conference on Predictive Models in Software Engineering (PROMISE), ACM, 2013. ISBN: 978-1-4503-2016-0, https://doi.org/10.1145/2499393.2499395.
[12] N. Bettenburg, M. Nagappan, A.E. Hassan, Think locally, act globally: improving defect and effort prediction models, in: Proceedings of the Ninth IEEE Working Conference on Mining Software Repositories (MSR), IEEE Computer Society, 2012. https://doi.org/10.1109/MSR.2012.6224300. ISSN 2160-1852.

[13] T. Menzies, A. Butcher, D. Cok, A. Marcus, L. Layman, F. Shull, B. Turhan, T. Zimmermann, Local versus global lessons for defect prediction and effort estimation, IEEE Trans. Softw. Eng. 39 (6) (2013) 822–834, https://doi.org/10.1109/TSE.2012.83. ISSN 0098-5589.

[14] G. Scanniello, C. Gravino, A. Marcus, T. Menzies, Class level fault prediction using software clustering, in: Proceedings of the 28th IEEE/ACM International Conference on Automated Software Engineering (ASE), IEEE Computer Society, 2013.

[15] S. Herbold, A. Trautsch, J. Grabowski, Global vs local models for cross-project defect prediction, Empir. Softw. Eng. 22 (4) (2017) 1866–1902. ISSN: 1573-7616, https://doi.org/10.1007/s10664-016-9468-y.

[16] S. Herbold, sherbold/replication-kit-tse-2017-comment-scottknottesd: release of the replication kit, 2017. https://doi.org/10.5281/zenodo.438025.

[17] S. Herbold, CrossPare: a tool for benchmarking cross-project defect predictions, in: The Fourth International Workshop on Software Mining (SoftMine), 2015.

[18] P. Leitner, How much of a problem are broken links in SE research? 2017, https://medium.com/@xLeitix/how-much-of-a-problem-are-broken-links-in-se-research-cedfdce3d030.

[19] S. Herbold, Replication kit for global vs local models for cross-project defect prediction, 2016. dl.handle.net/21.11101/0000-0001-3C55-D.

[20] CERN, Zenodo, 2017, https://zenodo.org.

[21] CERN, 2017, https://home.cern/.

[22] Inc Github, GitHub, 2017, https://github.com.

[23] S. Herbold, sherbold/replication-kit-emse-2016-local-models: release of the final replication kit, 2017, https://doi.org/10.5281/zenodo.321369. https://doi.org/10.5281/zenodo.321369.

[24] S. Herbold, sherbold/replication-kit-tse-2017-benchmark: release of the replication kit, 2017, https://doi.org/10.5281/zenodo.581178.

[25] H. Altinger, S. Herbold, F. Schneemann, J. Grabowski, F. Wotawa, Performance tuning for automotive software fault prediction, in: 2017 IEEE 24th International Conference on Software Analysis, Evolution and Reengineering (SANER), 2017, pp. 526–530, https://doi.org/10.1109/SANER.2017.7884667.

[26] S. Herbold, H. Altinger, sherbold/replication-kit-saner-2017: release of the final replication kit for SANER 2017. 2017, https://doi.org/10.5281/zenodo.321370.

[27] J.O. Coplien, Why most unit testing is waste, 2014, http://rbcs-us.com/documents/Why-Most-Unit-Testing-is-Waste.pdf (accessed 26.11.2018).

[28] F. Trautsch, S. Herbold, P. Makedonski, J. Grabowski, Adressing problems with external validity of repository mining studies through a smart data platform, in: Proceedings of the 13th International Conference on Mining Software Repositories, MSR '16, ACM, New York, NY, USA, ISBN: 978-1-4503-4186-8, 2016. pp. 97–108, https://doi.org/10.1145/2901739.2901753.

[29] F. Trautsch, J. Grabowski, Are there any unit tests? An empirical study on unit testing in open source python projects, in: 2017 IEEE International Conference on Software Testing, Verification and Validation (ICST), 2017, pp. 207–218, https://doi.org/10.1109/ICST.2017.26.

[30] Inc Github, GitHub Documentation, 2017, https://git-scm.com/docs/git-rebase.

[31] R.D. Cosmo, S. Zacchiroli, Software heritage: why and how to preserve software source code, in: iPRES 2017: 14th International Conference on Digital Preservation, 2017, https://hal.archives-ouvertes.fr/hal-01590958;https://hal.archives-ouvertes.fr/hal-01590958/file/ipres-2017-software-heritage.pdf. Kyoto, Japan.

[32] V. Honsel, S. Herbold, J. Grabowski, Hidden Markov models for the prediction of developer involvement dynamics and workload, in: Proceedings of the 12th International Conference on Predictive Models and Data Analytics in Software Engineering, PROMISE, ACM, New York, NY, USA, ISBN: 978-1-4503-4772-3, 2016, pp. 8:1–8:10, https://doi.org/10.1145/2972958.2972960.

[33] V. Honsel, Replication Kit, 2016, https://filepool.informatik.uni-goettingen.de/publication/swe/2016/vh-material-promise2016.zip.

[34] V. Herbold, S. Herbold, sherbold/replication-kit-promise-2016: release of the final replication kit for PROMISE 2016, 2017, https://doi.org/10.5281/zenodo.376505.

[35] A. Engelfriet, Is it legal for GHTorrent to aggregate Github user data?, 2016, https://ictrecht.nl/2016/02/28/is-it-legal-for-ghtorrent-to-aggregate-github-user-data/.

[36] P. Harms, Automated Field Usability Evaluation Using Generated Task Trees (Ph.D. thesis), 2016, http://hdl.handle.net/11858/00-1735-0000-0028-8684-1.

[37] H. Holderied, Data set of the thesis "evaluation of interaction concepts in virtual reality applications" 2017, https://doi.org/10.5281/zenodo.571138.

[38] P. Harms, Replication Kits for the papers "automated usability evaluation of virtual reality applications" and "VR interaction modalities for the evaluation of technical device prototypes" submitted to the CHI 2018, 2017, https://doi.org/10.5281/zenodo.894173.

[39] F. Trautsch, S. Herbold, P. Makedonski, J. Grabowski, Addressing problems with replicability and validity of repository mining studies through a smart data platform, Empir. Softw. Eng. 23 (2) (2017) 1036–1083.

[40] M. Zaharia, M. Chowdhury, M.J. Franklin, S. Shenker, I. Stoica, Spark: cluster computing with working sets, in: Proceedings of the Second USENIX Conference on Hot Topics in Cloud Computing (HotCloud), 2010.

[41] MongoDB Inc., MongoDB, https://www.mongodb.org/, n.d. (accessed 26.11.2018).

[42] SBST contest—java unit testing at the class level, 2017, http://sbstcontest.dsic.upv.es/.

[43] S. Herbold, P. Harms, AutoQUEST - automated quality engineering of event-driven software, in: Proceedings of the IEEE Sixth International Conference on Software Testing, Verification and Validation Workshops (ICSTW), 2013, https://doi.org/10.1109/ICSTW.2013.23.

[44] https://www.openhub.net/p/auto_quest, 2017.

[45] A. Roychoudhury, A. Zeller, ISSTA 2016 REVIEWING, 2016, https://issta2016.cispa.saarland/wp-content/uploads/2015/05/ISSTA16-Report.pdf.

ABOUT THE AUTHORS

Steffen Herbold manages a Professorship on Data Science at the Institute of Computer Science of the University of Goettingen. He received his doctorate in 2012 from the University of Goettingen. Dr. Herbold research interests is the application of data science methods, with a focus on empirical software engineering. His research includes work on defect prediction, software repository mining, software testing, as well as scalable analytics in cloud environments.

Fabian Trautsch is a PhD student at the University of Goettingen. He received the BSc degree in applied computer science from the University of Goettingen in 2013 and the subsequent MSc degree in 2015. Since then he is working as a PhD student in the Software Engineering for Distributed Systems group at the Institute of Computer Science of the University of Goettingen. His research interests include mining software repositories, software evolution, software testing, and empirical software engineering.

Dr. Patrick Harms leads the research focus "Usability Engineering, AR, MR, VR" in the research group Software Engineering for Distributed Systems at the Institute of Computer Science of the University of Goettingen. He finished his PhD on "Automated Field Usability Evaluation Using Generated Task Trees" in 2015. Currently, he transfers this research to Augmented and Virtual Reality (AR/VR) and assesses the applicability of AR/VR for usability evaluations of virtual prototypes of technical consumer products.

Verena Herbold is a PhD student at the University of Goettingen. She received a Diploma in Math from the University of Hannover in 2011. Since 2013 she is working as a PhD student in the Software Engineering for Distributed Systems group at the Institute of Computer Science of the University of Goettingen. Her research interests include mining software repositories, software evolution, software testing, and empirical software engineering.

Jens Grabowski is professor at the University of Goettingen. He is vice director of the Institute of Computer Science and is heading the Software Engineering for Distributed Systems Group at the Institute. Prof. Grabowski is one of the developers of the standardized testing languages TTCN-3 and UML Testing Profile. The current research interests of Prof. Grabowski are directed toward model-based development and testing, managed software evolution, and empirical software engineering.